Environmental Pollution and Remediation Techniques

Environmental Pollution and Remediation Techniques

Edited by Elliot Franco

SYRAWOOD
PUBLISHING HOUSE

New York

Published by Syrawood Publishing House,
750 Third Avenue, 9th Floor,
New York, NY 10017, USA
www.syrawoodpublishinghouse.com

Environmental Pollution and Remediation Techniques
Edited by Elliot Franco

International Standard Book Number: 978-1-64740-001-9 (Hardback)

Cataloging-in-Publication Data

Environmental pollution and remediation techniques / edited by Elliot Franco.
    p. cm.
Includes bibliographical references and index.
ISBN 978-1-64740-001-9
1. Pollution. 2. Pollution prevention. 3. Environmental protection.
4. Environmental engineering. I. Franco, Elliot.
TD174 .E58 2020
363.73--dc23

TABLE OF CONTENTS

Permissions

List of Contributors

Index

PREFACE

This book has been an outcome of determined endeavour from a group of educationists in the field. The primary objective was to involve a broad spectrum of professionals from diverse cultural background involved in the field for developing new researches. The book not only targets students but also scholars pursuing higher research for further enhancement of the theoretical and practical applications of the subject.

The contamination and degradation of the biological and physical components of the environment to an extent where normal environmental processes are adversely affected is termed as environmental pollution. Some of its prominent causes are carbon dioxide emissions, greenhouse gas emissions, reduction in biodiversity and acid rain. Pollutants are naturally occurring substances which are present in excessive amounts or concentrations. The removal of contaminants or pollution from the environmental media such as groundwater, surface water and soil is termed as environmental remediation. There are numerous technologies which are used for remediation such as thermal desorption, excavation, solidification and stabilization, in situ oxidization, and soil vapor extraction. This book unravels the recent studies related to environmental pollution. Most of the topics introduced herein cover new techniques for the remediation of pollution. Researchers and students in this field will be assisted by this book.

It was an honour to edit such a profound book and also a challenging task to compile and examine all the relevant data for accuracy and originality. I wish to acknowledge the efforts of the contributors for submitting such brilliant and diverse chapters in the field and for endlessly working for the completion of the book. Last, but not the least; I thank my family for being a constant source of support in all my research endeavours.

Editor

Adsorption and Desorption of Emerging Water Contaminants on Activated Carbon Fabrics

Sandrine Delpeux-Ouldriane[1], Mickaël Gineys[1], Nathalie Cohaut[1], François Béguin[1],
Sylvain Masson[2], Laurence Reinert[2], Laurent Duclaux[2]
[1]CNRS-Université d'Orléans, ICMN
1B Rue de la Férollerie, Orléans, France, 45071
delpeux@cnrs-orleans.fr
[2]Université Savoie Mont Blanc, LCME
Chambéry, France, 73000
laurent.duclaux@univ-savoie.fr

Abstract - *Nowadays, a wide variety of organic contaminants is present at trace concentrations in wastewater effluents. In order to mitigate these pollution problems, the implementation of the REACH European regulation has defined lists of targeted pollutants to be eliminated selectively in water. This therefore implies the development of innovative and more efficient remediation techniques. In this regard, adsorption processes can be successfully used to achieve the removal of organic compounds in waste water treatment processes, especially at low pollutant concentration. Activated carbons possess a highly developed porosity and thus demonstrate high adsorption capacities. More specifically, carbon cloths show high adsorption rates, ease of handling, good mechanical integrity and regeneration potential. When loaded with pollutants, these materials can be regenerated using electrochemical polarization.*

Keywords: Nanoporous carbons, activated carbon cloths, adsorption, micropollutants, emerging contaminants, regeneration, electrochemistry.

1. Introduction

A wide variety of organic compounds that are used in domestic, agricultural and industrial applications are present at trace concentrations in wastewater effluents [1, 2]. These micropollutants include personal-care products, plasticizers, reproductive hormones, pesticides and pharmaceuticals [3]. The adsorption properties of activated carbons offer great potential for water purification, particularly in the case of tertiary treatments as they are the most prevalent and competitive adsorbents, especially at low pollutant concentration. However, the major disadvantage encountered is their short lifetime due to the low and expensive regeneration capacities. Generally, the loaded carbon adsorbent can be regenerated *ex-situ* through high energy-consuming processes, like thermal treatments or steaming.

Specifically, as compared to powder or granules, activated carbon cloths (ACC) show numerous advantages, thanks their ease of handling, high mechanical integrity and regeneration potential. Additionally, due to their microtexture and their small fiber diameters (around 10 µm), they are ideal candidates for adsorption purposes as they show minimal diffusion limitation and greater adsorption rates towards noxious organic pollutants [4].

In the present work, the adsorption properties of some micropollutants and emerging pollutants, especially pharmaceutical residues, were investigated using an activated carbon cloth. Electrochemical polarization was applied to achieve the reversible desorption of adsorbed species and the regeneration of the adsorbent porosity [5, 6].

The involved mechanisms were carefully examined and correlated to the nanoporous texture of the adsorbent by taking into account the adsorbate speciation and physico-chemical properties. Results show that the reversible electrochemical desorption of the induced charged molecules offers great promise. Such systems could find a place of choice in industrial processes for tertiary treatment and for the treatment of hospital effluents, but also for underground water exploitation [7].

2. Experimental

2. 1. Adsorbent

An activated carbon cloth possessing a highly developed surface area was used (Table 1). Its porous network consisted of a large amount of supermicropores (0.7-2 nm) and ultramicropores (< 0.7 nm). The pore size distribution was very narrow and centered at 1-1.2 nm.

Table 1. Microtextural adsorbent characteristics.

S_{BET} (m^2/g)	V_{MICRO} N_2, DFT (cm^3/g)	V_{MESO} N_2, DFT (cm^3/g)	V_{MICRO} N_2, DR (cm^3/g)	V_{MICRO} CO_2, DR (cm^3/g)	V_{TOTAL} N_2 (cm^3/g)
1175	0.59	0.09	0.57	0.57	0.68

This material was rather hydrophobic and contained 95 % (mass %) of carbon and 1.2% of oxygen. It presented a basic zero charge pH (8.9) and a low amount of oxygenated surface groups (0.2 mmol/g).The diameters of the fibers were in the range of ten micrometers and micropores were directly accessible all along the fibers (Figure 1).

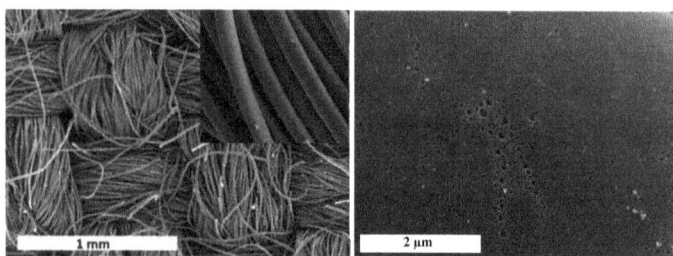

Figure 1. SEM images of the activated carbon cloth.

2. 2. Targeted Pollutants

Seven organic molecules, diclofenac (DFN), caffeine (CAF), acetaminophen (also called paracetamol POL), bisphenol A (BPA), carbamazepine (CBZ), pentachlorophenol (PCP) and ofloxacin (OFX) were selected (Figure 2).

These contaminants possess different physico-chemical properties: water solubility, polarity, pK_A values, and molecular volumes (Table 2).

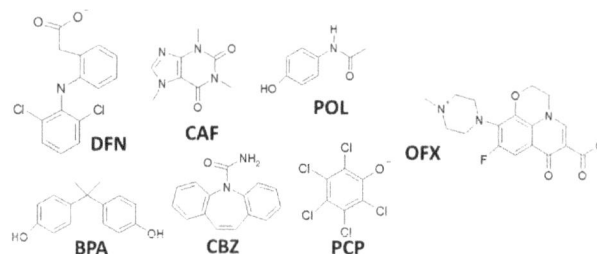

Figure 2. Adsorbates formulae and speciations at pH 7.5.

Table 2. Pollutants characteristics at pH7.5.

	pK_{A1} / pK_{A2}	V ($Å^3$)	s (mmol/L)	Log K_{ow}
POL	9.5	138	74.0	1.08
PCP	5.0	144	5.7	1.90
CAF	0.6	340	225.2	-0.79
CBZ	2.3/13.9	400	0.2	3.22
BPA	9.8	460	0.7	4.32
DFN	4.0	580	154.1	0.55
OFX	5.5/8.2	850	144.5	-0.93

2. 3. Analytical Detection and Quantification

Reverse phase HPLC was used to perform the detection and the quantification during adsorption kinetics, in binary or complex mixture and isotherm adsorption. The chromatographic separation was performed in the reversed phase mode using a Hypersil Gold C_{18} column at 25°C (100 × 2.1 mm with a particle size of 3 μm). The eluents were water (A) at pH 2.9 through acidification by orthophosphoric acid 0.01 % (v/v) and acetonitrile (B). The following multi-step linear gradient was applied: from 10 % B to 80 % B in 25.45 min (slope of 2.75 mL/min), followed by a plateau for 2 min then a decrease from 80 % B to 10 % B in one minute and a final plateau of 3 min at the initial conditions. The flow rate was set to 0.25 ml/min and the volume of injection to 50 μL [8].

2. 4. Adsorption Measurements

All the kinetics studies were conducted in a phosphate buffer at pH 7.5. In the case of binary mixtures, a total concentration of 10^{-4} mol/L was set (5.10^{-5} mol/L per each pollutant). For the complex mixture, a total concentration of 10^{-4}mol/L was set ($1.43.10^{-5}$ mol/L per each pollutant). The kinetics data were fitted with the pseudo second order model.

The adsorption isotherms were studied at 13, 25 and 40°C for DFN, CAF, CBZ, BPA and OFX. Experimentally, stoppered vials containing disks of carbon adsorbent (12 mg) were placed in 50 mL pollutant solution at various concentrations (from about 10^{-5} to 10^{-3} mol/L) and stirred until equilibrium was reached. The equilibrium time was strongly dependent of the pollutants nature. For some molecules, seven days were sufficient, except for OFX (21 days), DFN (14 days) and PCP (10 days). The solutions prepared at a given concentration were exactly the same for each of the three temperatures in order to obtain precise and comparable results. The solutions were filtered on 0.45 μm filter membranes prior to HPLC analysis.

2. 5. Regeneration Under Polarization

Regeneration was performed using a classical three electrode system connected to a galvanostat/potentiostat (VMP-1, Biologic) in a 0.01 mol.L^{-1} Na$_2$SO$_4$ electrolyte (pH = 5.9 and σ = 2.5 mS.cm^{-1}). After adsorption in open circuit voltage (OCV), the loaded ACC disk was washed with ultra-pure water (σ_{water} = 0,055 μS.cm^{-1}) and attached to a current collector for the electro-desorption experiments. A counter electrode and Hg/Hg$_2$SO$_4$ (E = 0.649 V vs NHE) as reference electrode were used. Negative currents (-10 mA) were applied to the activated carbon cloth electrode [7, 9].

3. Results and Discussion

3. 1. Adsorption Isotherms

Isotherms were studied at 298 K (not shown) in order to calculate the Gibbs free energy variation ΔG^0_{298K} at a maximum uptake for the adsorption of five molecules on the carbon fabric (Table 3). All the isotherms except that of OFX were simulated using the Langmuir-Freundlich equation Eq. 1 as:

$$Q_e = Q_{max} \times (K_{lf}\, C_e)^n / (1 + (K_{lf}\, C_e)^n) \tag{1}$$

where Q_e is the adsorption uptake at equilibrium (mmol/g), C_e is the concentration at equilibrium (mmol/L), k_{lf} is the Langmuir-Freundlich constant (L/mmol), Q_{max} is the maximum uptake (mmol/g) and n is the Langmuir-Freundlich exponent.

The OFX and BPA isotherms were better simulated by a Langmuir model for which the n exponent values were equal to one in the previous equation (Table 2).

The Gibbs free energy variations ΔG^0_{298K} were determined from Eq. 2 as:

$$\Delta G^0_T = -RT \ln(K) \tag{2}$$

where R is the ideal gas constant, T the temperature (K) and K the equilibrium constant (estimated from the relation $K = C_{ads}/C_e$, where C_{ads} is the concentration adsorbed at equilibrium in mmol/L).

In order to compare the ΔG^0_{298K} values of the different adsorbates, they were calculated at the Q_e uptake value equal to 80% of Q_{max} for each isotherm. Negative and small ΔG^0 showed a spontaneous and physical adsorption process. ΔG^0 values are discussed in section 3.2 where it is shown that the thermodynamic process is the dominant process for long adsorption time kinetics.

Table 3. Gibbs free energy for different micropollutants at a maximum uptake and Langmuir-Freundlich parameters of the isotherms at 298 K.

	OFX	DFN	BPA	CBZ	CAF
ΔG^0 (kJ/mol)	-17.5	-14.6	-13.5	-10.2	-0.7
Q_{max} (mmol/g)	0.7	1.35	1.72	1.59	1.9
Q_{max} (mg/g)	253	400	393	376	369
K_{lf} (L/mmol)	14930	7051	9612	2828	820
n	1	1.12	1	1.15	0.85

3. 2. Adsorption Kinetics

In order to understand the adsorption mechanisms and to determine the key parameters governing adsorption, kinetics studies in binary or complex mixtures were performed. In some cases, competition effects were clearly visible as for example between DFN and CAF and between DFN and POL. Two regimes were detected: during the first five days the speciation and hydrodynamic volume of the adsorbates controlled the adsorption kinetics, with anionic adsorbates being adsorbed more slowly because of electrostatic repulsions. In a second step for longer times, polarity and therefore water solubility governed the adsorption uptake, with the less soluble and most hydrophobic DFN becoming easier to adsorb onto the hydrophobic carbon adsorbent surface (Figure 3 and Figure 4).

When approaching equilibrium, the adsorption of DFN occurred to the detriment of CAF or POL molecules, thus demonstrating that these molecules were in competition for some adsorption sites and that when DFN entered the pores, POL or CAF were removed from the pores. The molecules of CAF with a weak Gibbs energy of adsorption (ΔG^0_{abs}) were desorbed from some of their adsorption sites and replaced by the competing contaminants (DFN) interacting at lower energy with the ACC.

The adsorption behavior of DFN in the different binary systems showed that its adsorption capacity and adsorption kinetic were affected to a different extent. Firstly, one must consider the large size of the DFN molecule, in the range of 580 Å³, as compared to the smaller molecules such as POL, PCP and CAF. More precisely, adsorbed DFN uptake was lowered especially when DFN was co-adsorbed with large species such as OFX and BPA (Figure 5 and Table 4).

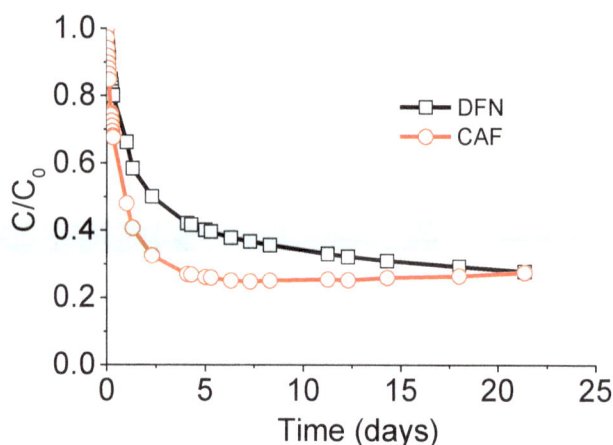

Figure 3. Adsorption of DFN and CAF (kinetics data).

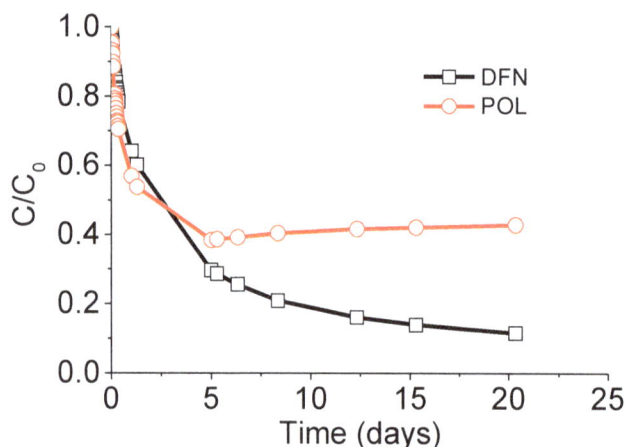

Figure 4. Adsorption of DFN and POL (kinetics data).

For co-adsorbates having the smallest hydrodynamic volumes, such as POL or CAF, the amounts of DFN adsorbed were less affected and high adsorption uptakes were measured.

These results showed the major role of adsorbate size during adsorption processes. For two adsorbates having different volumes, each of them could easily find an adsorption site in the porosity. For molecules that

are large compared to the small size of the pores, it became more complicated, and adsorbates were in competition to reach the bigger micropores (Figure 6).

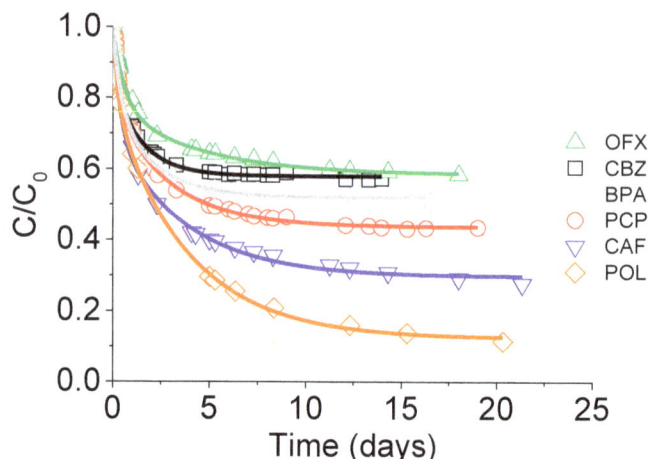

Figure 5. Adsorption kinetics of DFN in binary systems depending on the co-adsorbate.

Table 4. DFN adsorption characteristics.

Co-adsorbate	Q_{ads} DFN mmol/g	Q_{ads} Co-adsorbate mmol/g
POL	1.03	0.70
PCP	0.69	1.07
CAF	0.82	0.90
CBZ	0.51	0.82
BPA	0.60	0.96
OFX	0.49	0.27

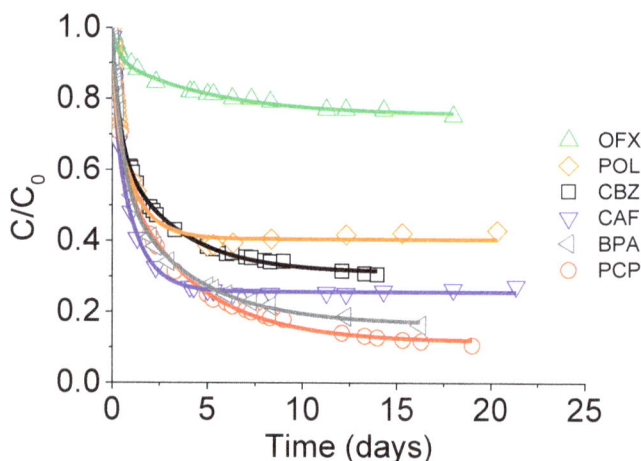

Figure 6. Adsorption kinetics of the co-adsorbate in binary systems with DFN.

When the adsorption kinetics were studied in a complex mixture containing the seven adsorbates, the same tendency was observed. The adsorbate size appeared to be the main parameter controlling the adsorption kinetics. The highest adsorption capacities were observed for the smallest organic molecules such as PCP and POL, whereas OFX and DFN which are large anionic molecules showed the slowest adsorption and the smallest uptake (Figure 7). The adsorption capacities could thus be directly correlated to the volume of the adsorbates.

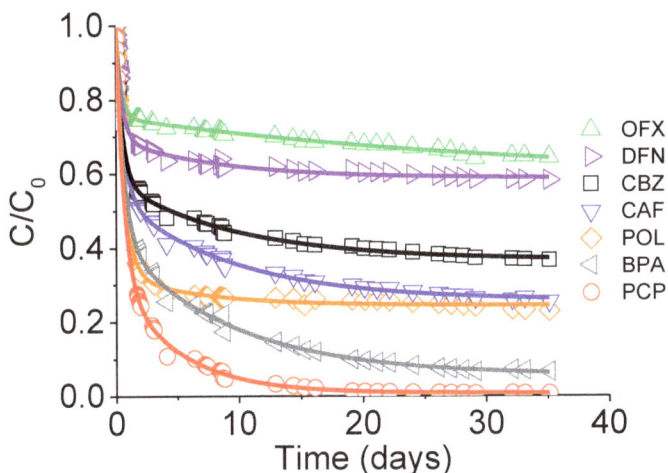

Figure 7. Adsorption kinetics of each pollutant in the complex mixture.

In order to complete the work, additional binary systems were studied with OFX in binary mixture and five co-adsorbates (PCP, BPA, DFN, CAF and CBZ). Like DFN, OFX was poorly adsorbed in the presence of PCP and BPA in the mixture (Table 5).

Table 5. OFX adsorption characteristics.

Co-adsorbate	Q_{ads} OFX mmol/g	Q_{ads} Co-adsorbate mmol/g
DFN	0.11	0.40
BPA	0.11	0.75
CBZ	0.26	1.02
CAF	0.41	0.95
PCP	0.09	0.94

OFX adsorption uptake at equilibrium was higher in the mixture with small molecules such as CAF and CBZ. Moreover, CAF single adsorption showed greater ΔG^0 than other molecules because of their small affinity with the carbon. The size effect was confirmed to be a key parameter driving the adsorption process in a binary system. Large molecules such as DFN and BPA diffused slowly into the pores because of steric hindrance whereas small molecules such as PCP and CAF diffused more rapidly and deeply inside the pores (Figure 8).

3. 3. Regeneration Under Polarization

After loading activated carbon cloths with organic contaminants, an attempt at regeneration was conducted under cathodic polarization of the carbon cloth. Depending on the nature on the adsorbate, it was possible to perform the reversible desorption of the adsorbed species. Desorption was assumed to be performed through electrostatic repulsions occurring between the negatively charged carbon surface and the dissociated organic molecules, reinforced by the presence of the electrical field. High regeneration efficiency of about 60 % was observed in the case of POL whereas for BPA (8.1 %) or DFN (11.5 %) desorption was more difficult (Figure 9).

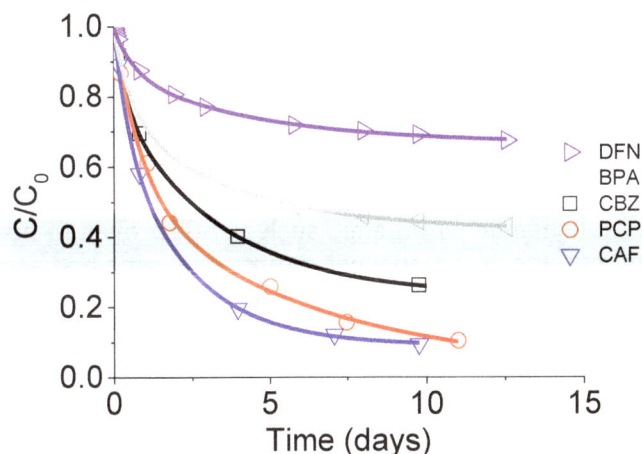

Figure 8. Adsorption kinetics of the co-adsorbates in binary systems with OFX.

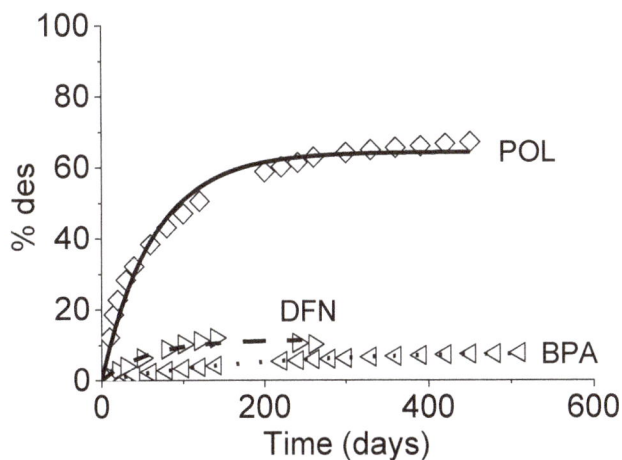

Figure 9. Regeneration percentages (%) as a function of polarization time for POL, BPA and DFN.

It was assumed that for small adsorbates as POL reversible desorption became more easy, especially because of a better diffusion of the molecule inside the narrow pores (< 0.7nm). The incomplete regeneration was explained by either some steric blockage, especially for large molecules (DFN and BPA), inside the narrow pores or the trapping of the molecule in high energetics sites.

4. Conclusion

Activated carbon cloths are one of the most prevalent adsorbents able to trap emerging water contaminants. High adsorption capacities ranging from 250 to 400 mg/g were reached. The pollutants are mainly adsorbed in the narrow porosity by π-π interactions. In two or seven-components systems, the adsorption kinetics of pollutants (CBZ, OFX, CAF, BPA, DFN, PCP and POL) studied at 10^{-4} mol.L^{-1} on a microporous activated carbon cloth was found to be related to the molecular volume. In contrast, the solubility and the pollutant polarity were involved to a lesser extent. Small neutral molecules indeed showed the best adsorption capacities. Knowledge of the thermodynamic parameters such as ΔG^0 proved to be a useful tool to assess to what extent a molecule had been well adsorbed in a binary mixture at equilibrium. Furthermore, competition effects were highlighted. Molecules showing less negative Gibbs adsorption energy variation (ΔG°_{ads}) were desorbed from some of their adsorption sites and replaced by the competing contaminants adsorbed with lower Gibbs energy variation. The adsorbent was partially regenerated through cathodic polarization. Steric blockages by large molecule sometimes take place and make the regeneration of the material more difficult to achieve.

Acknowledgements

The authors thank the French ANR for financial support.

References

[1] M. Huerta-Fontela, M. T. Galceran, J. Martin-Alonso and F. Ventura, "Occurrence of psychoactive stimulatory drugs in wastewaters in north-eastern Spain," *Science of the Total Environment*, vol. 397, pp. 31-40, 2008.

[2] C. Stavrakakis, R. Colin, C. Faur, V. Héquet and P. Le Cloirec, "Analysis and behaviour of endocrine disrupting compounds in wastewater treatment plant," *European Journal of Water Quality*, vol. 39, no. 2, pp. 145-156, 2008.

[3] S. Snyder, C. Lue-Hing, J. Cotruvo, J. E. Drewes, A. Eaton, R. C. Pleus and D. Schlenk, *Pharmaceuticals in the Water Environment,* Report from NACWA and Association of Metropolitan Water Agencies: 2010.

[4] H. Guedidi, L. Reinert, J. M. Lévêque, Y. Soneda, N. Bellakhal and L. Duclaux, "Adsorption of ibuprofen from aqueous solution on chemically surface-modified activated carbon cloths," *Carbon*, vol. 54, pp. 432-443, 2013.

[5] C. O. Ania and F. Béguin, "Mechanism of adsorption and electrosorption of bentazone on activated carbon cloth in aqueous solutions," *Water Research*, vol. 41, pp. 3372-3380, 2007.

[6] S. Delpeux-Ouldriane, N. Cohaut and F. Béguin, "Electrochemical Removal of Ionogenic Pesticides Adsorbed on Activated Carbon Textiles," in *Proceedings of the Annual World Conference on Carbon*, Biarritz, France, 2009. vol. 3, pp. 1777-1784.

[7] S. Delpeux-Ouldriane and F. Béguin, "On Carbon Capture Reversible Assets," French Patent 1257895, August 20, 2012.

[8] M. Gineys, T. Kirner, N. Cohaut, F. Béguin and S. Delpeux-Ouldriane, "Simultaneous determination of pharmaceutical and pesticides compounds by reversed phase high pressure liquid chromatography," *Journal of Chromatography and Separation Techniques,* vol. 6, no. 6, 2015.

[9] S. Delpeux-Ouldriane, M. Gineys, N. Cohaut and F. Béguin, "The role played by local pH and pore size distribution in the electrochemical regeneration of carbon fabrics loaded with bentazon," *Carbon,* vol. 94, pp. 816-825, 2015.

2

Effects of Operation Parameters on Heavy Metallic Ion Removal from Mine Waste by Natural Zeolite

Amanda L. Ciosek, Grace K. Luk
Department of Civil Engineering, Faculty of Engineering and Architectural Science, Ryerson University
350 Victoria Street, Toronto, Ontario, Canada M5B2K3
amanda.alaica@ryerson.ca; gluk@ryerson.ca

Abstract – *This study investigates the effects of particle size (0.420-1.1410 mm), dosage (40, 80 g/L), influent concentration (total 10 meq/L, 400 mg/L), contact time (5-180, 270, 360 min), set-temperature (20-32°C), and heat pre-treatment (200, 400, 600 °C) of natural zeolite on the removal efficiency of heavy metallic ions (HMIs); lead (Pb^{2+}), copper (Cu^{2+}), iron (Fe^{3+}), nickel (Ni^{2+}), and zinc (Zn^{2+}). The sorption process is performed in batch mode with a 100 mL aqueous solution, acidified to a pH level of 2 with concentrated nitric (HNO_3) acid. For all experimental parameter conditions examined, the removal efficiency order follows: $Pb^{2+}>>Fe^{3+}>Cu^{2+}>Zn^{2+}>Ni^{2+}$; the zeolite mineral exhibits the greatest preference towards the Pb^{2+} ion in all parameter trends. Overall, the removal efficiency is increased with decreasing particle size, as well as increasing dosage, contact time, and set-temperature. The operation is influenced by the studied parameters in the order of: influent concentration > heat pre-treatment level > dosage > particle size > contact time > set-temperature.*

Keywords: Natural Zeolite, Heavy Metallic Ions, Sorption Capacity, Removal Efficiency.

1. Introduction

Waterways are prone to acid mine drainage (AMD) contamination caused by the discharge of mineral mining and processing effluent [1,2]. Characterized by low pH levels and the presence of heavy metallic ions (HMIs) and other toxic elements [2], AMD significantly threatens our health and environment, causing various diseases and disorders [3-5]. The process of sorption has attained the interest of the mining industry as an industrial wastewater treatment method [6,7]. The uptake of HMIs is attributed to both adsorption (on the surface of the sorbents' micropores) and ion-exchange (through the sorbents' framework pores and channels) mechanisms [8]; referred to as sorption as a unified treatment process [9,10].

Natural zeolites have progressed among researchers' interests [6,11]. The mineral's structure is comprised of three independent components [11-13]: (1) hydro-aluminosilicate crystalline structure of SiO_4 and AlO_4 tetrahydras linked by oxygen atoms, (2) interconnected void spaces in a framework containing exchangeable cations, and (3) zeolitic water present at 10-20% of the dehydrated phase of the natural zeolites' structure. Its open, homogenous microporous negatively charged three-dimensional framework of voids and channels [11] enables the exchange with cations present when in solution [1,14]. Clinoptilolite, a globally abundant and well-documented form of zeolite [13], is used in this research. One of the most significant properties of zeolite is its high cation exchange capacity (CEC), and it is considered as a strong candidate for the removal of wastewater contaminants [15].

The industry holds great interest in the physico-chemical influential factors that dictate sorption efficiency of zeolite; which include particle size, initial concentration, pH level, and contact time. A smaller *particle size* of the sorbent material provides greater contact surface area, which improves the performance of the sorption process [5,16]; which may be attributed to diffusion as the rate-limiting step of the overall ion-exchange mechanism in the sorption process [16]. The effect of the *dosage* (solid-mass-to-solution-volume) on the uptake of HMIs is well-established. An increase in dosage translates to an increase in the rate of uptake; although the amount sorbed per unit mass decreases, there is a higher availability of active sorption sites which sorb more HMIs from the solution [1].

The *initial concentration* of the ions influences the removal efficiency due to the availability of functional groups on the specific surface to bind with the HMIs. This is primarily the case at higher concentrations, demonstrating a higher overall uptake given that the concentration difference is the driving force to overcome mass transfer resistance to metal ion transport between the solution and the sorbent surface [5]. The *pH level* influences the dissociation of the sorbent and solution chemistry, and affects the surface charge of the sorbents and degree of ionization of different pollutants [5]. This influence of acidity is particularly the case for HMIs that are in a rather low preference by zeolite; the initial pH must be attentively selected to ensure a balance among all ionic species. The goal is to avoid precipitation; for once precipitated, the ions of interest cannot be sorbed [16].

The state of equilibrium is altered throughout the sorption process. Room temperature is preferred for analysis, although higher thermal treatment temperatures are assumed to enhance sorption capacity with increased surface activities and solute kinetic energy [1,5], by removing the 'zeolitic water' present in the framework [1]; however, the dehydration of zeolite is an endothermic process, thereby causing 'activation' of the material [12] to a certain threshold, after which may lead to the structural collapse of the mineral [1].

The *contact time* is an important factor in the relationship of pollutants and sorbents. The rapid uptake of pollutants and equilibrium is established in a specific and limited period, which demonstrates efficiency of the sorbent for treatment. The mechanism study conducted by Sprynskyy et al. [4] states that the sorption of HMIs by natural zeolite is a heterogeneous process with three distinct stages: (1) rapid uptake within the first 30 min of contact, (2) inversion due to desorption prevalence, and (3) slower increase in uptake. In the kinetic studies conducted by Motsi et al. [1], the initial stage of rapid adsorption occurs within the first 40 contact min; when all of the adsorption sites are available for cations to interact, and when the concentration difference between the influent stock and sorbent–solution interface is very high. Inglezakis et al. [14] tributes this period to ion-exchange in the micropores on the zeolite particles' surface. The predominance of desorption is most likely caused by

slower diffusion of exchangeable cations within the internal zeolite crystalline structure, and consequently these preferred ions occupy the available exchange positions on the zeolite surface. During the third stage, the gradual deceleration of sorption in the micropores is caused by poor access as well as by more intensive sorption in comparison to the particles' surface. All of these factors are significant towards establishing the performance of any sorbent material [5].

The composition of AMD is uniquely complex and contains numerous contaminants, which include heavy metals and other pollutants, and the presence of these in solution affect the overall removal potential [2,9]. The existence of HMIs in AMD is mine-specific, and the concentrations fluctuate extensively [17]. This is evident in the vast variations of expected HMI levels, such as: copper, iron, zinc, aluminum, and manganese at 0.17, 0.82, 101.2, 22.6, and 10.7 mg/L, respectively [18]; copper, iron, zinc, aluminum, manganese, arsenic, and cadmium at 12, 200, 85, 15, 15, 9 and 1 mg/L, respectively [1]; or lead, copper, iron, zinc, nickel, aluminum, manganese, arsenic and cobalt at 0.045, 5.4, 4.9, 11.5, 0.145, 32.8, 8.1, 0.004 and 0.269 mg/L, respectively [19]. In addition to HMIs, other AMD constituents, such as the variations in minerals, micro-organisms, and (weather and seasonal) temperatures, all influence the quality and quantity of AMD [19]. A majority of previous research on sorption capacity of zeolite has investigated synthetic simple solute solutions spiked in single-component systems [20], and have demonstrated greater removal performance compared to investigating actual AMD [1,20]. However, there is still limited knowledge of the sorption capacity by zeolite for heavy metals and the associated mechanisms when in various multi-component systems [14,21]. The synthetic simple heavy metal solution permits the analysis of the effects of the selected operation parameters in a controlled environment for improved quantification, and identification of the important trends in this study.

The authors have designed a four-phase research project, which investigates: (1) the effects of preliminary parameters and operative conditions (particle size, sorbent-to-sorbate dosage, influent concentration, contact time, set-temperature, and heat pre-treatment), (2) HMIs component system combinations and selectivity order with a focus on its effects on the removal of lead (Pb^{2+}) [22], (3) kinetic modelling trends [23], and (4) the design of a packed, fixed-bed, dual-column sorption treatment system [24].

The study presented in this paper refers to the first phase. In feasible treatments of industrial waste, it is essential to classify the degree of influence of each operational parameter on the overall system performance [17]. Therefore, the objective of this present study is to assess the sorption capacity of natural zeolite for the removal of five fundamental HMIs, specifically lead (Pb^{2+}), copper (Cu^{2+}), iron (Fe^{3+}), nickel (Ni^{2+}), and zinc (Zn^{2+}) [18,25], combined in various component systems. The operative conditions of zeolite particle size and dosage, HMI influent concentration, contact time, set-temperature and heat pre-treatment level are all investigated. This is of great importance, in order to harness the full potential of zeolite in tertiary treatment processes.

2. Methodology

2. 1. Materials and Equipment

2. 1. 1. Heavy Metallic Ion Influent Concentration

The removal efficiency order indicates the variation of the selectivity for each HMI [16]. Overall, this selectivity or preference of zeolite for one cation compared to another [26] is stronger for the counter-ion of higher valence, increasing with dilution of solution and strongest with ion-exchange of high internal molality [16]. Therefore, comparative analysis of various HMIs should be conducted at the same normality and temperature [16]; as executed in this study. The synthetic ion solutions are prepared from analytical grade nitrate salts of $Pb(NO_3)_2$ (CAS No. 10099-74-8), $Cu(NO_3)_2 \cdot 3H_2O$ (CAS No. 10031-43-3), $Fe(NO_3)_3 \cdot 9H_2O$ (CAS No. 7782-61-8), $Ni(NO_3)_2 \cdot 6H_2O$ (CAS No. 13478-00-7), and $Zn(NO_3)_2 \cdot 6H_2O$ (CAS No. 10196-18-6), respectively; dissolved in deionized distilled water. The metals are combined to maintain a total normality of 0.01N (10 meq/L) [14,16] in the following systems:

- single-component system–10 meq/L per metal (lead [Pb], copper [Cu], iron [Fe], nickel [Ni], zinc [Zn]),;
- dual-component system–5.0 meq/L per metal (lead-copper [Pb-Cu], lead-iron [Pb-Fe]),;
- triple-component system–3.3 meq/L per metal [T] (lead, copper and iron), and;
- multi-component system–2.0 meq/L per metal [M] (all five metals).

The corresponding HMI concentrations are approximately 1036 mg/L for Pb^{2+}, 318 mg/L for Cu^{2+}, 186 mg/L for Fe^{3+}, 293 mg/L for Ni^{2+}, and 327 mg/L for Zn^{2+}. In addition to maintaining a total 10 meq/L concentration, the study is also conducted at 400 mg/L initial concentration for each HMI, based on the median range of conversion from meq/L to mg/L concentrations for a majority of the HMI investigated.

The Canada-Wide Survey of Acid Mine Drainage [17] reports a seasonal average of a majority of the mines surveyed to have documented pH values ranging from 2 to 5. Consequently, the influent stock is acidified with concentrated nitric acid (HNO_3) (CAS No. 7697-37-2) to a pH level of less than 2 [27]. This study is conducted in the conservative manner, with a majority of the pH values documented to be below this reported average and within comparability. It is important to note that the mechanism of HMI uptake by natural zeolites is influenced by the pH level; shifting from ion exchange/adsorption in the acidic region to adsorption/complexation and possible precipitation in the basic region [28]. The neutralization increases the pH level to reach a threshold of solubility of the metal hydroxides; the removal may be partially attributed to precipitation (or sorption/co-precipitation) rather than just sorption [29]. With the use of highly soluble nitrate salts and by maintaining very low pH levels in the batch-mode configuration of all the experiments, the formation of inorganic ligands (such as OH^-) is prevented and the precipitation of the HMIs is avoided [6,16]; under the testing conditions of this study.

Based on the dosage of zeolite mass to a selected 100 mL volume of aqueous solution, the HMI uptake is calculated by the following relationship [16]:

$$q_t = \frac{V \times (C_O - C_t)}{M} \quad (1)$$

where q_t is the HMI adsorbed at time t (in meq/g or mq/g), C_O and C_t are the initial and final HMI concentrations in solution (in meq/L or mg/L) after time t, V is the solution volume (in L), and M is the zeolite mass (in g).

2. 1. 2. Natural Zeolite Mineral

This study employs a natural zeolite mineral sample composed primarily of 85-95% clinoptilolite (CAS No. 12173-10-3) and is sourced by a deposit located in Preston, Idaho [30]. This sample holds a cation exchange capacity of 180-220 meq/100 g, a pH level ranging from 7-8.64. It has a maximum water retention and an overall specific surface area of 55 wt% and 24.9 m^2/g, respectively. The zeolite mineral sample is applied in its natural state, without any chemical modifications, to minimize associated costs and environmental impacts of the process investigated in this study. The particle size of the raw mineral sample ranges from 1.41 mm (pass No. 14) to 0.420 mm (retain No. 40). This sample is divided into sizes A $(d_{p,A})$ (1.190-1.410mm), B $(d_{p,B})$ (0.707-0.841 mm), and C $(d_{p,C})$ (0.420-0.595 mm) with standard mesh sieves and a mechanical shaker (Model No. Humboldt H4330; CAT No. G118-H-4330). Size D $(d_{p,D})$ (0.841-1.19 mm) has also been selected, ranging between A (pass No. 16) and B (retain No. 20). This additional size range holds the greatest percent yield within the +14-40 source and also, being a broader, coarser size range, is of interest to this study. Overall, these four divisions are selected to provide a distinct variance, based on the approximate distribution of the +14-40 source, as displayed in **Table 1**. The particles are put through a cleaning cycle, which involves thoroughly rinsing in deionized distilled water to remove residual debris and dust, and drying at 80 $\pm$ 3°C for 24 hr (Isotemp® Oven Model 630G; Serial No. 30300047; CAT No. 13-246-630G; 115 V; 6.5 A; 60 Hz; Fisher Scientific, USA) to remove residual moisture [31].

Table 1. Preliminary Distribution of Zeolite Supply.

	Test Sample Size (g)	1006.60	
	Sieve Gradations	**Sample Distribution**	
		(g)	(%)
	#14 Retain	76.5	7.6
A	#14 Pass | #16 Retain	199.9	19.9
D	#16 Pass | #18 Retain	181.1	18.0
D	#18 Pass | #20 Retain	150.9	15.0
B	#20 Pass | #25 Retain	119.1	11.8
	#25 Pass | #30 Retain	94.0	9.3
C	#30 Pass | #35 Retain	68.1	6.8
	#35 Pass | #40 Retain	48.3	4.8
	#40 Pass (PAN)	57.9	5.8
	SUM	995.8	98.9
	LOST	10.8	1.1

2. 1. 3. Analytical Equipment

Quantitative observations are conducted by analyzing the HMI concentrations in their aqueous phase with Inductively Coupled Plasma – Atomic Emission Spectroscopy (ICP-AES) technology (Optima 7300 DV, Part No. N0770796, Serial No. 077C8071802, Firmware Version 1.0.1.0079, Perkin Elmer Inc.), with corresponding WinLab32 Software (Version 4.0.0.0305). This spectrometry technique is considered to have true multi-element performance with

exceptional sample throughput, and with a very wide range of analytical signal intensity [32]. The primary wavelengths of each HMI element analyte targeted are 327.393 (Cu), 238.204 (Fe), 231.604 (Ni), 220.353 (Pb), and 206.200 (Zn), respectively; selected on the basis that these wavelengths have the strongest emission and provide the best quantifiable detection limits. With the plasma setting in radial view (to concentrations of greater than 1 mg/L), auto sampling of 45 seconds normal time at a rate of 1.5 mL/min, and a processing setting of 3 to 5 points per peak with 2 point spectral corrections are applied.

The calibration curve is generated through 'linear calculated intercept' by applying a stock blank and a multi-element Quality Control Standard 4 with 1, 10, 50, 90, and 100 mg/L concentrations (as per Standard Methods Part 3000) [27]. Based on the corresponding influent concentrations in mg/L, the samples are diluted with deionized distilled water, by zero to four pre-determined 50% steps, in order to be within this calibration range. The sorbed amount of HMI is calculated based on the initial concentration and its 0.45 μm filtered supernatant concentration.

Triplicate readings and their mean concentrations in calibration units are generated in mg/L by the ICP-AES software. Three major check parameters are selected to evaluate the calibration quality during each ICP-AES analytical session. The triplicate concentration of the median 50-mg/L standard detects an average of 51.37 mg/L, which is within 5% of the known value. The percent relative standard deviation (%RSD) of this selected standard reports an average of 0.4944% (well within the ≤3% limit) and the correlation coefficient of each HMI analyte primary wavelength generates an average of 0.9997, which is very close to unity. These check parameters indicate that the calibration is of a reasonable level of accuracy and reliability [33], such that there is acceptable error associated with the experimental data.

2. 2. Experimental Design

All analyses are conducted in batch mode, combining the HMI solutions of various component systems to 100 mL (total 10 meq/L initial concentration) with 4 g of the zeolite sorbent mineral. The mixture is agitated on a bench-top orbital shaker with triple-eccentric drive (MaxQ™ 4450, CAT No. 11-675-202, ThermoFisher Scientific) at a controlled condition of 400 r/min set at 22°C, for a period ranging from 5 to 180 min. In all experiments, after reaction,

HMI sorbate and zeolite sorbent are separated through a 0.45 μm syringe filter. The hydrothermal pre-treatment is conducted by placing the cleaned zeolite into a pre-heated muffle furnace (NEY M-525 SII; Serial No. AKN 9403-108; 120 V; 50/60 Hz; 12.5 A; 1500 W; Barkmeyer Division, USA) at the three selected temperatures of 200∘C, 400∘C, and 600∘C [1], for 1-hr. Table 2 summarizes the parameters investigated to determine their influence of the overall removal of the selected HMIs.

Table 2. Operation Parameters and Conditions.

Parameter	Conditions
Particle Size	Single-Component Systems: [Pb], [Cu], [Fe], [Ni], [Zn] **A** *1.140-1.190mm (pass No. 14, retain No. 16)* **B** *0.707-0.841mm (pass No. 20, retain No. 25)* **C** *0.420-0.595mm (pass No. 30, retain No. 40)*
Dosage	Single-Component Systems: [Pb], [Cu], [Fe], [Ni], [Zn] Particle Size: **D** *0.841-1.19 mm* *Dosage: 4 g/100 mL, 8 g/100 mL*
Influent Concentration	Systems: [Pb], [T], [M] Particle Size: **D** *Concentrations: total 10 meq/L, 400 mg/L*
Contact Time	Systems: [Pb], [Pb-Cu], [Pb-Fe], [T], [M] Particle Size D: 0.841-1.19 mm *Contact Time: 180, 270, 360 min*
Set-Temperature	Systems: [T], [M] Particle Size: **D** Contact Time: 180 min *Set Temperature: 20∘C, 24∘C, 28∘C, 32∘C*
Heat Pre-Treatment	Systems: [Pb], [T], [M] Particle Size: **D** *Heat Pre-Treatment: 200∘C, 400∘C, 600∘C*

3. Results and Discussion
3. 1. Particle Size and Dosage

The particle size and dosage parameters are significant to this study, as well as to the industry that adopt sorption as a treatment method. Figure 1 displays the uptake of each HMI at 180 min of contact with zeolite. As expected, with a reduction in the particle size (d_p) from A to C, the uptake and percent removal increases. This trend is most prevalent for the HMI Pb^{2+}, with a 45.63% decrease in concentration or a 15.18% increase in uptake from $d_{p,A}$ (0.1872 meq/g) to $d_{p,B}$ (0.2157 meq/g). However, this trend is not as prevalent from $d_{p,B}$ to $d_{p,C}$, with only a 3.98% in improved HMI uptake. This may be due to the greater particle size gradation range between $d_{p,A}$ and $d_{p,B}$ specifically, as well as a 40.47% decrease in nominal geometric mean diameter of 1.30 mm $(d_{p,A})$ to 0.77 mm $(d_{p,B})$. Based on the sieve distribution presented in Table 1, an average

of 10% per mesh range was detected for particle sizes B and C. In order to eliminate skater/variability, and to maintain a controlled environment, the particle size selected to observe the other experimental parameters is between A and B, denoted hereon in as size D $(d_{p,D})$. Based on these initial observations in the removal trends by particle size, the $d_{p,D}$ is considered a more feasible and conservative range moving forward; with a nominal geometric mean diameter of 1.00 mm [34].

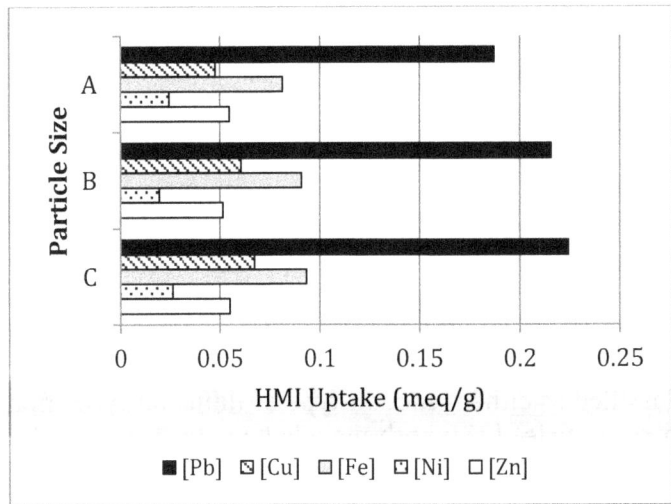

Figure 1. HMI Uptake based on Particle Size Parameter.

Figure 2 displays the overall percent removal of each HMI (in single-component solutions) at 180 min of contact with natural zeolite by increasing the zeolite sorbent dosage from 4 g to 8 g, for every 100 mL of HMI sorbate volume. As illustrated, when the dosage increases (doubled), the percent removal increases substantially; which is attributed to higher site uptake availability [35]. At a contact time of 180 min, the HMI effluent concentration is reduced for Cu^{2+} at 19.91%, Fe^{3+} at 35.93%, and significantly for Pb^{2+} at 82.37%. Additionally the overall removal efficiency of the selected $d_{p,D}$ falls within the range achieved of $d_{p,A}$ and $d_{p,B}$; demonstrating experimental continuity.

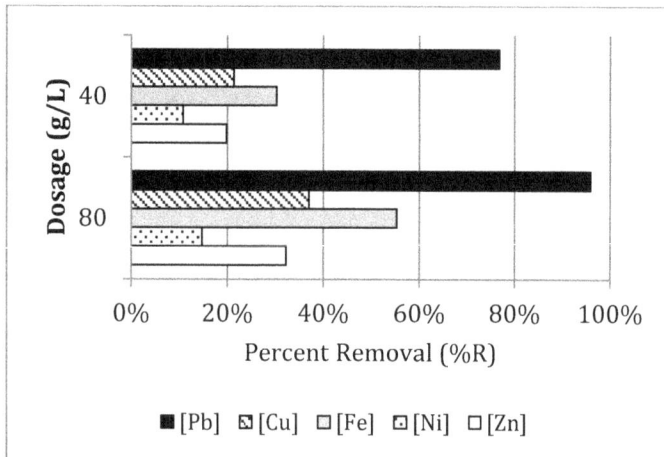

Figure 2. HMI Percent Removal based on Dosage Parameter.

Kinetic modelling is a powerful tool to assess the performance of sorbent materials and to comprehend the fundamental mechanisms involved in the sorption process. The sorption rate depends on the amount of ions on the sorbent surface at time t and what is sorbed when an equilibrium state is reached. The models are classified as either reaction-type or diffusion-type (film, intra-particle) [35]; both models have been thoroughly investigated and have demonstrated strong correlation [35-38].

The reaction-type known as the pseudo-second-order (PSO) kinetic model has well-demonstrated this rate process of various contaminants, including metal ions and organic substances in an aqueous state [36,39]. This model implies that the rate-limiting step is by chemical adsorption (chemisorption). It is represented in Eq. (2) and by applying the boundary conditions of $t = 0 \rightarrow q_t = 0$ and $t = t \rightarrow q_t = q_t$, its linearized form is presented in Eq. (3) [4,7,36,38]:

$$\frac{dq_t}{dt} = k_2(q_e - q_t)^2 \qquad (2)$$

$$\frac{t}{q_t} = \frac{t}{q_e} + \frac{1}{k_2 q_e^2} \qquad (3)$$

where $h = k_2 q_e^2$ is the initial sorption rate (in meq/g·min) as t approaches zero [38], and k_2 is the PSO rate constant (in g/meq·min). These constants are determined by a plot of the linearized form of t/q_t versus t [36,38]. The PSO rate constants and correlation coefficients are summarized in Table 3 and Table 4 for the particle size and dosage parameters, respectively. Based on the linearized form of Eq. (3), the slope (m) and y-intercept (b) values are interpreted to determine

the theoretical sorption at equilibrium (q_e in meq/g). The experimental sorption at 180 min (q_{180} in meq/g) of contact is also presented.

As demonstrated by the coefficients (CC), a strong correlation is established for all HMIs for both parameters. For all HMIs on average, the particle size q_{180} reaches the theoretical q_e uptake of 92.27% for $d_{p,A}$, 86.93% for $d_{p,B}$, and 93.45% for $d_{p,C}$; the dosage q_{180} reaches the theoretical q_e uptake on average of 84.27% for dosage 40 g/L and 80.72% for dosage 80 g/L. The particle size uptake rate trends in Table 3 are systematically consistent; with the $q_{e,[Pb]}$ within 5% of the theoretical maximum 0.25 meq/g threshold for total HMIs.

Table 3. PSO – Particle Size Data.

System	q_{180}	CC	m	b	q_e
Size A					
[Pb]	0.1872	0.9840	4.374	216.46	0.2286
[Cu]	0.0476	0.8193	12.780	1445.40	0.0782
[Fe]	0.0813	0.9741	11.002	372.35	0.0909
[Ni]	0.0245	0.9141	31.057	969.75	0.0322
[Zn]	0.0548	0.7413	9.569	1420.10	0.1045
Size B					
[Pb]	0.2157	0.9970	3.856	147.15	0.2594
[Cu]	0.0607	0.9866	13.611	626.99	0.0735
[Fe]	0.0908	0.9934	10.190	202.53	0.0981
[Ni]	0.0196	0.9872	47.121	298.41	0.0212
[Zn]	0.0514	0.9623	16.317	795.49	0.0613
Size C					
[Pb]	0.2242	0.9964	3.783	110.79	0.2644
[Cu]	0.0674	0.9700	13.942	373.99	0.0717
[Fe]	0.0933	0.9976	10.067	155.73	0.0993
[Ni]	0.0263	0.9942	37.924	274.89	0.0264
[Zn]	0.0550	0.9646	17.211	430.21	0.0581

Table 4. PSO – Dosage Data.

System	q_{180}	CC	m	b	q_e
Dosage 40					
[Pb]	0.1919	0.9926	4.098	217.01	0.2440
[Cu]	0.0533	0.9291	15.750	836.09	0.0635
[Fe]	0.0757	0.9708	11.872	419.08	0.0842
[Ni]	0.0268	0.9806	34.919	739.14	0.0286
[Zn]	0.0494	0.9147	15.237	1106.10	0.0656
Dosage 80					
[Pb]	0.1198	0.9986	7.343	174.00	0.1362
[Cu]	0.0463	0.9821	17.211	926.12	0.0581
[Fe]	0.0691	0.9899	12.893	344.90	0.0776
[Ni]	0.0184	0.9967	50.300	655.14	0.0199
[Zn]	0.0403	0.6369	13.507	1771.00	0.0740

The dosage level is not directly proportional to the sorption removal efficiency. The removal efficiency of Pb^{2+} improves from 76.82% to 95.91%; however, the q_{180} uptake has decreased from 0.1919 to 0.1198 meq/g, and the theoretically anticipated q_e uptake at equilibrium decreases from 0.2440 to 0.1362 meq/g, comparing dosage 40 g/L to 80 g/L, respectively. This may be attributed to the very rapid uptake of the first stage of sorption. The two HMIs preferred by zeolite in this study exhibit a faster initial sorption rate (h); for Pb^{2+} the rate increases from 0.0046 to 0.0057 meq/g·min, and for Fe^{3+} this rate increases from 0.0024 to 0.0029 meq/g·min; comparing dosage 40 g/L to 80 g/L, respectively. This finding in correlation with the lower overall expected uptake at equilibrium demonstrates that the Dosage 80 (8 g/100 mL) has reached its threshold of available active sorption sites. A higher removal at a faster rate comes at a cost of consuming more zeolite material; with the Dosage 40 (4 g/100 mL) considered more economically feasible.

In accordance with the fundamental principles of sorption (adsorption and ion-exchange), when intra-particle diffusion (IPD) as considered as the rate-limiting step, the sorption rate is proportional to $\overline{D}/d_p^2$; where $\overline{D}$ is the diffusion coefficient of a specific HMI. Since the d_p should not affect either the equilibrium state or the $\overline{D}$, higher sorption rates should be observed for smaller particle sizes. However, smaller particle sizes may exhibit lower rates, due to lower effective $\overline{D}$ values, caused by structural problems or pore clogging [21]. It is important to note that the natural (as-received) zeolite mineral sample is put through a systematic cleaning cycle, thoroughly washing before use. Therefore, pore clogging is not expected to affect the diffusion coefficients which are considered to be constant regardless of particle size. Then, with intra-particle diffusion considered as the controlling step, the exchange rate should be increased by decreasing particle size [21]; as demonstrated.

Based on the trends observed, the ideal levels of these two operation parameters (particle size $d_{p,D}$ and 4g/L dosage) are selected moving forward in this study.

3. 2. Influent Concentration

In addition to maintaining a total 10 meq/L initial concentration, this component of the study is also conducted at 400 mg/L for each HMI, based on the median range of conversion from meq/L to mg/L concentrations for a majority of the HMI investigated throughout this research endeavour; for single-lead [Pb], triple- [T], and multi- [M] component system combinations (Table 5).

The difference in the removal of each HMI investigated when the influent concentration is set to meq/L versus mg/L is evident. The trends detected are consistent with the literature; the amount in mg of Pb^{2+} ions available for uptake by zeolite decreases, theoretically from 1036 mg/L to 400 mg/L and the amount of the other four ions (in mg) has increased with this conversion of influent concentration. Oter and Akcay [35] demonstrated consistent findings, as the initial concentration increases, the amount of sorbed HMI increases, while the percent of sorbed HMI decreases for all ions.

Table 5. The HMI Removal Variation by Influent Concentration.

System	HMI	Total 10 meq/L			400 mg/L per HMI	
		q_{180}		%R	q_{180}	%R
		mg/g	meq/g		mg/g	
[Pb]	Pb^{2+}	23.31	**0.192**	77	9.36	95
[T]	Pb^{2+}	9.01	0.075	90	7.35	80
	Cu^{2+}	0.64	0.016	19	1.04	11
	Fe^{3+}	0.85	0.041	50	1.73	18
	TOTAL	–	**0.132**	–	–	–
[M]	Pb^{2+}	5.52	0.047	94	7.62	80
	Cu^{2+}	0.41	0.011	22	0.75	8.4
	Fe^{3+}	0.58	0.028	56	1.51	16
	Ni^{2+}	0.15	0.005	9.1	0.18	1.8
	Zn^{2+}	0.30	0.008	16	0.41	4.6
	TOTAL	–	**0.099**	–	–	–

Inglezakis et al. [16] demonstrates that dilution leads to an increase in the volume of treated solution to breakthrough (5-10% of the influent concentration) in continuous column configuration; the magnitude of which depends on the specific metal exchanged. This finding can be attributed to the increase of selectivity in the ion-exchange mechanism of sorption by dilution. The valences of the exchanging cations have a strong effect on ion-exchange at equilibrium, and consequently on the removal efficiency. This attribute is referred to as the "concentration-valency effect". It is theoretically recognized that when the exchanging ions are not of equal valence, the equilibrium is a function of the total concentration; at higher concentrations, this process prefers the uptake of the lower charged cations and subsequently excludes higher charged cations from the sorbent [16]. The cations present in the sorbent have valences that differ from those in solution.

Consequently, as the dilution increases, the selectivity of the sorbent for the ion with a higher valence also increases. Accordingly, comparative analysis of various metal ions should be conducted at the same normality and temperature, in order to minimize the changes observed in isotherm configuration with dilution [16].

3. 3. Contact Time and Set-Temperature

With [Pb], [T], and [M] component system combinations at total 10 meq/L influent concentration: (1) the contact time is extrapolated from 3 hrs to 4.5 and 6 hrs (Table 6), and (2) the set-temperature is evaluated to an adjusted range of 20 to 32°C at 180 contact min (Table 7).

When the uptake data (in meq/g) of Table 5 is compared to the contact time observations in Table 6 at 180 contact min, the removal efficiency is similarly on trend. Only a 5.94% average percent difference in the uptake of total HMIs of [Pb], [T], and [M] is detected. When this same comparison is made with 20°C uptake of [T] and [M] data of Table 7 (a temperature below the controlled 22°C), only a 4.85% average percent difference in the uptake of total HMIs is detected. These observations demonstrate continuity and repeatability of the experimental procedure.

To visualize the influence of both operating parameters, Figure 3 and Figure 4 display the total HMI uptake (meq/g) with respect to extrapolated contact time and set-temperature, respectively.

Table 6. The HMI Removal Variation by Contact Time at 22°C Set-Temperature.

System	HMI	Contact Time (mins)					
		180		270		360	
		q_{180}	%R	q_{270}	%R	q_{360}	%R
[Pb]	Pb^{2+}	**0.211**	84	**0.223**	90	**0.230**	92
[Pb-Cu]	Pb^{2+}	0.116	93	0.119	95	0.120	97
	Cu^{2+}	0.025	20	0.031	25	0.034	27
	TOTAL	**0.141**	–	**0.150**	–	**0.155**	–
[Pb-Fe]	Pb^{2+}	0.109	87	0.114	92	0.118	94
	Fe^{3+}	0.054	43	0.060	48	0.064	52
	TOTAL	**0.163**	–	**0.174**	–	**0.182**	–
[T]	Pb^{2+}	0.076	92	0.079	95	0.080	96
	Cu^{2+}	0.018	21	0.022	27	0.025	30
	Fe^{3+}	0.041	49	0.046	55	0.049	59
	TOTAL	**0.135**	–	**0.147**	–	**0.153**	–
[M]	Pb^{2+}	0.047	95	0.048	97	0.049	97
	Cu^{2+}	0.013	26	0.015	30	0.016	33
	Fe^{3+}	0.029	59	0.031	63	0.033	67
	Ni^{2+}	0.005	9.6	0.005	9.8	0.005	10
	Zn^{2+}	0.011	22	0.013	25	0.014	28
	TOTAL	**0.106**	–	**0.112**	–	**0.117**	–

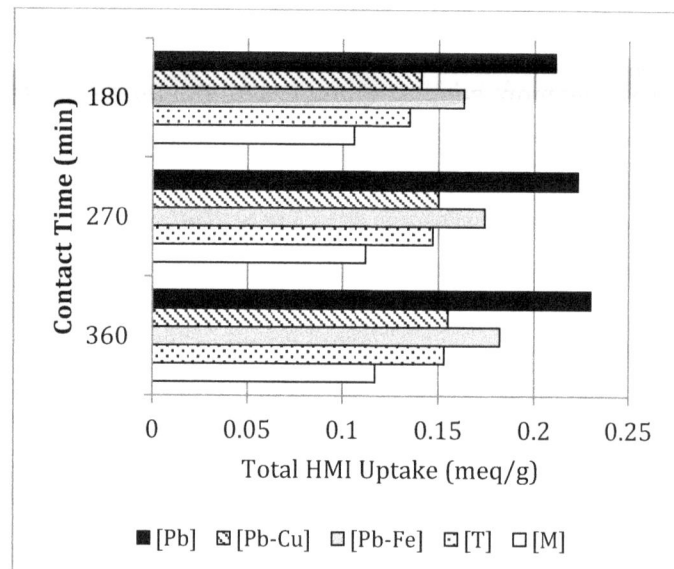

Figure 3. Total HMI Uptake based on Contact Time Parameter.

Table 7. The HMI Removal Variation by Set-Temperature at 180 Contact Min.

	HMI	Set-Temperature (°C)							
		20		24		28		32	
		q_{180}	%R	q_{180}	%R	q_{180}	%R	q_{180}	%R
[T]	Pb^{2+}	0.075	91	0.076	92	0.076	92	0.077	93
	Cu^{2+}	0.018	22	0.019	23	0.020	24	0.022	27
	Fe^{3+}	0.041	50	0.041	49	0.043	52	0.045	54
	TOTAL	**0.135**	–	**0.137**	–	**0.140**	–	**0.144**	–
[M]	Pb^{2+}	0.047	95	0.048	95	0.048	95	0.048	96
	Cu^{2+}	0.013	25	0.014	28	0.014	28	0.015	30
	Fe^{3+}	0.030	60	0.030	60	0.031	62	0.031	63
	Ni^{2+}	0.006	13	0.007	13	0.007	13	0.007	13
	Zn^{2+}	0.011	22	0.012	25	0.013	25	0.014	27
	TOTAL	**0.107**	–	**0.111**	–	**0.112**	–	**0.114**	–

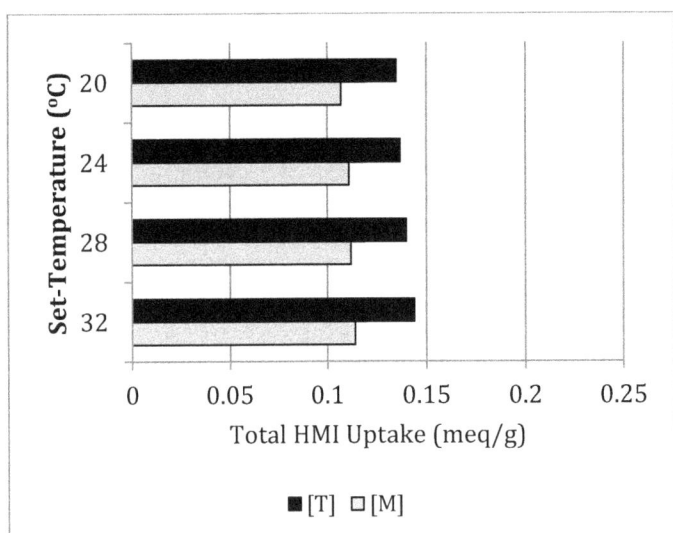

Figure 4. Total HMI Uptake based on Set-Temperature Parameter.

A greater increase in uptake of the total HMIs occurs from 180 to 270 min by, on average 0.010 meq/g compared to 0.006 meq/g from 270 to 360 min. This demonstrates the expected proportionality between uptake and contact time. For Pb^{2+}, the theoretical $q_{e,[Pb]}$ of 0.2440 meq/g by the PSO model (Table 4) is experimentally supported by the q_{360} of 0.230 meq/g. Research conducted by Oter and Akcay [35] demonstrates that equilibrium is attained for Cu^{2+} and Zn^{2+} in approximately 6 contact hr, while more rapidly for Pb^{2+} and Ni^{2+} at only 1 contact hr. As such, the uptake threshold quickly approaches equilibrium at 360 min of contact.

The results in Table 7 are on trend, with the direct proportionality between the systematic increase in set-temperature and uptake. This supports the fact that the sorbent's structure and surface functional groups are influenced by temperature between 20-35°C, observed by the overall sorption capacity [4]. However, the impact of set-temperature is not as significant within this selected range of study conditions.

3. 4. Heat Pre-Treatment

The hydrothermal stability of zeolites establishes the operational lifetime of a material, as well as degradation and regeneration conditions [39]. It is a measure of the structural changes that occur when exposed to water vapour at high temperatures and pressures [40]. This characteristic depends primarily on the type of zeolite, the silica/aluminum ratio, as well as the divalent/monovalent ratio and nature of cations

entering the framework [39,41]. The dehydration process is related to the considerable energy required to break bonds holding water molecules in the intra-crystalline channels of zeolite, as well as to overcome the energy barrier with diffusion of water molecules in the channels of the framework. The structural changes that occur are influenced by the degree of participation of water molecules in the energy balance of zeolite. Therefore, water molecules positioned in cavities and channels of the zeolite framework contribute to the compensation of the non-uniformly distributed charge of the silicate framework and cations. When water is separated from the crystalline lattice, the charge distribution breaks down. This leads to a deformation of the framework and variation in the mobile cations' positions [39].

Motsi et al. [1] investigates the uptake efficiency of natural zeolite for the heavy metallic ions (HMIs) Fe^{3+}, Cu^{2+}, Mn^{2+} and Zn^{2+}. The effects of heat pre-treatment are examined with the exposure to a muffle furnace at 200°C, 400°C, and 800°C for 30 mins. The pre-treated zeolite is then in contact with the HMIs in single-component solutions for 6 hrs. It is observed that the specific surface area is improved when treated at 200°C. An increase in both the adsorption rate and capacity due to this thermal treatment is caused by the removal of water from internal channels, which leave them vacant [1]. However, this trend is minimized beyond this temperature threshold. The structure collapses and the porosity inevitably decreases. The rate of adsorption by calcined zeolite is faster compared to untreated zeolite, but the efficiency decreases for zeolite exposed to very high temperatures.

The dehydration of zeolite occurs at a temperature that significantly exceeds the boiling point of water. A considerable amount of water is removed continuously and reversibly, both partially and completely [39,41], when exposed to heat from air at room temperature. When exposed to heat at approximately 350-400°C [12,41,43], the water is eliminated, and the cations fall back into positions on the inner surface of channels and central cavities of the zeolite structure. Dehydration of zeolite is an endothermic process, thereby causing 'activation' of the material [12] between 250-400°C [40] at approximately 350°C [41]; with a structural stability of up to 750°C [41,44]. Research has also revealed that the relationship between the dehydration mechanism of zeolite and positions occupied by aluminum and cations in its structure have an effect on the thermal stability.

Thermal treatment of zeolite between 500-600°C causes the loss of one H_2O for every two tetrahedral aluminum atoms. This temperature range instigates a loss of oxygen atoms in framework, producing structural vacancies [42]. Beyond this thermal threshold, the crystalline structure breaks down and the clinoptilolite becomes an amorphous solid [40,41].

Langelia et al. [42] investigates three thermal behavioural types of zeolites. This work emphasizes that reversible dehydration with minimal framework contraction would be observed upon heating up to approximately 230°C (Type-1) and 280°C (Type-2), while irreversible structural changes hinders rehydration at a range of 230-260°C (Type-1) and 280-400°C (Type-2). Also, heat pre-treatment greater than 450°C (Type-1) and 550°C (Type-2) causes a thermally induced collapse of the zeolite structure. Behavioural Type-3 exhibits continuous reversible dehydration with only very small structural contraction; the framework is not destroyed at an exposure of up to 750°C. High aluminum and alkaline-earth contents give rise to Type-1. An increase in silicon and/or alkaline-earth cations leads to a progressive change in thermal behavior in the order of Type-1 to -2 to -3. The study presented in this paper is also comparable with the findings of Langelia et al. [42], as the temperature levels analyzed exhibit Type-2 behaviour [45].

3. 4. 1. Heavy Metallic Ion Pre-Treatment Trends

Table 8 provides the HMI uptake at 180 min (q_{180} in meq/g) of contact for non-heated and heat-pre-treated zeolite in the triple-[T] and multi-[M] component systems. With each heat pre-treatment level, the same trend is maintained among the various component systems. Once again, for all operation parameters investigated in this study, the zeolite exhibits the highest affinity and favoured uptake for that of the Pb^{2+} ion [4,14,16,21,22,46] followed by Fe^{2+} and Cu^{2+}, with a lower affinity to Zn^{2+} then Ni^{2+}. A significant loss in crystallinity and hence catalytic activity are common with this pre-treatment process [47]. Dehydration temperature as well as micropore volume and transitional porosity development are directly proportional [40]. It is important to increase surface area, porosity and sorption capacities of natural zeolites without crystallinity loss [47]. The percent removal of the Pb^{2+} ion in [M] is 93.97%, while only 56.70% in [M-600]. The percent removal of the total HMIs reduces from 16.47% to 3.68% going from non-heat-pre-treated to 600°C exposure. This demonstrates

the extreme temperature effects on the zeolite's sorption capacity to the HMIs of interest.

Table 8. HMI Uptake by Heat Pre-Treatment Level.

Heat Level	Non-Heated		200°C		400°C		600°C	
System	[T]	[M]	[T]	[M]	[T]	[M]	[T]	[M]
HMI								
Pb^{2+}	0.075	0.047	0.073	0.047	0.057	0.040	0.038	0.02
Cu^{2+}	0.016	0.011	0.015	0.011	0.008	0.006	0.006	0.00
Fe^{3+}	0.041	0.028	0.036	0.025	0.024	0.017	0.026	0.01
Ni^{2+}	–	0.005	–	0.004	–	0.004	–	0.00
Zn^{2+}	–	0.008	–	0.008	–	0.004	–	0.00
TOTAL HMI	0.132	0.099	0.124	0.094	0.088	0.071	0.070	0.05

Figure 5 displays the effects of each heat pre-treatment level, with respect to the percentage of non-heated zeolite uptake. Evidently, the presence of each HMI in solution impacts the uptake of the other; as seen by the interference of the Ni^{2+} and Zn^{2+} ions in the [M] component system uptake of the Cu^{2+}, Fe^{3+}, and Pb^{2+} ions associated with the [T] system. Comparing [T] to [M], the uptake of the Pb^{2+} ion is reduced by 36.2%, 30.4% and 25.9% at the heat pre-treatment levels of 200°C, 400°C, and 600°C, respectively, and 37.5% without heat pre-treatment. When heat-pre-treated to 600°C, the total HMI uptake is reduced by approximately 47% in both systems.

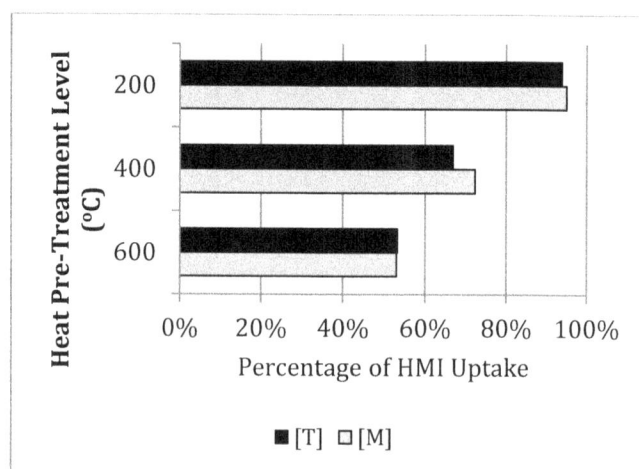

Figure 5. Percentage of Total HMI Uptake Compared to Non-Heated Level.

Figure 6 displays the total HMI uptake over the 3-hr contact period for each heat pre-treatment level, for the [T] (6a.) and [M] (6b.) component systems, respectively. This study is consistent with the three distinct stages discussed by Sprynskyy et al. [4]. As expected, there is a slightly greater uptake in the [T] over time; attributed to the interference of the additional two HMIs in the [M]. The rate of uptake for both component systems is not significantly affected by the 200°C heat exposure. The first 45-min period is very similar for [M-200], compared to the non-heated. Consistent with the findings of Motsi et al. [1], a substantial reduction of HMI uptake occurs at the 400°C threshold.

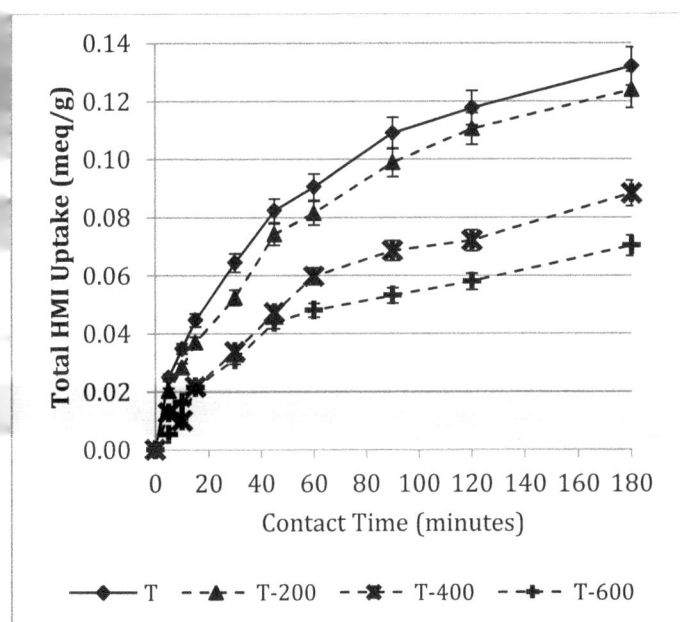

a. Triple-Component System

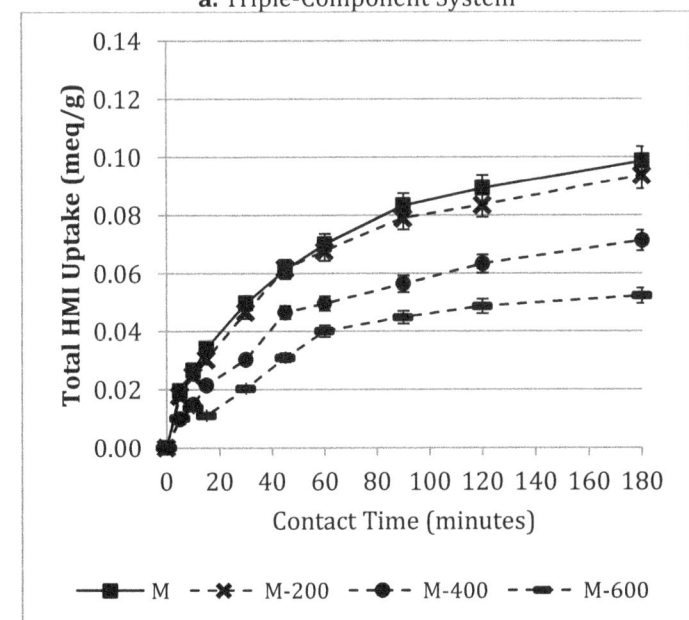

b. Multi-Component System
Figure 6. Heat Pre-Treatment Variation of Total HMI Uptake over Time (adapted from [45]).

3. 4. 2. Qualitative Pre-Treatment Trends

Qualitative analysis of the natural zeolite is conducted to observe the surface topography over time, by a high-resolution Scanning Electron Microscopy (SEM) technology (6380LV, JEOL, USA), equipped with Oxford energy dispersive X-ray spectroscopy (EDS) and electron backscatter diffraction (EBSD) capacity.

Once the cleaning cycle of the raw zeolite sample is complete, the as-received pale green colour is sustained. Following the progressive heat pre-treatment exposure, this colour transitions to a pink, pale pink, then light brown colour [45]. Figure 7 provides SEM images obtained by the high-resolution microscope, taken at ×5000 magnification (5 μm scale bar). Subtle physical changes of the surface structure are observed when comparing the raw granules (shown in 7a.) to those exhibited to the cleaning cycle (shown in 7b.). The images of the zeolite exposed to heat pre-treatment of 200°C, 400°C, and 600°C are shown in 7c., 7d., and 7e., respectively. To point out once again, a substantial reduction of HMI uptake occurs within the **400-600°C temperature range of this study [1]. The non-heated uptake in [T] and [M] is achieved by 93.9%** and 95.0% in [T-200] and [M-200], respectively. This is qualitatively observed in Figure 7b. and 7c., with the visual similarity.

As seen in Table 8, the sorption capacity is significantly compromised at the 400°C and 600°C exposures; which is supported by the lack of textural complexity in Figure 7d. and 7e., respectively. It is visually evident that the raw sample possesses textured granularity and significant detail, which is subsequently diminished with heat-pre-treatment towards the inter-granular spaces and mineral crevasses. This provides additional knowledge into how the structure of the zeolite mineral has been modified [45]. However, the process of dehydration requires a considerable amount of energy, which practically outweighs the interest to 'activate' the structure of the mineral sample. Both the quantitative and qualitative observations demonstrate that there is no economic benefit to the hydrothermal pre-treatment of the zeolite mineral, under the testing conditions.

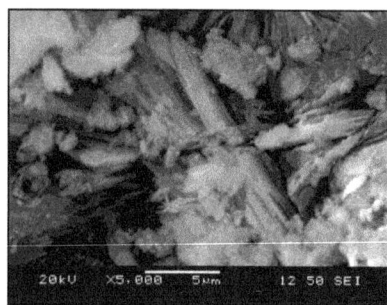

a. Natural Zeolite – As Received

b. Cleaning Cycle

c. Heat Pre-treated at 200°C

d. Heat Pre-treated at 400°C

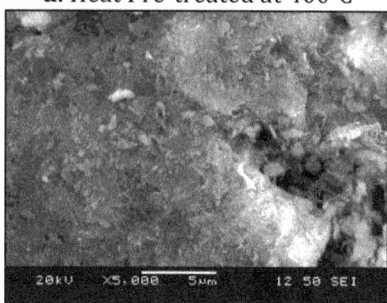

e. Heat Pre-treated at 600°C

Figure 7. SEM Images of Natural Zeolite Exposure to Heat-Pre-treatment (adapted from [45]).

Innovative treatment technologies are a challenge for all related industry, which include but are not limited to high associated economic costs and pollutant specific methods. Consequently, the conversion of inorganic ion exchangers into hybrid fibrous or nanoscale ion exchangers is considered to be the latest development of the water treatment industry. These materials are gaining attention, as they demonstrate a high efficiency and rate of sorption with short diffusion path towards environmental pollutants. Among metal-containing nanoparticles, carbonaceous materials and dendrimers, zeolites are considered as one of the most progressive functional and nano-sized materials of the millennium. The prospects of this mineral are promising, and its unique position is attributed to its sorption properties particularly through their surface treatment. Nanoscale science and engineering developments are providing extraordinary opportunities to develop more cost effective and environmentally acceptable water purification processes [48].

3. 5. Acidity Observations

It is important to note that the pH level of every ICP-AES sample (stock solution and sorbent-solution contact) is measured for all operating parameters investigated; utilizing the accumet Basic AB15 pH Meter (Fisher Scientific; CAT No. 13 636 AB15). This is conducted after batch mode contact, and before dilution (following 24 hr storage) in preparation for calibration. The set-up of the initial stock pH level is kept consistent throughout all experiments; an average of 1.85 is detected. The average maximum and minimum values between the batch bottle and stored supernatant samples are summarized in Table 9.

Table 9. The Average pH Level Data.

Observation	pH Sample	
	Batch Contact	Storage Filter
Maximum	2.18	2.14
Minimum	1.90	1.89

This data is based on a Dosage 40 ratio (4 g/100 mL); excluding the pH observations for the Dosage 80 (8 g/100 mL) parameter. At 45 min of contact for all HMIs at Dosage 80 conditions, the effluent becomes more basic to reach a pH level of just above 2.20. Overall, the samples collected show an average pH level maximum and minimum of 2.39 and 1.91, respectively. This level is still quite acidic, and is well within the typical 2 to 5 range as investigated in the Canada-Wide AMD Survey [17]. This brings attention to the fact that the H^+ ions are in competition with the HMI of interest [15,16]. The doubled dosage provides greater active sites availability for sorption to occur. The decision to proceed with particle size D ($d_{p,D}$) as the controlled parameter in the analysis is justified, given that the pH level was kept consistent to the completion of this study.

4. Conclusions

The removal efficiency of heavy metallic ions by natural zeolite increases by:

- Decreasing particle size and influent concentration;
- Increasing dosage level, contact time and set-temperature, and by;
- Maintaining the heat-pre-treatment within limits of the activation threshold.

The efficiency is improved depending on the specific metal and the parameter observed. Since the heavy metals selected in this present study possess different chemical and physical properties influenced in the same manner, it can be qualitatively speculated that other heavy metals would be equally influenced [46]. The removal efficiency order (or selectivity series) is consistent for all examined experimental conditions: $Pb^{2+} >> Fe^{3+} > Cu^{2+} > Zn^{2+} > Ni^{2+}$ [22,45].

The sorption uptake of HMIs by natural zeolite is complex, due to the aqueous chemistry of the elements and the nature of the sorbent mineral [35]. However, this research provides a greater insight into how the presence of multiple metallic ions and various operative parameters impact the overall removal efficiency, and indicates how the sorption properties of zeolite influence the overall selectivity trends. This is a significant contribution to the current knowledge-base, and how these parameters impact the natural mineral batch mode, for the scale-up to continuous column design and configuration for industrial treatment applications.

References

[1] T. Motsi, N. A. Rowson, M. J. H. Simmons, "Adsorption of heavy metals from acid mine drainage by natural zeolite," *International Journal of Mineral Processing*, vol. 92, no. 1-2, pp. 42-48, 2009. DOI: 10.1016/j.minpro.2009.02.005.

[2] A. Akcil, S. Koldas, "Acid Mine Drainage (AMD): Causes, treatment and case studies," *Journal of Cleaner Production*, vol. 14, pp. 1139-1145, 2006. DOI: 10.1016/j.jclepro.2004.09.006.

[3] E. Erdem, N. Karapinar, R. Donat, "The removal of heavy metal cations by natural zeolites," *Journal of Colloid and Interface Science*, vol. 280. pp. 309-314, 2004. DOI: 10.1016/j.jcis.2004.08.028.

[4] M. Sprynskyy, B. Buszewski, A. P. Terzyk, J. Namiesnik, "Study of the selection mechanism of heavy metal (Pb^{2+}, Cu^{2+}, Ni^{2+}, and Cd^{2+}) adsorption on clinoptilolite," *Journal of Colloid Interface Science*, vol. 304, pp. 21-28, 2006. DOI: 10.1016/j.jcis.2006.07.068.

[5] M. A. Acheampong, R. J. W. Meulepas, P. N. L. Lens, "Removal of heavy metals and cyanide from gold mine wastewater," *Journal of Chemical Technology and Biotechnology*, vol. 85, pp. 590-613, 2010. DOI: 10.1002/jctb.2358.

[6] C. Wang, J. Li, X. Sun, L. Wang, X. Sun, "Evaluation of zeolites synthesized from fly ash as potential adsorbents for wastewater containing heavy metals," *Journal of Environmental Sciences*, vol. 21, pp. 127-136, 2009. DOI: 10.1016/S1001-074260022-X.

[7] T. Motsi, N. A. Rowson, M. J. H. Simmons, "Kinetic studies of the removal of heavy metals from acid mine drainage by natural zeolite," *International Journal of Mineral Processing*, vol. 101, pp. 42-49, 2011. DOI: 10.1016/j.minpro.2011.07.004.

[8] L. Curkovic, S. Cerjan-Stefanovic, T. Filipan, "Metal ion exchange by natural and modified zeolites," *Water Research*, vol. 31, no. 6, pp. 1379-1382, 1997. DOI: 10.1016/S0043-135400411-3.

[9] F. Helfferich, *Ion Exchange. Series in Advanced Chemistry*. McGraw-Hill, New York, USA, 1962, pp. 95-322.

[10] V. J. Inglezakis, S. G. Poulopoulos, "Adsorption and Ion-Exchange (Kinetics)," *In Adsorption, Ion Exchange and Catalysis, Design of Operations and Environmental Applications*, 1st ed., Elsevier Science: Amsterdam, The Netherlands, 2006, pp. 262-266. ISBN-13: 978-0-444-52783-7.

[11] G. V. Tsitsishvili, "Perspectives of Natural Zeolite Applications," *Occurrences, Properties and Utilization of Natural Zeolites*, Akademiai Kiado: Budapest, Hungary, 1988, pp. 367-393.

[12] F. A. Mumpton, J. R. Boles, E. M. Flanigen, A. J. Gude, R. A. Sheppard, R. L. Hay, R. C. Surdam, *Mineralogy and Geology of Natural Zeolites*. Washington, D.C.: Mineralogical Society of America, 1977, pp. 165. ISBN: 0939950049.

[13] S. Wang, Y. Peng, "Natural zeolites as effective adsorbents in water and wastewater treatment," *Chemical Engineering Journal*, vol. 156, pp. 11-24, 2010. DOI: 10.1016/j.cej.2009.10.029.

[14] V. J. Inglezakis, M. D. Loizidou, H. P. Grigoropoulou, "Equilibrium and kinetic ion exchange studies of Pb^{2+}, Cr^{3+}, Fe^{3+} and Cu^{2+} on natural clinoptilolite," *Water Research,* vol. 36, pp. 2784-2792, 2002. DOI: 10.1016/S0043-135400504-8.

[15] B. Ersoy, M. S. Celik, "Electrokinetic properties of clinoptilolite with mono- and multivalent electrolytes," *Microporous and Mesoporous Materials*, vol. 55, pp. 305-312, 2002. DOI: 10.1016/S1387-181100433-X.

[16] V. J. Inglezakis, M. D. Loizidou, H. P. Grigoropoulou, "Ion exchange of Pb^{2+}, Cu^{2+}, Fe^{3+}, and Cr^{3+} on natural clinoptilolite: selectivity determination and influence of acidity on metal uptake," *Journal of Colloid Interface Science,* vol. 261, pp. 49-54, 2003. DOI: 10.1016/S0021-979700244-8.

[17] L. J. Wilson. (2014, October 30). "Canada-Wide Survey of Acid Mine Drainage Characteristics. Project Report 3.22.1-Job No. 50788," Mineral Sciences Laboratories Division Report MSL 94–32 (CR), Ontario Ministry of Northern Development and Mines. Mine Environment Neutral Drainage (MEND) Program, Canada, 1994. [Online]. Available: http://mend-nedem.org/wp-content/uploads/2013/01/3.22.1.pdf

[18] H. Cui, L. Y. Li, J. R. Grace, "Exploration of remediation of acid rock drainage with clinoptilolite as sorbent in a slurry bubble column for both heavy metal capture and regeneration," *Water Research*, pp. 3359-3366, 2006. DOI: 10.1016/j.watres.2006.07.028.

[19] T. Motsi, "Remediation of Acid Mine Drainage using Natural Zeolite," Ph.D. Thesis, School of Chemical Engineering, The University of Birmingham, United Kingdom, 2010.

[20] W. Xu, L. Y. Li, J. R. Grace, G. Hebrard, "Acid rock drainage treatment by clinoptilolite with slurry bubble column: Sustainable zinc removal with regeneration of clinoptilolite," *Applied Clay Science,* vol. 80-81, pp. 31-37, 2013. DOI: 10.1016/j.clay.2013.05.009.

[21] V. J. Inglezakis, H. Grigoropoulou, "Effects of operating conditions on the removal of heavy metals by zeolite in fixed bed reactors," *Journal of*

Hazardous Materials, vol. 112, pp. 37-43, 2004. DOI: 10.1016/j.jhazmat.2004.02.052.

[22] A. L. Ciosek, G. K. Luk, "Lead Removal from Mine Tailings with Multiple Metallic Ions," *International Journal of Water and Wastewater Treatment*, vol. 3, pp. 1-9, 2017. DOI: 10.16966/2381-5299.134.

[23] A. L. Ciosek, G. K. Luk, "Kinetic Modelling of the Removal of Multiple Heavy Metallic Ions in Mine Waste by Natural Zeolite Sorption," *Special Issue: Treatment of Wastewater and Drinking Water through Advanced Technologies, In Water*, vol. 9, no. 7, pp. 482, 2017. DOI: 10.3390/w9070482.

[24] A. L. Ciosek, G. K. Luk, "An Innovative Dual-Column System for Heavy Metallic Ion Sorption by Natural Zeolite," *Special Issue: Wastewater Treatment and Reuse Technologies, In Applied Sciences*, vol. 7, no. 8, pp. 795, 2017. DOI: 10.3390/app7080795.

[25] Canadian Minister of Justice (CMJ). *Metal Mining Effluent Regulations. Consolidation SOR/2002-222.* 2014. [Online]. Available: http://laws-lois.justice.gc.ca

[26] H. V. Bekkum, E. M. Flanigen, J. C. Jansen, "Ion-Exchange in Zeolites" *In Introduction to Zeolite Science and Practice—Studies in Surface Science and Catalysis*, 1st ed., Elsevier Science: Zeist, The Netherlands, vol. 58, 1991, pp. 359-390.

[27] E. W. Rice, R. B. Baird, A. D. Eaton, L. S. Clesceri, "Part 1000-Introduction, Part 3000-METALS," In *Standard Methods for the Examination of Water and Wastewater*, 22nd ed., APHA, AWWA, WEF: Washington DC, USA, 2012, pp. 1.1-68; 3.1-112, ISSN: 978-087553-013-0.

[28] M. Minceva, R. Fajgar, L. Markovska, V. Meshko, "Comparative Study of Zn2+, Cd2+, and Pb2+ Removal From Water Solution Using Natural Clinoptilolitic Zeolite and Commercial Granulated Activated Carbon: Equilibrium of Adsorption," *Separation Science and Technology*, vol. 43, pp. 2117-2143, 2008. DOI: 10.1080/01496390801941174.

[29] U. Wingenfelder, C. Hansen, G. Furrer, R. Schulin, "Removal of Heavy Metals from Mine Waters from Natural Zeolites," *Environmental Science and Technology*, vol. 39, no 12, pp. 4606-4613, 2005. DOI: 10.1021/es048482s.

[30] Bear River Zeolite Co. Inc. (2012 and 2017, September 1). *Zeolite-Specifications and MSDS.* [Online]. Available: http://www.bearriverzeolite.com

[31] V. J. Inglezakis, C. D. Papadeas, M. D. Loizidou, H. P. Grigoropoulou, "Effects of Pretreatment on

Physical and Ion Exchange Properties of Natural Clinoptilolite," *Environmental Technology*, vol. 22, pp. 75-82, 2001. DOI: 10.1080/09593332208618308.

[32] Perkin Elmer Inc., *Atomic Spectroscopy—A Guide to Selecting the Appropriate Technique and System: World Leader in AA, ICP-OES, and ICP-MS*, Perkin Elmer Inc., Waltham MA, USA, 2011.

[33] Perkin Elmer Inc., *WinLab32 for ICP—Instrument Control Software, version 5.0*, Perkin Elmer Inc., Waltham MA, USA, 2010.

[34] J. Mullin, "Physical and thermal properties," In *Crystallization*, 4th ed., Read Educational and Professional Publishing Ltd., Woburn, MA, USA, pp. 76-77, 2001. IBSN: 0-7506-4833-3.

[35] O. Oter, H. Akcay, "Use of Natural Clinoptilolite to Improve Water Quality: Sorption and Selectivity Studies of Lead(II), Copper(II), Zinc(II), and Nickel(II)," *Water Environment Research*, vol. 79, no. 3. pp. 329-335, 2007. DOI: 10.2175/106143006X111880.

[36] H. Qiu, L. Lv, B. C. Pan, Q. J. Zhang, W. M. Zhang, Q. X. Zhang, "Critical review in adsorption kinetic models," *Journal of Zhejiang University Science A*, vol. 10, pp. 716-724, 2009. DOI: 10.1631/jzus.A0820524.

[37] N. Bektas, S. Kara, "Removal of lead from aqueous solutions by natural clinoptilolite: Equilibrium and kinetic studies," *Separation and Purification Technology*, vol. 39, pp. 189-200, 2004. DOI: 10.1016/j.seppur.2003.12.001.

[38] Y. Ho, A. E. Ofomaja, "Pseudo-second-order model for lead ion sorption from aqueous solutions onto palm kernel fiber," *Journal of Hazardous Materials*, vol. 129, pp. 137-142, 2005. DOI: 10.1016/j.jhazmat.2005.08.020.

[39] M. Jovanovic, N. Rajic, B. Obradovic, "Novel kinetic model of the removal of divalent heavy metal ions from aqueous solutions by natural clinoptilolite," *Journal of Hazardous Materials*, vol. 233, pp. 57-64, 2012. DOI: 10.1016/j.jhazmat.2012.06.052.

[40] M. N. Kostandyan, S. G. Babayan, M. A. Balayan, "Effect of heat treatment on the structural characteristics and sorption properties of clinoptilolite," *Inorganic Materials*, vol. 18, no. 10, 0020-1685, pp. 1498 -1501, 1982.

[41] D. W. Breck, *Zeolite Molecular Sieves: Structure, Chemistry, and Use*. John Wiley & Sons, New York, 1974.

[42] A. Langelia, M. Pansini, G. Cerri, P. Cappellietti, M. De Gennaro, "Thermal Behavior of Natural and Cation-Exchanged Clinoptilolite from Sardinia (Italy)," *Clays and Clay Minerals*, vol. 51, no. 6, pp. 625-633, 2003. DOI: 10.1346/CCMN.2003.0510605.

[43] E. Yörükoğulları, G. Yılmaz, S. Dikmen, "Thermal treatment of zeolitic tuff," *Journal of Thermal Analysis and Calorimetry*, vol. 100, no. 3, pp. 925-928, 2010. DOI: 10.1007/s10973-009-0503-8.

[44] K. Margeta, N. Zabukovec Logar, M. Šiljeg; A. Farkaš, "Natural Zeolites in Water Treatment—How Effective Is Their Use," *InTech, Water Treatment* W. Elshorbagy, Ed., pp. 81-112, 2013. DOI: 10.5772/50738.

[45] A. L. Ciosek, G. K. Luk, "Effects of Heat Pre-Treatment on Metallic Ion Sorption by Natural Zeolite," *Technical Session: Industrial Treatment A, Water Environment Association of Ontario Technical Symposium and OPCEA Exhibition*, Ottawa, Canada, 2017. Paper No: 2017-010.

[46] S. K. Ouki, M. Kavannagh, "Performance of natural zeolites for the treatment of mixed metal contaminated effluents," *Waste Management and Research*, vol. 15, no. 4, pp. 383-394, 1997. DOI: 10.1006/wmre.1996.0094.

[47] D. B. Akkoca, M. Yilgin, M. Ural, H. Akcin, A. Mergen, "Hydrothermal and Thermal Treatment of Natural Clinoptilolite Zeolite from Bigadiç, Turkey: An Experimental Study," *Geochemistry International*, vol. 51, no. 6, pp. 495-504, 2013. DOI: 10.1134/S0016702913040022.

[48] E. Chmielewská, L. Sabová, K. Jesenák, "Study of adsorption phenomena ongoing onto clinoptilolite with the immobilized interfaces," *Journal of Thermal Analysis and Calorimetry*, vol. 92, no. 2, pp. 567-571, 2008. DOI: 10.1007/s10973-006-8315-6.

3

Immobilisation of Metals in Contaminated Landfill Material Using Orthophosphate and Silica Amendments

Danielle Camenzuli[1], Damian B. Gore[1], Scott C. Stark[2]
[1] Macquarie University, Department of Environmental Sciences,
North Ryde 2109, Sydney, Australia.
Danielle.Camenzuli@mq.edu.au; Damian.Gore@mq.edu.au
[2]Australian Antarctic Division,
201 Channel Highway, Kingston 7050, Hobart, Australia.
Scott.Stark@aad.gov.au

Abstract- Immobilization and encapsulation of contaminants using silica and orthophosphate based chemical treatments are emerging technologies applicable to the management of metal contaminated soil. While the efficacy of orthophosphate treatment is well documented, there is a paucity of research on the application of silica or coupled orthophosphate and silica chemical treatments to metal contaminated soil. This paper presents a pilot scale bench study on the use of silica and coupled orthophosphate-silica treatments for the immobilization of metal contaminants in soil material obtained from the Thala Valley landfill, East Antarctica, which in places has petroleum hydrocarbons mixed with metal-contaminated sediment. The performance of the treatments trialed was assessed by the concentrations of copper, zinc, arsenic and lead released using the Toxicity Characteristic Leaching Procedure. The results of this pilot study demonstrate that the orthophosphate-silica treatment was the most effective and reduced leachable copper, zinc and lead by 95%, 96% and 99%, respectively, relative to the experimental controls. Further development of this technique will require additional research evaluating its long-term performance under a range of environmental conditions. Studies investigating potential adverse effects of silica and orthophosphate-silica treatments are also necessary, to demonstrate the environmental risk and efficacy of these remediation technologies.

Keywords: Contamination; Remediation; Silica; Orthophosphate; Landfill; Chemical Fixation

1. Introduction

Human activities such as mining, smelting, manufacturing and agriculture have resulted in a broad distribution of metal contaminated sites across the globe (Järup 2003). The legacy and exposure effects of contaminants such as copper, cadmium and lead on the environment and human health are well documented (Järup 2003; Poland et al., 2003; Taylor et al., 2010, 2013), and the management of these contaminated sites remains an ongoing global problem. Efficient management of metal contaminated sites requires the development of cost-effective techniques that are applicable under a range of environmental conditions, and do not engender environmental harm. Technologies being developed to manage metal contaminated land more efficiently include phytoremediation, bioremediation and electrokinetic remediation; however, the efficacy of these techniques relies heavily on favorable environmental conditions (Martin and Ruby 2004). Silica and orthophosphate based immobilization treatments also demonstrate potential for a range of contaminated sites and environments (Mitchell et al., 2000a, b; White et al., 2012).

Orthophosphate immobilization has been trialed successfully in laboratory experiments conducted under cold (2 °C) conditions and through multiple freeze-thaw cycles (Hafsteinsdóttir 2011, 2013; White et al., 2012). However, two concerns remain with orthophosphate treatments. Firstly, over dosage with reagents may lead to orthophosphate contamination of the environment, with the attendant risks of ecosystem disturbance including eutrophication. Secondly, if organic contaminants such as petroleum hydrocarbons have been co-disposed with inorganic waste, organic coatings on contaminants may hinder or even prevent effective reaction with orthophosphate. In this situation, silica amendments offer an alternative approach to contaminant immobilization. Silica treatment can work via direct reaction with contaminants and by microencapsulation, allowing the immobilization of hydrocarbon-coated contaminants. Therefore, its use alone or coupled with orthophosphate, may be far more efficient than treatment with orthophosphate only. A slight excess of silica in the environment is not of great concern in most earth surface environments where silicate rocks and sediments abound, and there is little foreseeable risk of ecosystem disturbance with the dosages to be applied using this method.

Treatment of contaminated soil with orthophosphate or silica requires the application of a powder or solution to contaminated soil (Arocha et al., 1996; Mbhele 2007). Once mixed with soil and water, orthophosphate and silica treatments react with metal cations including Mg^{2+}, Ca^{2+}, Fe^{3+}, Cu^{2+}, Zn^{2+} and Pb^{2+} to immobilize metals and precipitate insoluble metal-orthophosphate (White et al. 2012) or metal-silicate (Mitchell et al., 2000a; b; Abdel-Hamid et al., 2012). Contaminants are considered immobilized, and thus less hazardous (Mitchell et al., 2000a, b; Mbhele 2007), once they are transformed into inert or sparingly soluble forms and their leachability is reduced significantly (McDowell 1994; Zhu et al., 2004; Sonmez and Pierzynski, 2005).

While the potential of orthophosphate treatments for remediating metal contaminated soil is well documented (Zhu et al., 2004; Sonmez and Pierzynski 2005; White et al., 2012; Camenzuli et al., 2014), there is a paucity of studies reporting on the efficacy of silica treatments at metal contaminated sites (Camenzuli and Gore 2013). Furthermore, we are not aware of any published studies on the use of coupled orthophosphate-silica treatments for metal immobilization. Therefore, the purpose of this pilot study is to investigate the potential of two silica treatments and one orthophosphate-silica treatment for immobilizing Cu, Zn, As and Pb in contaminated soil. This study will provide a platform for subsequent investigations of

these treatments on a wider range of contaminated materials and contaminants, and under varying environmental conditions.

2. Materials and Methods

2.1. Experimental Design

The contaminated soil used in this study was sourced in January 2008 from a stockpile at the Thala Valley waste disposal site at Casey Station, East Antarctica (Stark et al., 2006). Concentrations of leachable Cu, Zn, As and Pb in the soil averaged 1.3 ± 1.0, 2.6 ± 1.8, <0.01 and 3.0 ± 3.8 mg/L, respectively (Thums et al., 2010).

Three treatments and one untreated control (Table 1) were applied in-duplicate at room temperature to columns loaded with 1 kg of Thala Valley soil sieved to <2 mm using a stainless steel mesh. The columns used in this study adopted a design similar to that described by Vandiviere and Evangelou (1998). The silica treatments were applied by mixing calcium carbonate powder with soil, followed by the application of sodium metasilicate dissolved in Type I (ASTM 2011) reagent water (Milli-Q). The phosphate-silica treatment contained Triple Super Phosphate (TSP), calcined magnesia ('Qmag'), calcium carbonate applied as powder, followed by the application of a sodium metasilicate solution (Table 1). X-ray diffractometry showed Qmag to consist mainly of anhydrite ($CaSO_4$) and quartz (SiO_2), with minor calcium montomorillonite ($Ca_{0.2}(Al, Mg)_2Si_4O_{10}(OH).2H_2O$), bassanite ($CaSO_4.0.5H_2O$) and trace amounts of other silicates.

The experiment was performed at room temperature, with each column leached daily with 100 ml of Milli-Q water for 10 days. Leachate samples were collected and stored unacidified (to prevent formation of a silica gel) for later analysis, if required. Six months after application of the silica treatments, two soil samples were obtained from each column. These 16 samples, along with 21 samples of untreated soil collected at the start of the experiment, were extracted according to the Toxicity Characteristic Leaching Procedure (TCLP; US EPA Method 1311; US EPA 1992) and analysed at the Australian Antarctic Division for metals using Inductively Coupled Plasma Optical Emission Spectrometry (ICP-OES). Four composite soil samples from the landfill were also analyzed for total metals by hot nitric acid digestion and ICP-OES at Analytical Services Tasmania.

Table. 1. Composition of silica and silica-phosphate treatments applied to Thala Valley landfill material.

Treatment ID	TSP (g)	Qmag (g)	Sodium metasilicate pentahydrate (g)	Calcium carbonate (g)	Solution volume (ml)
Control	0	0	0	0	210
Silica 1	0	0	70	40	210
Silica 2	0	0	100	40	300
Phosphate-silica	30	20	70	30	210

2.2. Analysis of Leachable Metals by Toxicity Characteristic Leaching Procedure (TCLP)

The TCLP simulates contaminant leachability from soil under landfill conditions, and can be used to estimate contaminant mobility or classify contaminated soil for disposal (Scott et al., 2005, 2007). To simplify this procedure, we adopted a scaled-down version of US EPA Method 1311, employing smaller quantities of soil and leachate.

TCLP extractions were performed on the <2.0 mm soil with analytical grade reagents and Type I deionized water. The pH of the soil averaged 8.2 ± 1.2 (n=12). Soil samples (2.5 g) were weighed into 50 ml polypropylene tubes (Sarstedt), mixed with 45 ml 0.10 M sodium acetate (pH 4.9) prepared from glacial acetic acid, 1 M NaOH and deionized water, and extracted for 18 h at 20 ± 1 °C on a rotary sample tumbler. Following filtration using 0.45 μm cellulose acetate syringe filters (Sartorius), the extracts were acidified to pH <2 with concentrated HNO_3 and analysed with a Varian 720-ES ICP-OES using standard operating conditions for the analytes. Analytical duplicates returned relative standard deviations for all elements of <0.5%. Analyte recovery was measured using matrix spikes and was >94% for all elements.

3. Results and Discussion

Total concentrations of Cu, Zn, As and Pb in the soil averaged 114 ± 88, 190 ± 100 and 3 ± 1 and 210 ± 170 mg/kg, respectively. The TCLP results (Table 2, Figure 1) demonstrate that the phosphate-silica treatment was the most effective at reducing the leachable concentrations of metals from the landfill material. The orthophosphate-silica treatment reduced Cu, Zn and Pb by 94%, 96% and 99%, respectively, relative to the experimental controls. The silica treatments also reduced Cu and Zn relative to the controls. Silica 2 treatment was the only treatment which did not reduce Pb relative to the controls (Figure 1). Arsenic concentrations were increased by all the treatments (Figure 1), which we attribute to competitive phosphate-arsenate (PO_4^{3-}-AsO_4^{3-}) and silicate-arsenate interactions in the soil (Peryea 1991; Luxton et al., 2006; Burton and Johnston 2012). The mobilizing effect of phosphate on As represents a major shortcoming of phosphate based remedial techniques (Peryea 1991; Munksgaard and Lottermoser 2013).

Table. 2. TCLP results from experimental controls and treated landfill material (mg/L).

	Cu	Zn	As	Pb
Control				
Mean	0.72	2.69	0.01	1.35
Range	0.48-1.02	2.24-3.52	0.01-0.02	0.63-2.68
Standard Deviation	0.25	0.57	0.00	0.95
Silica 1				
Mean	0.55	1.65	0.02	0.63
Range	0.37-0.71	1.52-1.90	0.03-0.03	0.04-1.57
Standard Deviation	0.13	0.18	0.00	0.51
Silica 2				
Mean	0.61	1.57	0.02	1.97
Range	0.49-0.83	1.50-1.62	0.018-0.026	1.30-2.34
Standard Deviation	0.15	0.05	0.00	0.56
Phosphate-silica				
Mean	0.04	0.12	0.03	0.02
Range	0.04-0.07	0.10-0.21	0.031-0.042	0.02-0.02
Standard Deviation	0.01	0.04	0.01	0.00

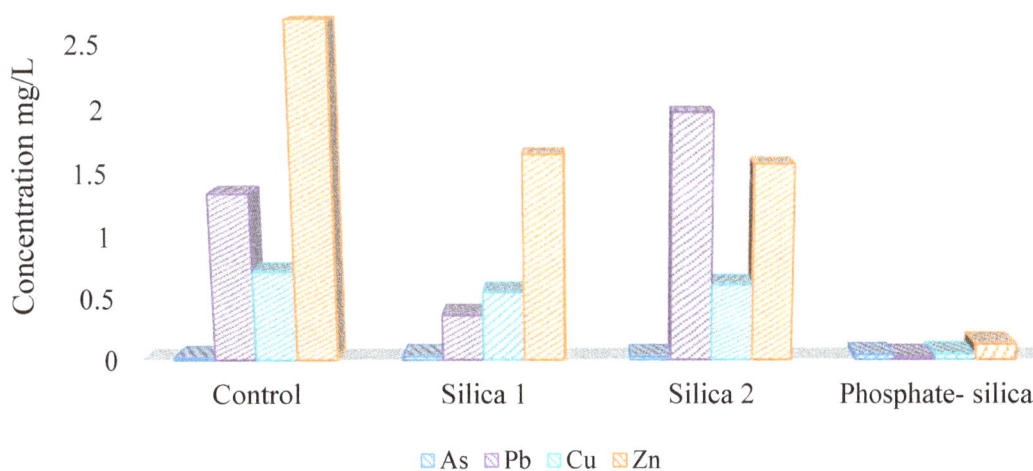

Figure 1. Mean TCLP results from treated and untreated landfill material (mg/L).

These results demonstrate the potential of silica and phosphate-silica treatments as a technique for remediating metal contaminated land; however, a paucity of research offers only a limited understanding of this approach. The small dataset presented in this paper is based on a small pilot scale experiment with few replicates. This leaves several uncertainties surrounding this technique which warrant further investigation. Of critical importance are long-term studies which evaluate the performance of silica and phosphate-silica treatments under a range of environmental conditions, including the effects of soil pH, oxidation-reduction potential, soil character and temperature on treatment performance, and the stability of minerals and other compounds formed during treatment. Understanding the effect of pH on treatment performance is essential for several reasons. The pH of soil can affect the solubility of treatments (particularly phosphorus based treatments) which can compromise mineral stability (Ma et al., 1993). Secondly, phosphorus release is enhanced in both low and high pH environments and this has important implications for eutrophication (Hafsteinsdóttir et al. 2014). Excessive applications of phosphorus can also acidify the soil and mobilise metals (Hafsteinsdóttir et al. 2014). Although the mobility of Cd, Cu, Sr and Zn in soil decreases with increasing pH, alkaline soil environments can be problematic for soil health, microbial activity and soil structure. Therefore, it is important to balance acidity-alkalinity when applying silica or phosphate-silica mix treatments (Camenzuli and Gore 2013). Furthermore, a study which compares coupled orthophosphate-silica treatments with orthophosphate-only based treatments is essential to distinguish the remedial effects of orthophosphate from silica, and ideally would examine a wider range of analytes than that presented here. Any potential adverse environmental effects associated with these treatments should also be investigated.

4. Conclusion

This pilot study investigated the potential of silica treatments and a coupled orthophosphate-silica treatment as a remediation strategy for metal contaminated soil. Results for contaminant leachability six months after treatment indicate that the coupled orthophosphate-silica treatment was most effective for Cu, Zn and Pb, but mobilized As. However, this study is not comprehensive enough to validate the safety of this technology. Since uncertainty about the safety of this technique under a range of conditions remains, further studies are required before *in situ* or on-site application to soil contaminated with metals and metalloids, particularly arsenic, is attempted. These should investigate the long-term effectiveness of the treatments under different environmental conditions. Coupled orthophosphate-silica treatments may be a promising technique for metal contaminated sites if future research is able to validate its long-term safety and reliability.

Acknowledgements

The authors wish to thank the Australian Antarctic Division for financial support (AAS4029), and Geoff Stevens, Kathryn Mumford, Tom Statham, Benjamin Freidman, and Chad Sanders for helpful discussions.

References

[1] Abdel-Hamid M. A., Kamel M. M., Moussa E. M. M., Rafaie H. A., "In-situ Immobilization Remediation of Soils Polluted with Lead, Cadmium and Nickel," *Global J. of Environmental Res.*, vol. 6, pp. 1-10, 2012.

[2] Arocha M. A., McCoy B. J., Jackman A. P., "VOC Immobilization in Soil by Adsorption, Absorption and Encapsulation," *J. of Hazardous Materials*, vol. 51, pp. 131-149, 1996.

[3] ASTM (American Society for Testing and Materials), "Standard Specification for Reagent Water," ASTM D1193 – 06, 2011.

[4] Burton E. D., Johnston S. G., "Impact of Silica on the Reductive Transformation of Schwertmannite and the Mobilization of Arsenic," *Geochemica et Cosmochimica Acta*, vol. 96, nol. 1, pp. 134-153, 2012.

[5] Camenzuli D., Freidman B. L., Statham T. M, Mumford K. A., Gore. D. B., "On-site and in Situ Remediation Technologies Applicable to Metal Contaminated Sites in Antarctica and the Arctic: A review," *Polar Research*, vol. 33, 21522, 2014.

[6] Camenzuli D., Gore D. B., "Immobilization and Encapsulation of Contaminants Using Silica Treatments: A review," *Remediation*, vol. 24, pp. 49-67, 2013.

[7] Hafsteinsdóttir E. G., White D. A., Gore D. B., Stark S. C., "Products and Stability of Phosphate Reactions with Lead Under Freeze-Thaw Cycling in Simple Systems," *Environmental Pollution*, vol. 159, pp. 3496-3503, 2011.

[8] Hafsteinsdóttir E. G., White D. A., Gore D. B., "Effects of Freeze-Thaw Cycling on Metal-Phosphate Formation and Stability in Single and Multi-Metal Systems," *Environmental Pollution*, vol. 175, pp. 168-177, 2013.

[9] Hafsteinsdóttir E. G., Fryirs K. A., Stark S. C., Gore D. B., "Remediation of Metal-Contaminated Soil in Polar Environments: Phosphate Fixation at Casey Station, Antarctica," *Appl. Geochemistry*, vol. 51, pp. 33-43, 2014.

[10] Järup L., "Hazards of Heavy Metal Contamination," *Brit. Medical Bulletin*, vol. 68, pp. 167-182, 2003.

[11] Luxton T. P., Tadanier C. J., Eick M. J., "Mobilization of Arsenite by Competitive Interaction with Silicic Acid," *Soil Sci. Soc. of Amer.*, 2006.

[12] Ma Q. Y., Traina S. J., Logan T. J., Ryan J. A., "In-situ Lead Immobilization by Apatite," *Environmental Sci. and Technol.*, vol. 27, pp. 1803-1810, 1993.

[13] Martin T. A, Ruby M. V., "Review of In Situ Remediation Technologies for Lead, Zinc, and Cadmium in Soil," *Remediation*, vol. 14, pp. 35-53, 2004.

[14] Mbhele P. P. "Remediation of Soil and Water Contaminated by Heavy Metals and Hydrocarbons Using Silica Encapsulation," PhD Thesis, University of Witwatersrand, South Africa, 2007.

[15] McDowell T., "Siallon: The Microencapsulation of Hydrocarbons Within a Silica Shell," in *Process Engineering for Pollution Control and Waste Minimization*, Wise DL, Trantolo DJ ed., Marcel-Dekker Inc., 1994, ch. 19.

[16] Mitchell P., Rybock J. T., Anderson A. L., "Silica Micro Encapsulation: An Innovative Commercial Technology for the Treatment of Metal and Radionuclide Contamination in Water and Soil," in *Proc. of the 6th Symposium on Environmental Issues and Waste Management in Energy and Mineral Prod.*, Calgary, Canada, 2000a.

[17] Mitchell P., Anderson A., Potter C., "Protection of **Ecosystem and Human Health via Silica Micro** Encapsulation of Heavy Metals," in *Proc. of 7th Int. Mine Water Assoc. Congr.*, Ustron, Poland, 2000b.

[18] Munksgaard N. C., Lottermoser B. G., "Phosphate Amendment of Metalliferous Tailings, Cannington Ag-Pb-Zn Mine, Australia: Implications for the Capping of Tailings Storage Facilities," *Environmental Earth Sci.*, vol. 68, pp. 33-44, 2013.

[19] Peryea F. J., "Phosphate-Induced Release of Arsenic from Soils Contaminated with Lead Arsenate," *Soil Sci. Soc. of Amer. J.*, vol. 55, pp. 1301-1306, 1991.

[20] Poland J. S., Riddle M. J., Zeeb B. A., "Contaminants in the Arctic and the Antarctic: A Comparison of Sources, Impacts, and Remediation Options," *Polar Rec.*, vol. 39, pp. 369-383, 2003.

[21] Scott J., Beydoun D., Amal R., Low G., Cattle J., "Landfill Management, Leachate Generation, and Leach Testing of Solid Wastes in Australia and Overseas," *Critical Rev. in Environmental Sci. and Technol.*, vol. 35, pp. 239-332, 2005.

[22] Snape I., Riddle M. J., Stark J. S., Cole C. M., King C. K., Duquesne S., Gore D. B., "Management and Remediation of Contaminated Sites at Casey Station, Antarctica," *Polar Rec.*, vol. 37, pp. 199-214, 2001.

[23] Stark J. S., Snape I., Riddle M. J., "Abandoned Antarctic Waste Disposal Sites: Monitoring Remediation Outcomes and Limitations at Casey Station," *Ecological Manag. and Restoration*, vol. 7, no. 1, pp. 21-31, 2006.

[24] Stark S. C., Snape I., Graham N. J., Brennan J. C., Gore D. B., "Assessment of Metal Contamination Using X-ray Fluorescence Spectrometry and the Toxicity Characteristic Leaching Procedure (TCLP) During Remediation of a Waste Disposal Site in Antarctica," *J. of Environmental Monitoring*, vol. 10, pp. 60-70, 2008.

[25] Sonmez O., Pierzynski G. M., "Phosphorus and Manganese Oxides Effects on Soil Lead Bioaccessibility: PBET and TCLP," *Water, Air, and Soil Pollution*, vol. 166, pp. 3-16, 2005.

[26] Taylor M. P., Mackay A. K., Hudson-Edwards K. A., Holz E. "Soil Cd, Cu, Pb and Zn Contaminants Around Mount Isa City, Queensland, Australia: Potential Sources and Risks to Human Health," *Appl. Geochemistry*, vol. 25, pp. 841-855, 2010.

[27] Taylor M. P., Camenzuli D., Kristensen L. J., Forbes M., Zahran S., "Environmental Lead Exposure Risks Associated with Children's Outdoor **Playgrounds**," *Environmental Pollution*, vol. 178, pp. 447-554, 2013.

4

Investigation of Impact Factors on the Treatment of Oily Sludge using a Hybrid Ultrasonic and Fenton's Reaction Process

Ju Zhang[1], Jianbing Li[1*], Ronald W. Thring[1], Guangji Hu[1,] Lei Liu[2]
[1]Environmental Engineering Program, University of Northern British Columbia,
3333 University Way, Prince George, British Columbia, Canada V2N 4Z9
jzhang5@unbc.ca; jianbing.li@unbc.ca;thring@unbc.ca; hug@unbc.ca
[2]Department of Civil and Resource Engineering, Dalhousie University,
1360 Barrington St., Halifax, Nova Scotia, Canada B3H 4R2
lei.liu@dal.ca

Abstract - In this study, a hybrid ultrasonic and Fenton's reaction process (US/Fenton) was applied for oily sludge treatment. The impacts of four different factors on the reduction of total petroleum hydrocarbons (TPH) in oily sludge were investigated. These include the initial sludge content, the molar ratio of hydrogen peroxide to iron (H_2O_2/Fe^{2+}), the ultrasonic power, and the ultrasonic treatment duration. Taguchi experimental design method was used to arrange laboratory experiments. The results indicated that a TPH reduction rate of up to 88.1% was reached with an initial sludge content of 20 g/L, a H_2O_2/Fe^{2+} molar ratio of 4:1, an ultrasonic treatment time of 5 min, and an ultrasonic power of 60 W. The initial oily sludge content and ultrasonic treatment duration were found to be the most significant factors affecting the US/Fenton treatment of oily sludge.

Keywords: Advanced oxidation, Fenton's reaction, oily sludge, petroleum hydrocarbons, ultrasound.

1. Introduction

As one of the major wastes generated in the petroleum industry, oily sludge is a complex mixture consisting of water, inorganic solid particles, and various petroleum hydrocarbons (PHCs). In particular, it contains a large amount of heavy PHCs, such as long-chain alkanes and alkenes, polycyclic aromatic hydrocarbons, asphaltenes, and resins [1, 2]. Due to the complicated composition and high concentration of heavy PHCs, the direct disposal of oily sludge could pose serious threats to the environment, and thus it needs effective treatment. Among various technologies, advanced oxidation processes (AOPs) have been recognized potential treatment approaches to effectively degrade the recalcitrant compounds [3]. During AOP processes, a large amount of hydroxyl radicals ($\cdot OH$) can be generated through various methods (e.g., ultrasonic irradiation, ultraviolet radiation, photo-catalysis, ozonation, and/or Fenton's reaction) [4]. The hydroxyl radicals are strong and non-selective oxidants which can oxidize various recalcitrant organic compounds. The final products of reaction include carbon dioxide, short-chain organic compounds, and inorganic ions, which are usually less toxic and favourable for biodegradation [5]. Among various AOPs, ultrasonic irradiation (US) and Fenton's reaction have been widely applied to a variety of fields. Ultrasonic treatment can generate •OH radicals due to the acoustic cavitation which involves the formation and subsequent expansion of micro-bubbles under the periodic pressure variations [6]. The Fenton's reagents, usually hydrogen peroxide (H_2O_2) and ferrous (Fe^{2+})

materials, can react with each other to generate sufficient hydroxyl radicals, while H_2O_2 serves as an oxidizing agent and the ferrous (Fe^{2+}) compound works as a catalyst for reaction [7].

In general, there is a similarity between the pollutant oxidation mechanisms among various AOPs, and it is thus expected that the combination of individual AOPs might achieve better results as compared to a single AOP. In fact, the combination of ultrasonic irradiation and Fenton's reaction has recently received increasing attention. In the hybrid AOP process, hydroxyl radicals can be generated by the decomposition of H_2O_2 which also converts Fe^{2+} ions into Fe^{3+} ($Fe-OOH^{2+}$). Meanwhile, the application of ultrasonic energy could isolate Fe^{2+} from $Fe-OOH^{2+}$, and the isolated Fe^{2+} could in turn react with H_2O_2 to generate hydroxyl radicals. As a result, the iron catalysts could be regenerated during the hybrid process of combining ultrasonic irradiation with Fenton's reaction, and this hybrid method can be more effective in degrading recalcitrant compounds. For example, Neppolian et al. [8] reported that the degradation rate of para-chlorobenzoic acid (p-CBA) was 7.3 $\times 10^{-3}$ min^{-1} under the combined process of ultrasound and Fenton's reaction as compared to the value of 4.5 $\times 10^{-3}$ min^{-1} under the process of only using 20 kHz ultrasound. Sun et al. [9] investigated the combined ultrasonic and Fenton's reaction process for the decolorization of acid black 1 (AB1) solution, and found that the optimal concentration of Fe^{2+} was 0.025 mM when the concentration of H_2O_2 was 8.0 mM. Virkutyte et al. [10] examined the effect of ultrasonic and Fenton's reaction process on the degradation of naphthalene when the mineral iron in soil was used as the catalyst, and they observed that the optimal degradation efficiency of naphthalene reached 97% at an ultrasonic power of 400 W when 600 mg/L of H_2O_2 was added into the soil with an initial naphthalene concentration of 200 mg/kg. In spite of the advantages of oxidizing recalcitrant organic compounds, these reported AOP studies mainly focused on the degradation of individual contaminants; very few studies have applied the hybrid AOP process to treat a mixture of hazardous organic compounds such as oily sludge.

The objective of this study is then to investigate the application of a combined ultrasonic and Fenton's reaction process (US/Fenton) to treat refinery oily sludge. The Taguchi experimental design method was used to plan the laboratory experiments and to examine the effects of different factors on the treatment performance, which is indicated by the oxidation of petroleum hydrocarbons. These factors include the initial oily sludge content in the treatment system, the molar ratio of H_2O_2 to Fe^{2+}, the ultrasonic power, and the ultrasonic treatment duration. The results could provide a sound basis for developing more efficient and economically competitive methods for oily sludge treatment.

2. Materials and Methods
2. 1. Oily Sludge and Chemicals

$FeSO_4 \bullet 7H_2O$ (from Sigma) and H_2O_2 (30% w/w) solution were used as Fenton's reagents. The oily sludge was obtained from the crude oil tank bottom in an oil refinery plant in western Canada, and was kept at 4 °C in a capped stainless-steel bucket before use. Table 1 lists the summary of its characteristics. The TPH content was measured according to the method described in Zhang et al. [11], water content was measured using ASTM D1744 [12], the solid content was calculated based on the measurement of TPH and water, and metal elements were measured using Inductively Coupled Plasma (ICP) analysis based on the ASTM D5185 [13].

Table 1. Characteristics of oily sludge.

Parameter	Concentration	Parameter	Concentration
TPH (by mass)	61%	Barium(mg/kg)	2,136
Water content (by mass)	24%	Iron (mg/kg)	6,339
Solid content (by mass)	15%	Zinc (mg/kg)	209
Sodium (mg/kg)	76	Copper (mg/kg)	43
Potassium (mg/kg)	423	Lead (mg/kg)	19
Magnesium (mg/kg)	432	Chromium (mg/kg)	11
Aluminium (mg/kg)	999	Nickel (mg/kg)	9
Calcium (mg/kg)	1,145		

2.2. Experimental Design

The ultrasonic apparatus used in this study was a 20-kHz Misonix Sonicator 3000 generator with a titanium sonic probe, and the ultrasonic power could be adjusted from 0 to about 75 W. For the hybrid ultrasonic and Fenton's reaction process, the impacts of four different factors were investigated. They include the initial oily sludge content in the treatment system, the molar ratio of H_2O_2 to Fe^{2+} (H_2O_2/Fe^{2+}), the ultrasonic power, and the ultrasonic treatment duration. Each factor was examined at 3 levels, and a Taguchi orthogonal experimental design method [14] was used to arrange the experiments through a L27 array (Table 2). This method can allow for the examination of both the main effects of each factor and the interaction effects between factors. There were 27 experimental runs (L1-L27), and each run was replicated twice. For each test, a constant volume of H_2O_2 (i.e. 15 mL) was used, and a given amount of oily sludge was added into a 100-mL beaker with 10 mL of de-ionized water. A final total liquid volume of 25 mL was thus obtained after adding 15 mL of H_2O_2 solution,

which was gradually added to avoid rigorous reaction. The sludge amount for each test (e.g., 0.5 g, 1.0 g, 1.5 g) was determined by multiplying the specified sludge content in Table 2 by the liquid volume (i.e. 25 mL). The amount of $FeSO_4 \bullet 7H_2O$ added to the beaker was calculated based on the molar ratio of H_2O_2 /Fe^{2+} for each test (Table 2). During the experiment, the ultrasonic probe was placed under the liquid to start ultrasonic irradiation with an ultrasonic power as specified in Table 2. After initiating the ultrasonic treatment, the H_2O_2 solution (i.e. a total volume of 15 mL) was gradually added into the beaker at a rate of about 15, 5, and 3 mL/min for the 1-, 3-, and 5-min ultrasonic treatment durations (Table 2), respectively,

Table 1. L27 array obtained from Taguchi experimental design.

Experimental test #	Sludge content (g/L)	Molar ratio of H_2O_2 /Fe^{2+}	US time (min)	US power (W)
L 1	20 (level 1)	4:1 (level 1)	1 (level 1)	20 (level 1)
L 2	20 (level 1)	4:1 (level 1)	3 (level 2)	40 (level 2)
L 3	20 (level 1)	4:1 (level 1)	5 (level 3)	60 (level 3)
L 4	20 (level 1)	10:1 (level 2)	1 (level 1)	40 (level 2)
L 5	20 (level 1)	10:1 (level 2)	3 (level 2)	60 (level 3)
L 6	20 (level 1)	10:1 (level 2)	5 (level 3)	20 (level 1)
L 7	20 (level 1)	50:1 (level 3)	1 (level 1)	60 (level 3)
L 8	20 (level 1)	50:1 (level 3)	3 (level 2)	20 (level 1)
L 9	20 (level 1)	50:1 (level 3)	5 (level 3)	40 (level 2)
L 10	40 (level 2)	4:1 (level 1)	1 (level 1)	40 (level 2)
L 11	40 (level 2)	4:1 (level 1)	3 (level 2)	60 (level 3)
L 12	40 (level 2)	4:1 (level 1)	5 (level 3)	20 (level 1)
L 13	40 (level 2)	10:1 (level 2)	1 (level 1)	60 (level 3)
L 14	40 (level 2)	10:1 (level 2)	3 (level 2)	20 (level 1)
L 15	40 (level 2)	10:1 (level 2)	5 (level 3)	40 (level 2)
L 16	40 (level 2)	50:1 (level 3)	1 (level 1)	20 (level 1)
L 17	40 (level 2)	50:1 (level 3)	3 (level 2)	40 (level 2)
L 18	40 (level 2)	50:1 (level 3)	5 (level 3)	60 (level 3)
L 19	60 (level 3)	4:1 (level 1)	1 (level 1)	60 (level 3)
L 20	60 (level 3)	4:1 (level 1)	3 (level 2)	20 (level 1)
L 21	60 (level 3)	4:1 (level 1)	5 (level 3)	40 (level 2)
L 22	60 (level 3)	10:1 (level 2)	1 (level 1)	20 (level 1)
L 23	60 (level 3)	10:1 (level 2)	3 (level 2)	40 (level 2)
L 24	60 (level 3)	10:1 (level 2)	5 (level 3)	60 (level 3)
L 25	60 (level 3)	50:1 (level 3)	1 (level 1)	40 (level 2)
L 26	60 (level 3)	50:1 (level 3)	3 (level 2)	60 (level 3)
L 27	60 (level 3)	50:1 (level 3)	5 (level 3)	20 (level 1)

2.3. Sample Analysis

The sample in the beaker after US/Fenton treatment was transferred into a 50-mL tube for centrifugation for 30 min in order to separate the solid from liquid for the analysis of PHCs within the two phases [15]. After centrifugation, the supernatant was transferred into a separating funnel for liquid extraction, and the solid residue left in the centrifugation tube was used to extract PHCs. The liquid and solid extraction procedures as well the chemical solvents used can be found in Zhang et al. [11]. The extracted solution was then analyzed for PHCs using a Varian CP-3800 Gas Chromatograph with flame ionization (GC-FID). The GC analysis conditions and procedures can be found in Zhang et al. [11] The TPH reduction rate (η) was used to analyze the efficiency of the US/Fenton process, based on the measured TPH mass in the system before and after treatment (M_0 and M_A):

$$\eta = \frac{M_A - M_0}{M_0} \times 100\% \qquad (1)$$

2.4. Experimental Data Analysis

Statistical analyses including signal to noise (S/N) ratio analysis and ANOVA were used to analyze the experimental data. The S/N ratio was evaluated using the following equation [14]:

$$S/N = -10\log \sum_{i=1}^{n} \left(\frac{1}{y_i^2}\right)/n \qquad (2)$$

where S/N denotes the performance statistic, y_i denotes the observed data, n is the number of observations. The unit of S/N ratio is decibels (dB). The higher the S/N ratio, the better the result is.

3. Results and Discussion
3. 1. Degradation of PHCs in Sludge by using the Hybrid US/Fenton Process

The reduction of individual petroleum hydrocarbons (PHCs) fraction in oily sludge after US/Fenton treatment was examined (Fig. 1), where fraction 1 (F1), fraction 2 (F2), fraction 3 (F3) and fraction 4 (F4) was defined as the group of PHCs from C6 to C10, C10 to C16, C16 to C34, and C34 to C50, respectively. It can be found that the US/Fenton process achieved a F2 fraction reduction in the range of 43.3% to 93.1%, a F3 fraction reduction in the range of 33.3% to 86.9%, and a F4 fraction reduction in the range of 11.8% to 90.1%, respectively. The decomposition of

long-chain hydrocarbons (i.e. F4 fraction) was generally less than those of the other two PHCs fractions, indicating that short-chain PHCs are more prone to be destructed by this hybrid oxidation process. All of the highest PHCs reduction occurred for treatment L3, but the reduction of the F3 fraction was less than that of F4 fraction. This might be caused by the accumulation into F3 fraction of intermediate products from the decomposition of the heavier F4 fraction. The lowest F2 reduction (i.e. 43.3%), F3 reduction (i.e. 33.3%) and F4 reduction (i.e. 11.8%) occurred for treatment L22 (i.e. corresponding to the lowest ultrasonic power and treatment duration level), L27 (i.e. corresponding to the lowest ultrasonic power level), and L1 (i.e. corresponding to the lowest level for all the four factors), respectively. Fig. 2 presents the reduction results of TPH in oily sludge for all of the experiments. The TPH reduction rate was in the range of 51.9% to 88.1%, 42.3% to 83.9%, and 36.0% to 81.3% for experiments with low (i.e. test # L1 to L9), medium (i.e. test # L10 to L18), and high initial sludge contents (i.e. test # L19 to L27), respectively. It is obvious that the initial oily sludge content could affect the treatment performance of the hybrid US/Fenton process. Among all of the 27 experimental runs, the highest TPH reduction rate of 88.1% was observed for treatment L3, with an initial sludge content of 20 g/L, a molar ratio of H_2O_2/Fe^{2+} of 4:1, an ultrasonic treatment time of 5 min, and an ultrasonic power of 60W.

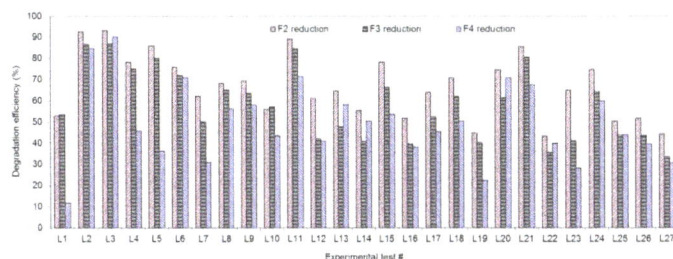

Figure 1. Reduction of PHCs fraction in oily sludge through US/Fenton process.

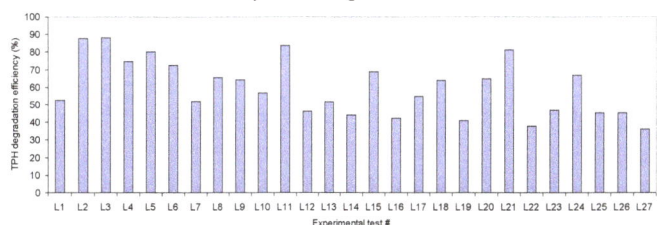

Fig. 2 Reduction of TPH in oily sludge through US/Fenton process.

3. 2. Impact of US/Fenton Operating Factors on the TPH Reduction

Fig. 3 presents the main effect plot which shows the contribution to S/N ratio change from variation in an impact factor from one level to another. It can be found that the S/N ratio decreased with the increase of initial oily sludge content and the increase of H_2O_2/Fe^{2+} ratio, but increased when the ultrasonic irradiation time was extended. The S/N ratio increased when the ultrasonic power increased from level 1 to level 2, but decreased when the ultrasonic power increased from level 2 to level 3. This indicates that further increase of ultrasonic power after level 2 didn't further promote the TPH reduction. The greater change in S/N ratio occurred when sludge content and ultrasonic treatment duration increased from level 1 to level 3, indicating that sludge content and ultrasonic treatment time were the most significant factors on TPH reduction. As shown in Fig.3, the optimal condition (when the highest S/N ratio occurs) for TPH reduction using US/Fenton process would be a sludge content of 20 g/L, a H_2O_2/Fe^{2+} molar ratio of 4:1, an ultrasonic irradiation time of 5 min, and an ultrasonic power of 40 W.

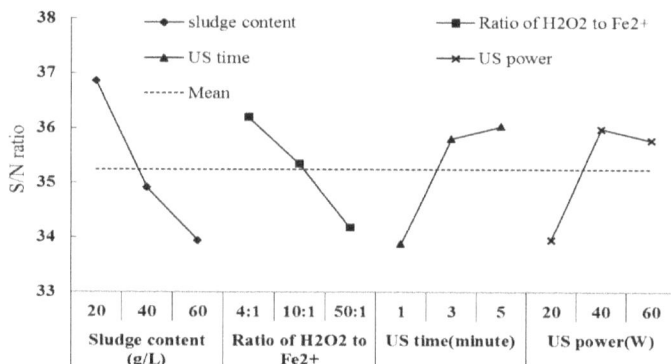

Figure 3. Main effects of factors on the reduction of TPH in oily sludge.

Fig. 4 presents the two-factor interaction effects on TPH reduction in oily sludge when using the US/Fenton process. In the interaction plot, the levels of one factor are set on the x-axis and a separate line stands for the mean S/N ratio of each level for the other factor. A larger vertical distance between the factor point and the mean S/N ratio line indicates a stronger interaction between that factor and the corresponding factor set on the x-axis. However, if the lines are parallel to each other, there is no interaction between them. Since each factor was examined with three levels in this study, three curves were displayed in each plot. The degree of interaction between the factors depends on the departure of a curve from the trend of another curve. It was observed in Fig. 4(c) that the interaction between H_2O_2/Fe^{2+} ratio and the ultrasonic treatment duration was greater than that between other factors, and a greater TPH reduction was obtained with a H_2O_2/Fe^{2+} ratio of 4:1 and an ultrasonic treatment time of 3 min.

In addition to S/N ratio analysis, the analysis of variance (ANOVA) was also carried out to verify the impacts of various factors and their interactions on the TPH reduction in oily sludge. The ANOVA was implemented by using MINITAB 16, and the results are shown in Table 3. Several parameters were generated during the ANOVA process, including the degree of freedom (DF), the sequential sums of squares (Seq SS), the adjusted sum of squares (Adj SS), and the adjusted means squares (Adj MS). F-test was performed with 95% confidence interval, and it was verified that oily sludge content and ultrasonic treatment duration had significant impacts on TPH reduction (e.g., their P values were less than 0.05 as shown in Table 3). The other two factors (with P values greater than 0.05) did not show significant impacts on the US/Fenton treatment performance. The ANOVA results also illustrated that a higher interaction existed between H_2O_2/Fe^{2+} ratio and ultrasonic treatment duration (with P value of 0.332), but there was no significant interaction between the factors to affect TPH degradation.

Table 3. ANOVA for TPH reduction using US/Fenton process.

Sources	DF	Seq SS	Adj SS	Adj MS	F	P
Sludge content	2	39.646	39.646	19.823	8.34	0.019
H_2O_2/Fe^{2+} ratio	2	18.395	18.395	9.1975	3.87	0.083
Ultrasonic treatment duration (US time)	2	24.982	24.982	12.491	5.25	0.048
Ultrasonic power (US power)	2	22.422	22.422	11.211	4.72	0.059
Sludge content*H_2O_2/Fe^{2+} ratio [a]	4	5.597	5.597	1.3994	0.59	0.684
Sludge content*US time [a]	4	2.595	2.595	0.6488	0.27	0.885
H_2O_2/Fe^{2+} ratio*US time [a]	4	13.544	13.544	3.3861	1.42	0.332
Residual error	6	14.262	14.262	2.377		
Total	26	141.444				

[a]* denotes the interaction between two factors

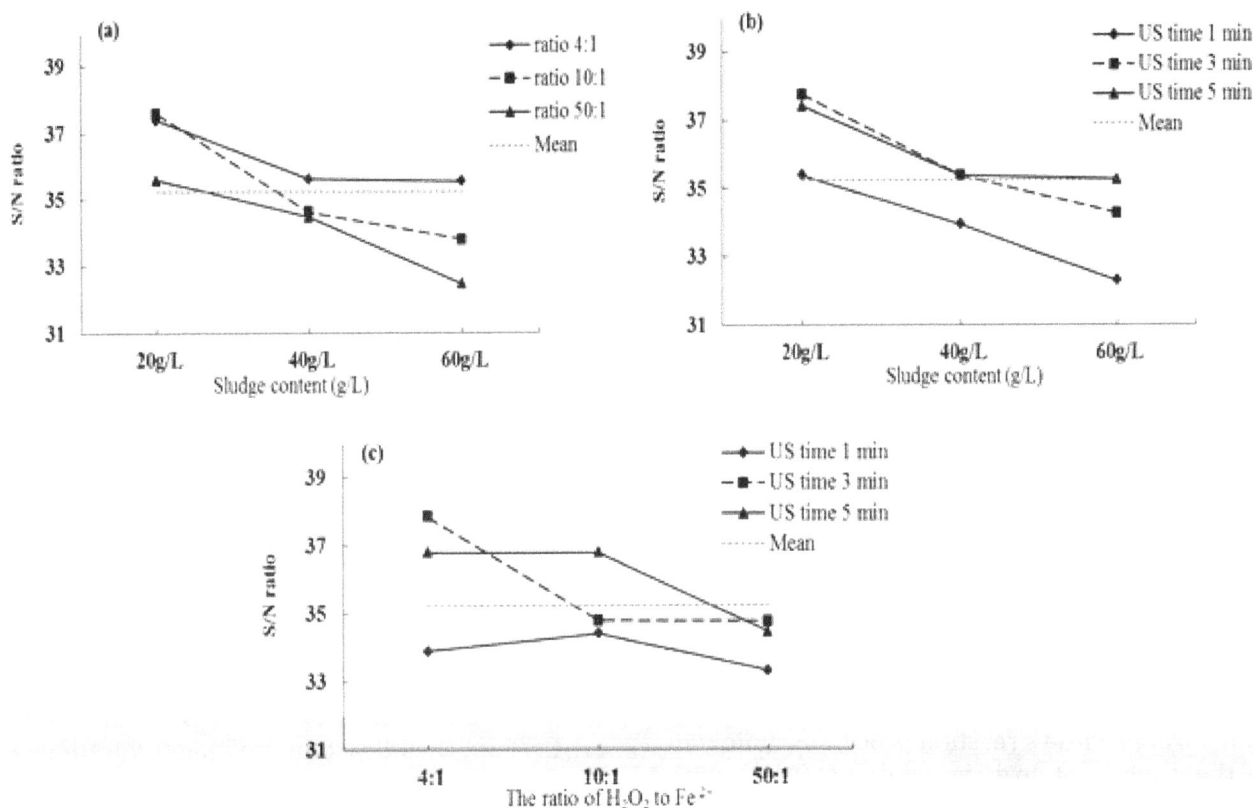

Figure 4. Interaction effects of factors on TPH reduction in oily sludge: (a) interaction between initial sludge content and H_2O_2/Fe^{2+} ratio, (b) interaction between initial sludge content and ultrasonic treatment duration, (c) interaction between H_2O_2/Fe^{2+} ratio and ultrasonic treatment duration.

3. 3. Discussion

As illustrated in Fig. 3 and further verified by ANOVA, the initial oily sludge content had a significant impact on the performance of the US/Fenton process. On one hand, when the initial oily sludge content was high, more PHCs would be in the form of a non-aqueous phase or attached to the solid particles. This would decrease the contact of hydroxyl radicals with PHCs, and thus the TPH reduction would decrease. When the initial sludge content was high in the US/Fenton treatment system, the viscosity of bulk liquid also increased. The increased viscosity could lead to the impedance of the formation and collapse of cavitation bubbles, and thus decrease the ultrasonic desorption of petroleum hydrocarbons [16]. For example, the TPH reduction rate ranged from 51.9% to 88.1% at level 1 (i.e. 20 g/L), from 42.3% to 83.9% at level 2 (i.e. 40 g/L), and from 36.0% to 81.3% at level 3 of sludge content (i.e. 60 g/L), respectively. This was consistent with many other studies. Virkutyte et al. [10] investigated the ultrasonic assisted oxidation of naphthalene-contaminated soil combining with a

Fenton-like process, and they observed a higher degradation efficiency (94-97%) with lower initial naphthalene concentration (200 mg/kg), but the efficiency decreased to 58-76% when the initial naphthalene concentration was doubled (i.e. 400 mg/kg). On the other hand, the yield of oxidation intermediates could increase when the initial sludge content increased, and the accumulation of intermediates might affect the degree of PHCs degradation. Lin et al. [17] examined the oxidation rate of azo dyes at high initial concentration under ultrasonic irradiation with Fenton-like reagents, and they found that a lower degradation efficiency was due to the formation of recalcitrant by-products. Since oily sludge is a mixture of many complex PHCs, the intermediates (e.g., carboxyl acids, alkene, ketones) could be accumulated when a large amount of PHCs were oxidized. The resistance to further decomposition might increase due to the accumulation of more oxidation intermediates. Consequently, a lower TPH reduction efficiency was observed in this study when the initial sludge content increased.

In spite of the above discussion, a relatively higher TPH reduction rate was still observed at a higher level of initial sludge content (i.e. 60 g/L). For example, TPH reduction reached 66.7% and 81.3% for experimental run L24 and L21, respectively (Fig. 2). This might indicate the complicated effects of other factors on the US/Fenton process or other interactions of factors. As the initial sludge content increased, more solid particles were brought into the treatment system, and they could also provide more interfacial areas for the formation of OH• radicals and thus the occurrence of free radical reactions [18]. Moreover, it was reported that many metal components could serve as catalysts to trigger the chain oxidation reactions similar to reactions associated with the Fenton process [18]. Many metals elements, such as Fe, Al, Ca, Cu, Zn, were found in oily sludge. These metals might serve as other sources of catalysts for Fenton-like reactions to improve the TPH reduction. In addition, ultrasonic treatment duration had a positive impact on TPH reduction when using the US/Fenton process. In general, the benefit of applying ultrasonic irradiation during Fenton's reaction process is mainly to enhance the contact of hydroxyl radicals (OH•) with PHCs compounds. The increase in ultrasonic treatment time could help the desorption of PHCs from solid particles and the dispersion of these hydrophobic compounds into the bulk liquid. In the meantime, the micro-jets generated from the heterogeneous sludge treatment system by ultrasonic irradiation could enhance the transfer of OH• radicals towards the solid-liquid interface where PHCs are attached. As a result, a higher TPH reduction rate could be achieved under US/Fenton's process when a longer ultrasonic treatment (i.e. 5 min) was applied.

Moreover, the abundant intermediates of Fe-OOH^{2+}, which are related to Fenton's reactions, can be decomposed into Fe^{2+} and hydroperoxyl (HO•$^{2+}$) by ultrasonic irradiation. With longer ultrasonic treatment duration, more Fe^{2+} could be regenerated to engage in the reactions with H$_2$O$_2$, leading to the production of more hydroxyl radicals. Another benefit from prolonged ultrasonic irradiation is that it might facilitate the decomposition and cleavage of more petroleum hydrocarbons. Although the ANOVA results did not confirm the significant impact of H$_2$O$_2$/Fe^{2+} ratio on TPH reduction when using US/Fenton's process, it is still worth further investigating the effect of this ratio. Many studies have examined the impact of the H$_2$O$_2$/Fe^{2+} ratio on wastewater treatment when using Fenton's reaction process alone, and reported a wide range of optimal ratios. Casero et al. [19] reported the optimal molar ratio was 5 to 40 by the use of Fenton's reagents to degrade aromatic amines in wastewater, and Tekin et al. [20] found that the molar ratio was between 150 and 250 when using Fenton's reaction process for pharmaceutical wastewater treatment. In this study, the highest TPH reduction rate was achieved when the H$_2$O$_2$/Fe^{2+} ratio was 4:1, and the existence of extra Fe^{2+} might play an important role in the oxidation of PHCs by reactions with intermediate radicals (carboxyl radicals) [21]. In fact, other studies have indicated that the mineralization of organic compounds can be increased with the increase of Fe^{2+} [22].

4. Conclusion

In this study, a hybrid ultrasonic and Fenton's reaction (US/Fenton) process was applied to treat refinery oily sludge. Four different factors were examined for their effects on the performance of the US/Fenton process. These factors include the initial oily sludge content in the treatment system, the molar ratio of H$_2$O$_2$/Fe^{2+}, the ultrasonic treatment duration, and the ultrasonic power. Taguchi experimental design method was used to arrange laboratory experiments for investigating the impact of these factors. It was found that a TPH reduction rate of up to 88.1% was reached with a sludge content of 20 g/L, a H$_2$O$_2$/Fe^{2+} ratio of 4:1, an ultrasonic power of 60 W, and an ultrasonic treatment duration of 5 min. The sludge content and ultrasonic treatment duration had significant impacts on the reduction of petroleum hydrocarbons, while the other two factors didn't show significant impacts. A higher TPH reduction was generally observed to be associated with a lower initial sludge content and a longer ultrasonic treatment duration. Although there was no significant interaction between factors with respect to TPH reduction, the interaction effect between ultrasonic treatment duration and H$_2$O$_2$/Fe^{2+} ratio was relatively higher than that between other factors. In summary, the combination of ultrasonic irradiation with Fenton's reaction can effectively reduce petroleum hydrocarbons in oily sludge within a short treatment period, and its treatment efficiency could be improved by appropriate combination of different factor levels.

References

[1]   D. Ramaswamy, D. D. Kar, S. De "A study on recovery of oil from sludge containing oil using froth flotation", Journal of Environmental Management, 85, 2007, 150-154.

[2] G. J. Hu, J.B. Li, G.M. Zeng "Recent development in the treatment of oily sludge from petroleum industry: a review", Journal of Hazardous Materials, 261, 2013, 470-490.

[3] R.P. Gogate, B.A. Pandit "A review of imperative technologies for wastewater treatment II: hybrid methods" Advances in Environmental Research, 8, 2004, 553-597.

[4] A.R. Torres, F. Abdelmalek, E. Combet, C. Petrier, C. Pulgarin "A comparative study of ultrasonic cavitation and Fenton's reagent for bisphenol A degradation in deionised and natural waters", Journal of Hazardous Materials, 16, 2007, 546-551.

[5] G.Y. Adewuyi "Sonochemistry in environmental remediation. 1 combinative and hybrid sonophotochemical oxidation processes for the treatment of pollutants in water" Environmental Science & Technology, 39, 2005, 3409-3420.

[6] M. Mohajerani, M. Mehrvar, F. Ein-Mozaffari "Recent achievements in combination of ultrasonolysis and other advanced oxidation processes for wastewater treatment" International Journal of Chemical Reactor Engineering, 8(R2), 2010, 1-78.

[7] J.J. Pignatello, E. Oliveros, A. MacKay "Advanced oxidation processes for organic contaminant destruction based on the Fenton reaction and related chemistry" Critical Reviews in Environmental Science & Technology, 36(1), 2006, 1-84.

[8] B. Neppolian, J. Park, H. Choi "Effect of Fenton-like oxidation on enhanced ocidative degradation of para-chlorobenzoic acid by ultrasonic irradiation" Ultrasonics Sonochemistry, 11, 2004, 273-279.

[9] J-H. Sun, S-P. Sun, J-Y. Sun, R-X, Sun, L-P. Qiao, H-Q. Guo, M-H. Fan "Degradation of azo dye acid black 1 using low concentration iron of Fenton process facilitated by ultrasonic irradiation" Ultrasonics Sonochemistry, 14, 2007, 761-766.

[10] J. Virkutyte, V. Vickackaite, A. Padarauskas "Sono-oxidation of soils: degradation of naphthalene by sono-Fenton-like process." Journal of Soils and Sediments, 10, 2010, 526-536.

[11] J. Zhang "Treatment of Refinery Oily Sludge using Ultrasound, Bio-surfactant, and Advanced Oxidation Processes" M.Sc. Thesis, University of Northern British Columbia, Prince George, Canada, 2013.

[12] ASTM "Standard test method for determination of water in liquid petroleum products" by Karl Fischer reagent, ASTM D1744, 1992.

[13] ASTM "Standard test method for determination of additive elements, wear metals, and contaminants in used lubricating oils and determination of selected elements in base oils" by Inductively Coupled Plasma Atomic Emission Spectrometry (ICP-AES), ASTM D5185, 2009.

[14] G. Taguchi "System of Experimental Design" White Plains, New York, 1987.

[15] Q.X. Huang, X. Han, F.Y. Mao, Y. Chi, J.H. Yan "A model for predicting solid particle behaviour in petroleum sludge during centrifugation" Fuel, 117, 30, 2014, 95-102.

[16] G.S. Gaikwad, B.A. Pandit "Ultrasound emulsification: effect of ultrasonic and physicochemical properties on dispersed phase volume and droplet size" Ultrasonics Sonochemistry, 15(4), 2008, 554-563.

[17] J.J. Lin, X.S. Zhao, D. Liu, Z.G. Yu, Y. Zhang, H. Xu "The decoloration and mineralization of azo dye C. I. Acid Red 14 by sonochemical process: rate improvement via Fenton's reactions" Journal of Hazardous Materials, 157, 2008, 541-546.

[18] Y.L. Pang, Z. Abdullah, S. Bhatia "Review on sonochemical methods in the presence of catalysts and chemical additives for treatment of organic pollutants in wastewater" Desalination, 277, 2011, 1-14.

[19] D.S. Casero, S. Rubio, D. Perez-Bendito "Chemical degradation of aromatic amines by Fenton's reagent" Water Research, 31, 1997, 1985-1995.

[20] H. Tekin, O. Bilkay, S.S. Ataberk, H.T. Balta " Use of Fenton oxidation to improve the biodegradability of a pharmaceutical wastewater" Journal of Hazardous Materials, 136, 2006, 258-265.

[21] C. Mansano-Weiss, H. Cohen, D. Meyerstein "Reactions of peroxyl radicals with $Fe(H_2O)_6^{2+}$" Journal of Inorganic Biochemistry, 91, 2002, 199-204.

[22] J.R. Watts, C.P. Stanton "Mineralization of sorbed and NAPL-phase hexadecane by catalyzed hydrogen peroxide", Water Research, 33(6), 1999, 1405-1414.

5

CO2 Sequestration by Direct Dry Gas-solid Contact of Serpentinite Mining Residues: A Solution for Industrial CO2 Emission

Sanoopkumar Puthiya Veetil[1*], Guy Mercier[1], Jean Francois Blais[1], Emmanuelle Cecchi[1], Sandra Kentish[2]

[1]INRS-ETE
490 rue de la couronne, Québec, Canada
sanoopkumar.puthiya_veetil@ete.inrs.ca; guy.mercier@ete.inrs.ca

[2]The University of Melbourne, Chemical and Biomolecular Engineering, VIC 3010, Australia
sandraek@unimelb.edu.au

Abstract – Direct dry gas-solid carbonation is a simple approach towards mineral carbon dioxide sequestration. The route theoretically implies the direct reaction of CO_2 with silicates of Calcium and Magnesium in dry condition to form stable, insoluble metal carbonates. The mining regions of southern Québec have a large deposit of serpentinite residues. The current study examines the suitability of serpentinite mining residues to use as feedstock material for mineral carbonation. The focus of the present work is to assess the CO_2 removal efficiency of the residue from a simulated flue gas mixture of a typical cement plant (18 Vol% CO_2). This approach avoids the requirement of separate CO_2 capture and pre-concentration prior to mineral carbonation. The reaction parameters considered are temperature, pressure and time. The optimization of parameters is carried out for the maximum CO_2 removal efficiency (%) from the feed gas. Operating condition for CO_2 removal is optimized at 258 °C, 5.6 barg ($pCO_2 \approx 1$) for 310 minutes with a removal efficiency of 37%. Preliminary analysis of reacted solid indicates carbonation is null at optimum condition, possibly a reversible adsorption might be responsible for the depletion of CO_2 from feed gas. The study also checks the importance of pre-treatment options such as grinding, magnetic separation and heat treatment on CO_2 removal. A separate optimization study is carried out for magnetic separation of serpentinite residue and the separation parameters are optimized at an initial pulp density of 40% and magnetic intensity of $7.5*10^{-3}$ T with about 70% of iron oxide removal from the initial feed.

Keywords: Serpentinite, mining residues, CO_2 sequestration, direct dry gas-solid.

1. Introduction

The augmentation of greenhouse gases such as carbon dioxide (CO_2) in the atmosphere has led to an increase in global temperature and changes in the climate. In order to mitigate the potentially devastating consequences of that phenomenon, the emissions of anthropogenic greenhouse gases especially CO_2 into the atmosphere should be reduced [1]. Carbon dioxide capture and storage (CCS) is a well-known option for mitigating the unwanted anthropogenic CO_2 emissions. Geological formation and ocean are the widely used sinks for carbon dioxide storage, but they are limited either due to the lack of permanence in storage or due to environmental issues associated with storage [2]. Mineral carbonation is the promising CCS option which guarantees the permanent storage of CO_2 sequestrated [3, 4]. This option mimics the process of natural silicate weathering in which CO_2 reacts with the divalent cation (Ca^{2+} or Mg^{2+}) of natural minerals to form metal carbonates [5]. The carbonates formed are environmentally benign and geologically stable [2-4, 6]. Mineral carbonation is now in the developing stage, but the cost and kinetics keep it laid-back from other CO_2 storage options [2]. Mafic and ultramafic rocks containing magnesium (Mg) and calcium (Ca) are generally used for mineral carbonation due to their relative abundance and admissible reactivity with CO_2 [5]. In comparison to other metal silicates, magnesium based minerals such as serpentine and olivine are most abundant in nature [7]. Therefore, these minerals are under research scrutiny to develop as a cheap raw material for mineral carbonation [8, 9]. Besides these natural forms, waste materials contained admissible concentration of Mg and Ca have also been employed as mineral carbonation feedstock. These include mineral tailings of asbestos, industrial waste like stainless steel slag, waste cement, fly and bottom ash from municipal solid waste incinerator etc. [1, 7].

Mineral carbonation could be done either directly or indirectly through aqueous or dry route. Direct gas-solid route is a straight forward approach towards mineral carbonation and exothermic in nature [7, 10, 11]. Since the carbonation rate of pure mineral under direct dry condition was found to be slow, many modifications were recently suggested. Most recent works have been reported using a multistep gas-solid carbonation of magnesium hydroxide [$Mg(OH)_2$] produced from serpentine [12-16]. Another recent approach has been reported with the addition of a small amount of water or water vapor for carbonation enhancement [17-21]. Most of the previous works on direct gas-solid route was mainly focused on the carbonation of solid material [13, 19, 21, 22] and less work have been reported on the CO_2 depletion potential of minerals [18, 20].

The present is a preliminary laboratory study conducted using a batch wise mode. The objective of the study is to check the feasibility of serpentinite mining residue as a CO_2 removal or carbonation material using direct dry gas-solid route. The mining residue used is from a chrysotile extraction mine in southern Québec. Direct dry gas-solid route was selected because of the simplicity and exclusion of chemical additives. The study adapted a new strategy of directly using a readily available mining residue for the capture and storage of CO_2 from an industrial flue gas. Instead of pure mineral, using a mine residue that contains various mineral phases might provide a better reaction and could avoid the mining and mineral purification requirements Direct use of a flue gas composition (18% CO_2) could avoid the separate steps of CO_2 capture and pre-concentration.

2. Experimental Section

Serpentinite Mining Residue (SMR) collected from Black Lake mine (Québec) was used for the present study. Collected samples were homogenised and then stored in separate sealed containers. The moisture content of the raw sample was measured by heating the mineral sample at 110 °C for around 24 h. The texture analysis of the sample was carried out by sieving through meshes with different cut sizes between 0.075 mm to 2 mm.

2.1. Material Characterization

Mineral phase of the residue was identified by X-ray diffraction (XRD) analysis using a Siemens D5000 diffractometer with Cu Kα radiation. Scans were taken for 2θ over 2° to 65° at 0.02°/s. Microscopic imaging and a semiquantitative analysis of the sample was done with a scanning electron microscope (SEM) equipped with energy dispersive X-ray spectroscopy (EDS) (Zeiss Evo 50 Smart SEM). Chemical composition of the mineral sample was determined by inductively coupled plasma atomic emission spectroscopy (ICP-AES) (Varian 725-ES, Model Vista-AX CCO, Palto Alto, CA, USA) after an alkaline fusion with lithium metaborate. The specific surface area and average pore diameter of the sample were determined by means of Brunauer–Emmett–Teller (BET) analyzer (BELSORP-max, BEL Japan Inc). For this, the samples were dried and analyzed for N_2 sorption and desorption isotherm at 77 K after degassing at 150 °C overnight and the residual pressure down to 10^{-5} Torr.

2.2. Experimental Procedure
2.2.1. Mineral Pre-treatment

The grinding of SMR was done in a shatter-box (BLEULER-NAEF shatter-box, model M04/06) to increase the surface area and thereby enhance the reaction rate. The particle size of the ground SMR was determined by a laser particle analyzer (Horiba laser particle size distribution analyzer LA-950). The others pre-treatments options adapted for the current study were thermal heat treatment and magnetic separation. Temperature for thermal treatment was adapted from the previous studies conducted with similar type materials [23-27]. For this, the ground SMR was heated to 650 °C for 30 min in an air muffle furnace (Thermolyne Furnatrol 133). Magnetic separation was carried out by using a wet high intensity magnetic separator (WHIMS -CARPCO model serial no. 221-02) in order to remove the magnetic impurities; especially compounds of iron (Fe) from SMR. This was carried out to avoid the formation of a passive layer of iron oxide such as hematite during heat treatment which negatively affects the carbonation process [27, 28]. Conducting magnetic separation prior mineral carbonation can also provide final carbonation products of less magnetic impurities and a separate stream of magnetic by-product with good potential market value. Optimization of parameters for magnetic separation of SMR was separately carried out and the non-magnetic sample at optimum condition was used for direct dry gas-solid experiments. Parameters considered for the magnetic separation study were magnetic intensity (T) and initial pulp density (%).

2.2.2. Direct Dry Gas-solid Reaction

Samples for direct dry mineral reaction were categorized into three types: (1) raw, (2) heat-treated, and (3) magnetically separated samples. All tests were conducted in a batch reactor designed by Parr (Parr 4560 Mini Bench Top Reactor) of capacity 300 mL. The representation of the experimental set-up is given in Figure 1.

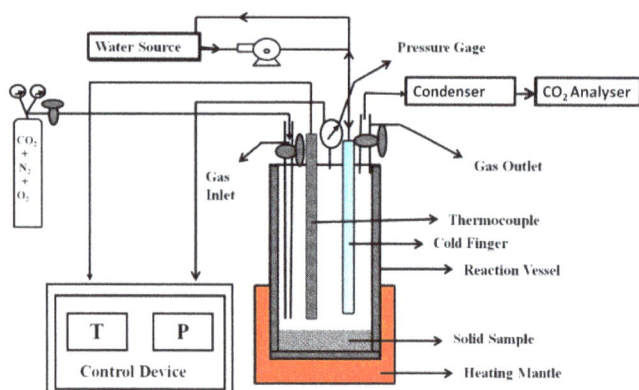

Figure 1. Diagrammatic representation of direct dry gas-solid carbonation.

Direct dry gas-solid reactions were carried out by contacting ground SMR with CO_2 gas mixture ($N_2/CO_2/O_2$-78/18/4% (v/v)) at varying temperature and pressure. Based on the previous studies on this route, the reaction parameters considered for optimization were temperature, pressure and time [11, 12, 19, 29]. The direct dry gas-solid experiments were conducted by loading a definite mass (g) of ground SMR into the reactor. Then, it was heated to the desired reaction temperature (°C). After reaching the reaction temperature, the air and moisture inside the reaction vessel was purged and filled with the gas mixture to a

desired pressure (bar). After a certain reaction time (min), the non-reacted gas was collected in a *Tedlar Bags* (3.8 L) equipped with an *on / off valve* and subjected to analysis. The concentration of CO_2 (Vol%), before and after reaction, was measured by means of a CO_2 analyzer (Quantec instrument model 906) and the mass of CO_2 in the inlet and outlet gas was calculated using the ideal gas equation (PV=nRT) by knowing the reaction temperature and CO_2 partial pressure.

The reaction deciding factor considered in the parameters optimization was the efficiency of CO_2 removal (%) (Equation 1). Initial experiments were conducted in conventional method (single variable at a time) using raw, heat-treated and nonmagnetic sample. This was carried out to get an approximate idea about the parameter influence on CO_2 removal. Then, the optimization was carried out with a statistical Box-Behnken model (Design Expert with ANOVA: Version 8.0.4, Stat-Ease Inc., Minneapolis, USA). The independent variables considered for the model were temperature, total pressure and time with corresponding ranges of 25 to 280 °C, 5 to 95 bar and 120 to 360 min respectively. The model designed 17 experiments and heat-treated sample of 25 g was used for the each experiment. Based on the experimental results obtained, the model predicted a most desirable interaction of parameters for the maximum response. Experimental validation of the model was carried out on the predicted optimized conditions. Comparative results were also generated with raw and nonmagnetic sample at optimum conditions.

$$CO_2 \text{ removal efficiency (\%)} = ((CO_{2in} - CO_{2out})/ CO_{2out}) \times 100 \qquad (1)$$

Where CO_{2in} is the mass of CO_2 in the inlet gas and CO_{2out} is the mass of CO_2 in the outlet gas. The solid samples at optimum result conditions were subjected to both elemental carbon analysis (Leco CHNS-932 auto analyzer) and XRD to assess the formation of carbonates.

3. Results and Discussion
3.1. Characterization of SMR

The average interstitial water content of SMR from Black Lake residue was about 5.3% (±0.5). The texture analysis of SMR shows about 38% of the sample was above 2 mm, 22.9% was between 1 to 2 mm and remaining 39.1% was below 1 mm. Since 60% of the sample was above 1 mm, grinding was adapted to reduce the size to a micro level. The mean size of the ground SMR given by laser particle analyzer was 75 μm with 90% distribution below 250 μm. The XRD gives major peaks for lizardite associated with minor components such as chrysotile, magnetite, brucite, talc and chlorite. This is in concurrence with mineral compositions reported for similar type materials near the present sampling station [17, 26, 30, 31]. The EDS analysis of the site within the SEM images confirmed the presence of Fe substituted lizardite and magnetite (Fe_3O_4). The chemical composition of SMR in each size fraction is presented in Table 1. The sample is mainly concentrated with magnesium and silicon with a considerable level of Fe. This confirms the presence of serpentine group mineral (lizardite). The presence of Fe would account for both magnetite and partial substituted Fe within the lizardite structure. The average percentage of MgO (42.5±0.2%) and SiO_2 (40±0.4%) in SMR are admissible range with known average value for lizardite [32]. The BET specific surface area and average pore diameter of the SMR were 11.5 m^2/g and 13.7 nm respectively.

Table 1. Particle size distribution and chemical composition of SMR from Black Lake mine.

Mesh size (mm)	Weight (%)	Fe₂O₃ (%)	MgO (%)	SiO₂ (%)	Ni (mg/kg)	Co (mg/kg)	Cr (mg/kg)
≥2	38.0	11.1	41.5	37.0	2409	63.9	2493
1-2	22.9	9.0	36.9	37.7	1702	107.6	2010
0.5-1	17.8	10.1	41.3	37.9	2017	74.8	2160
0.3-0.5	9.3	11.9	39.4	36.5	1895	63.3	1991
0.15-0.3	7.5	12.2	34.8	35.9	1696	108.6	2554
0.075-0.15	3.4	18.6	36.9	34.5	1968	73.4	2856
≤0.075	1.1	21.9	31.9	32.4	1739	135.3	2545

3.2. Magnetic Separation

The best conditions for magnetic separation of SMR from mine Black Lake was optimized at a magnetic intensity of 7.5*10⁻³ T and an initial pulp density of 40% with ground SMR of mean size 75 μm. The second pass of magnetic separation products at above mentioned conditions given a maximum non-magnetic mass recovery of 90.0% (±0.2). The magnetic separation reduces the iron oxide (as Fe_2O_3) concentration of SMR from 10.9% (±0.4) to 3.4% (±0.1). From the raw feed about 97% of magnesium has been recovered in non-magnetic fractions. The final recovered magnetic fraction is rich in iron oxide with 79.0% (±0.3) composition. All the above results are calculated based on the ICP-AES analysis (after alkaline fusion with lithium metaborate). In short, conducting magnetic separation at above conditions resulted in 70% of iron oxide impurities removal from feed SMR.

3.3. Direct Dry Gas-solid Reaction

The maximum CO_2 removal obtained using conventional single variable method was about 36.7% (±2) with heat treated sample at 200 °C, 5 bar (pCO₂ =0.90 bar) after 360 min (6 h) duration. The response surface methodology (RSM) graph obtained from Box-Behnken analysis is given in Figure ure 2. The result shows that the CO_2 removal percentage was increased with increases in temperature and decreases in total pressure. The Model Prob > F value less than 0.05; implies that the model is significant. The value of multiple regression coefficients, R^2 = 0.98, shows that the model could explain 98% of the response variability and which was in reasonable agreement with the adjusted R^2 value (0.94). The model predicted an optimum condition at 258 °C, 5.6 bar (pCO₂ ≈ 1) for 310 min (≈ 5 h) with 40% CO_2 removal and was validated experimentally with 37% (±0.6) CO_2 removal at these conditions. Direct dry reaction of raw and non-magnetic samples was also conducted at optimum condition (Box-Behnken) and obtained almost equal CO_2 removal. At the maximum removal efficiency (37%), 25 g SMR removed about 0.12 g (±0.01) of CO_2 from the feed gas with initial CO_2 concentration of 0.32 g CO_2. The elemental carbon and XRD analysis carried for reacted solid at optimum conditions did not show any variation from the original sample. This indicates CO_2 might be removed from the feed gas due to a reversible adsorption rather than carbonation.

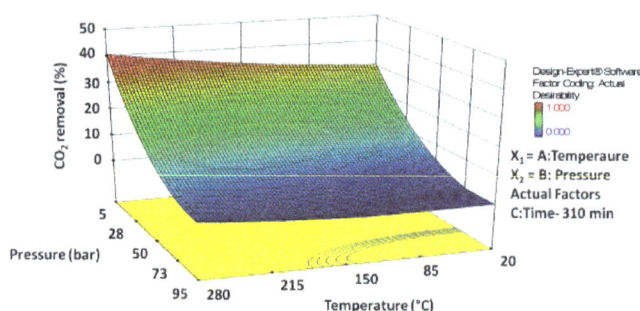

Figure 2. RSM graph showing the interaction of T and P on CO_2 removal efficiency.

The results furnished for CO_2 removal are only based on the CO_2 analyzer measurement. The current removal is too low for practical application. We suggest that a limited CO_2 removal occurred due to the poor gas-solid contact in the batch set-up. Experiments are required to be conducted with the presence of water vapor to enhance carbonation [17-21]. More process improvement is required to enhance the CO_2 removal efficiency of solid such as increasing the gas-solid interaction through the fluidization of the bed and increasing the surface area and pore size of solid through a series of grinding.

4. Conclusion

The present work highlighted a new approach towards direct gas-solid carbonation by incorporating a mine waste for the direct capture or sequestration of CO_2 from an industrial chimney. The usage of a residue rather than pure mineral can help in avoiding the mining and mineral purification cost requirement for mineral CO_2 sequestration. Beside this, the process will revalorize the waste residue into environmentally benign material that can be used for land filling and mine reclamation (only if carbonation occurs). Switching from 100% CO_2 to simulated flue gas composition of 18 Vol% CO_2 can reduce the separate CO_2 capture and pre-concentration cost. The main challenge of the route is achieving an admissible reaction rate with a low cost and energy consuming process.

The preliminary results obtained through the characterization and the parameter optimization for the mineral CO_2 sequestration of SMR shows that SMR can be selected as a feedstock for CO_2 sequestration. The reacted solid analysis indicates that the carbonation

under dry condition is negligible, but that the CO_2 is depleted from the feed gas possibly due to a reversible adsorption. Carbonation possibilities have to be assessed by the addition of water vapor. Studies are further required in a suitable design such as fluidized bed providing better gas-solid interaction to increase the reaction rate and kinetics. More extensive studies are required in this area to reach an admissible reaction rate and kinetics and thereby develop a promising process for an industrial application. The prime goal of the future work is to have an admissible carbonation reaction at low pressure and low temperature within the short reaction period.

Acknowledgements

The authors are grateful for financial support from the Fonds Québécois de la Recherche sur la Nature et les Technologies (FQRNT), Carbon Management Canada and the support from INRS-ETE, Quebec, Canada.

References

[1]   IPCC, "Synthesis Report of the IPCC on climate change," Valencia, Spain, 2007.

[2]   IPCC, "IPCC special report on carbon dioxide capture and storage," Cambridge University Press, Cambridge, United Kingdom and New York, NY, USA, 2005.

[3]   W. Seifritz, "CO_2 Disposal by mean of silicates," Nature, vol. 345, pp. 486-486, Jun 7 1990.

[4]   K. S. Lackner, C. H. Wendt, D. P. Butt, E. L. Joyce, and D. H. Sharp, "Carbon-dioxide disposal in carbonate minerals," Energy, vol. 20, pp. 1153-1170, Nov 1995.

[5]   W. J. J. Huijgen and R. N. J. Comans, "Carbon dioxide sequestration by mineral carbonation: Literature review," Energy Research Centre of the Netherlands, vol. ECN-C--03-016, 2003.

[6]   K. S. Lackner, "Climate change: A Guide to CO_2 Sequestration," Science, vol. 300, pp. 1677-1678, 2003.

[7]   W. J. J. Huijgen and R. N. J. Comans, "Carbon dioxide sequestration by mineral carbonation: Literature review update 2003-2004," Petten, The Netherlands, ECN-C--05-0222005.

[8]   H. Bearat, M. J. McKelvy, A. V. G. Chizmeshya, D. Gormley, R. Nunez, R. W. Carpenter, K. Squires and G.H. Wolf "Carbon sequestration via aqueous olivine mineral carbonation: Role of passivating layer formation," Environmental Science & Technology, vol. 40, pp. 4802-4808, Aug 1 2006.

[9] M. J. McKelvy, A. V. G. Chizmeshya, J. Diefenbacher, H. Béarat, and G. Wolf, "Exploration of the Role of Heat Activation in Enhancing Serpentine Carbon Sequestration Reactions," Environmental Science & Technology, vol. 38, pp. 6897-6903, Dec 1 2004.

[10] J. Sipilä, S. Teir, and R. Zevenhoven, "Carbon dioxide sequestration by mineral carbonation - Literature review update 2005-2007," Abo Akademi, 2008.

[11] R. Zevenhoven, S. Teir, and S. Eloneva, "Heat optimisation of a staged gas-solid mineral carbonation process for long-term CO_2 storage," Energy, vol. 33, pp. 362-370, Feb 2008.

[12] J. Fagerlund, E. Nduagu, I. Romão, and R. Zevenhoven, "A stepwise process for carbon dioxide sequestration using magnesium silicates," Frontiers of Chemical Engineering in China, vol. 4, pp. 133-141, 2010.

[13] J. Fagerlund, E. Nduagu, I. Romão, and R. Zevenhoven, "CO_2 fixation using magnesium silicate minerals part 1: Process description and performance," Energy, vol. 41, pp. 184-191, 2012.

[14] E. Nduagu, T. Björklöf, J. Fagerlund, J. Wärnå, H. Geerlings, and R. Zevenhoven, "Production of magnesium hydroxide from magnesium silicate for the purpose of CO_2 mineralisation–Part 1: Application to Finnish serpentinite," Minerals Engineering, vol. 30, pp. 75-86, 2012.

[15] R. Zevenhoven, J. Fagerlund, E. Nduagu, I. Romão, B. Jie, and J. Highfield, "Carbon Storage by Mineralisation (CSM): Serpentinite Rock Carbonation via $Mg(OH)_2$ Reaction Intermediate Without CO_2 Pre-separation," Energy Procedia, vol. 37, pp. 5945-5954, 2013.

[16] J. Fagerlund, E. Nduagu, and R. Zevenhoven, "Recent developments in the carbonation of serpentinite derived $Mg(OH)_2$ using a pressurized fluidized bed," Energy Procedia, vol. 4, pp. 4993-5000, 2011.

[17] J. Pronost, G. Beaudoin, J. Tremblay, F. Larachi, J. Duchesne, R. Hebert, M. Constantin, "Carbon Sequestration Kinetic and Storage Capacity of Ultramafic Mining Waste," Environmental Science & Technology, vol. 45, pp. 9413-9420, Nov 1 2011.

[18] H. F. Da Costa, M. Fan, and A. T. R., "Method to sequester CO_2 as mineral carbonate," U.S Patent, 20100221163A1, 2010.

[19] F. Larachi, I. Daldoul, and G. Beaudoin, "Fixation of CO_2 by chrysotile in low-pressure dry and moist carbonation: Ex-situ and in-situ characterizations," Geochimica Et Cosmochimica Acta, vol. 74, pp. 3051-3075, 2010.

[20] S. Kwon, M. Fan, H. F. DaCosta, and A. G. Russell, "Factors affecting the direct mineralization of CO_2 with olivine," Journal of Environmental Sciences, vol. 23, pp. 1233-1239, 2011.

[21] F. Larachi, J. P. Gravel, B. P. A. Grandjean, and G. Beaudoin, "Role of steam, hydrogen and pretreatment in chrysotile gas–solid carbonation: Opportunities for pre-combustion CO_2 capture," International Journal of Greenhouse Gas Control, vol. 6, pp. 69-76, 2012.

[22] R. Zevenhoven, J. Kohlmann, and B. A. Mukherjee, "Direct Dry Mineral Carbonation for CO_2 Emissions Reduction in Finland," in 27th International Technical Conference on Coal Utilization & Fuel Systems Clearwater (FL), USA, 2002.

[23] M. M. Maroto-Valer, D. J. Fauth, M. E. Kuchta, Y. Zhang, and J. M. Andresen, "Activation of magnesium rich minerals as carbonation feedstock materials for CO_2 sequestration," Fuel Processing Technology, vol. 86, pp. 1627-1645, Oct 2005.

[24] G. W. Brindley, "A structural study of the thermal transformation of serpentine minerals to forsterite," American Mineralogist, vol. 42, pp. 461-474, 1957.

[25] C. Jolicoeur and D. Duchesne, "Infrared and thermogravimetric studies of the thermal degradation of chrysotile asbestos fibres: evidence for matrix effects, Canadian Journal of Chemistry, v. 59, p. 1521-1526.," 1981.

[26] M. Nagamori, A. J. Plumpton, and R. Le Houillier, "Activation of magnesia in serpentine by calcination and the chemical utilization of asbestos tailings - a review," CIM Bulletin, vol. 73, pp. 144-156, 1980.

[27] W. O'connor, D. Dahlin, G. Rush, S. Gerdemann, L. Penner, and D. Nilsen, "Aqueous mineral carbonation, mineral availability, pretreatment, reaction parametrics, and process studies," in US DOE, DOE/ARC-TR-04-002, Albany Research Centre: Albany, OR, 2005., 2004.

[28] W. K. O'Connor, D. C. Dahlin, D. N. Nilsen, R. P. Walters, and P. C. Turner, "Carbon dioxide sequestration by direct aqueous mineral carbonation," in Proceedings of the 25th International Technical Conference on Coal

Utilization & Fuel Systems, Clear Water, Florida., 2000.

[29]   R. Baciocchi, A. Polettini, R. Pomi, V. Prigiobbe, V. N. Von Zedwitz, and A. Steinfeld, "CO_2 sequestration by direct gas-solid carbonation of air pollution control (APC) residues," Energy & Fuels, vol. 20, pp. 1933-1940, Sep 20 2006.

[30]   I. M. Power, S. A. Wilson, D. P. Small, G. M. Dipple, W. Wan, and G. Southam, "Microbially Mediated Mineral Carbonation: Roles of Phototrophy and Heterotrophy," Environmental Science & Technology, vol. 45, pp. 9061-9068, 2011.

[31]   A. L. Auzende, I. Daniel, B. Reynard, C. Lemaire, and F. Guyot, "High-pressure behaviour of serpentine minerals: a Raman spectroscopic study," Physics and Chemistry of Minerals, vol. 31, pp. 269-277, Jun 2004.

[32]   A to Z Listing of Minearls, [Online] Available at: http://webmineral.com/data/Lizardite.shtml. consulted 25 Feb. 2013.

6

Phytotreatment of Polychlorinated Biphenyls Contaminated Soil by *Chromolaena odorata* (L) King and Robinson

Raymond Oriebe Anyasi*, Harrison Ifeanyichuku Atagana
Department of Environmental Sciences
Institute for Science and Technology Education
University of South Africa, 1, Preller street, Muckleneuk Ridge, Pretoria, South Africa
41525981@mylife.unisa.ac.za

Abstract -The ability of *Chromolaena odorata* propagated by stem cuttings and grown for six weeks in the greenhouse to thrive in soil containing different concentrations of PCB congeners found in Aroclor 1254, and to possibly remediate such soil was studied under greenhouse conditions. *Chromolaena odorata* plants were transplanted into soil containing 100, 200, and 500 ppm of Aroclor in 1L pots. The experiments were watered daily at 70 % moisture field capacity. Parameters such as fully expanded leaves per plant, shoot length, leaf chlorophyll content as well as root length at harvest were measured. PCB was not phytotoxic to *C. odorata* growth but plants in the 500 ppm treatment only showed diminished growth at the sixth week. Percentage increases in height of plant were 45.9, 39.4 and 40.0 for 100, 200 and 500 ppm treatments respectively. Such decreases were observed in the leaf numbers, root length and leaf chlorophyll concentration. The control sample showed 48.3 % increase in plant height which was not significant from the treated samples, an indication that *C. odorata* could survive such PCB concentration and could be used to remediate contaminated soil. Mean total PCB absorbed by *C. odorata* plant was between 6.40 and 64.60 ppm per kilogram of soil, leading to percentage PCB absorption of 0.03 and 17.03 % per kilogram of contaminated soil. PCBs were found mostly in the root tissues of the plants, and the Bioaccumulation factor were between 0.006-0.38. Total PCB absorbed by the plant increases as the concentration of the compound is increased. With these high BAF ensured, *C. odorata* could serve as a promising candidate plant in phytoextraction of PCB from a PCB-contaminated soil.

Keywords: Phytoremediation, Bioremediation, Soil restoration, Polychlorinated biphenyls (PCB), Biological treatment, Aroclor.

1. Introduction

The unprecedented growth in agriculture, chemical industries, oil production, transportation, military activities and mining has contributed in the intensive generation of pollution to the environment (Graham and Ramsden, 2008). The concentrations of these anthropogenic toxic substances in the environment have risen beyond set limits; although quantification of such increases had been difficult to ascertain. However, annual estimation of the spread has been reported to be in billions of tons (USEPA, 1997). This sudden rise in waste generation leads to nature cycling and environmental degradation. Environmental degradation causes loss in biodiversity and the ecosystem which eventually impacts on human health if proper measures are not employed to address the consequences (Pilon-Smith, 2005; Mosaddegh *et al.*, 2014). There are different types of contaminants found in the environment. The most dangerous among them are those that have high capabilities to persist, bioaccumulate, and be toxic to man in the food chain e.g. polychlorinated biphenyls (PCBs). Polychlorinated biphenyls (PCBs) are a family of compounds produced commercially by direct chlorination of biphenyls. As a result of its dielectric nature, the compound is used for various activities, example as a component of the transformer oil. Through its use and incessant disposal,

PCB finds its way into the environment and its sink is the soil from where it contaminates other part of the environment (Graham and Ramsden, 2008).

Knowledge of the environmental occurrence of PCB emanated from the discovery of extremely high level of PCBs in a white-tailed sea eagle found dead in Stockholm archipelago reported by Jenson in 1966 (Andersson, 2000). Today, PCBs can be found in all environmental compartments including water, soil and air even in the Polar Regions. They spread into the environment from dumps, landfills, combustion process, and from their use in various open and close systems leading to their toxic effects in wildlife and human (Low et al., 2009). The effects of toxicity of PCB were brought to public awareness by the Yusho incident in Western Japan in 1968, where more than 1800 persons suffered from toxicity due to consumption of contaminated rice oil (Xu et al., 2010). Subsequently, the production of PCBs in Sweden and many other industrial countries have been strictly restricted since the 1970s. The most widely accepted method for the destruction of PCBs is incineration (Rodriguez and Lafuente, 2002). However, incineration is an expensive practice and often produces more toxic compounds as by-products (Andersson, 2000). Current chemical remediation techniques were developed as a result of the demerits of incineration.

The ability of PCBs to be degraded or be transformed in the environment depends on the degree of chlorination of the biphenyl molecule as well as isomeric substitution pattern (Bhandary, 2007). As a result biochemical abilities of microorganisms became one of the leading strategies in the biological treatment of PCB contaminated sites (Idris and Ahmed, 2003). This nascent shift into biological means of PCB remediation came into place because of the disadvantages of chemical and physical methods and the method is referred to as bioremediation. Bioremediation is the use of living organisms to reduce or eliminate environmental hazards resulting from accumulations of toxic chemicals or other hazardous waste (Lee, 2013). Bacteria are generally used for bioremediation, but fungi, algae and plants could also be used as is the case in this study. When plants are used in bioremediation, it is usually referred to as phytoremediation. Phytoremediation is the use of vegetation for in situ treatment of contaminants in soil and water bodies. It is a promising technique that is made up of different forms depending on the technique involved. Available research investigations into

phytotreatment of PCB contaminated soil used mostly food crops, with interests in the members of the Cucurbita family (Zeeb et al., 2006; Mattina et al., 2007; Ficko et al., 2011). These however will add pressure into the already crippling world food security since such plant should not be consumed for its adversity. For phytoremediation of organic contaminants to be effective there should be maximum possession of plants phytoremediation abilities as indicated in Ficko et al, (2011). Weeds have been shown to possess such abilities which include amongst other factors the ability to be propagated and cultivated with simple agronomic measures; they are relatively inexpensive, self-sustainable, have unique root system and also have the ability to grow in a diverse environment (Singh et al., 2009).

Chlomolaena odorata (L) R.M. King & H. Robinson (Asteraceae) is an invasive bushy shrub of Neotropical origin and has been described as one of the world's worst tropical weeds (Tanhan, 2011). The plant is a member of the tribe Eupatoreae in the sunflower family Asteraceae. Chromolaena odorata is described as a perfect competitor in its physiology; this means that it scavenges for available nutrients in the soil, as a result suppresses the growth of other plants even the weeds of its category (De Rouw, 1991).

Plants should be considered for phytoremediation studies if they possess a number of growth characteristics which include and not limited to plants ability to survive under stress, accumulates high biomass as well as the potential to dominate native vegetation at any new environment (Tanhan, 2011). Chromolaena odorata possesses most of these qualities, some of which are responsible for its success as an invasive plant in new environments. These factors therefore present Chromolaena odorata as a potential plant for phytoremediation of PCB-contaminated soil. This study involves an investigation of the capabilities of C. odorata plants to grow in Aroclor contaminated soil and reduce the concentration of the compound under greenhouse conditions.

2. Materials and Methods

Chromolaena odorata plants were transplanted directly into a 1 kg of soil containing different concentrations of Aroclor in PVC pots, noting the initial length and mature leaf per plant (MLPP). The experiment which was translated into treated soil with plants (T) made in three different concentrations of 100, 200 and 500 ppm and treated soil without plants (control=C), was monitored for six weeks at prevailing environmental conditions, maintaining other agronomic procedures. Measurements were made over time of plant growth parameters including length of plants, MLPP, leaf colour (Chlorophyll content) and root length which was measured on the day of harvest. Harvested plant samples were analyzed for total PCB (tPCB) using GC-MS while the residual soil was analyzed to ascertain the reduction rate of PCB at the end of the experiment. They were no application of inorganic manures to the soil mixes, but animal compost was used during soil preparation. After six weeks of experimentation, the soil and plants were harvested, measured and sampled for analysis

Initial length of plant was measured on the day of contamination using strings and meter rule and subsequent measurements were taken at weekly intervals. MLPP was also measured on Day one of contamination and subsequently at weekly intervals using manual counting of the leaves. The same measurement sequence was employed in leaf chlorophyll using Chlorophyll meter from the UNISA Unit for Horticulture (SPAD-502Plus Konica Minolta, Japan-). Root length was measure first on the day of contamination and finally on the day of harvest using strings and meter rule to get the initial and final measurements respectively. Soil and plant samples were thoroughly homogenized for analysis and sub-sampled for the determination of wet and dry weight ratio. The samples for biomass determination were dried at 50°C until constant mass using Lancon industrial oven (Labcon South Africa) with heating integration of 40-100°C and were measured to obtain the dry mass. The dried plant samples were then ground using commercial blender, sieved at 2 mm and were stored prior to extraction while the soil samples were ground using a commercial mortar and was sieved at 2 mm. The extraction process adopted was 'Soxhlet Extraction Method 3540' (US EPA, 1997), USEPA Method 3630B: for cleaning, and USEPA modified 8089/8081 method was used for the determination of total PCB. The analysis was conducted according to Anyasi (2012). The whole values presented from the analyses of samples were the mean values of three replicates. General linear model of analysis of variance (ANOVA) was used at P = 0.05 level of significance difference (SPSS) (version 11.0 for windows).

3. Results and Discussion

The properties of the soil in which the plant was grown indicated slightly acidic clayey sandy soil, other parameters indicated compatibility with optimal growth of plants [7ppm of TOC, N=0.03 %wt, P=9 ppm, K=15.5 ppm, Ca=83 ppm, Mg=1.2, and moisture content was 4.8 %]. The low value of total nitrogen in the soil compared to carbon and phosphorus indicated the need for a nitrogen impacted manure. However, the CNP ratio of the soil in relation to the manure was 233:1:300 and 1:2:1 respectively. There was progressive growth of C. odorata throughout the duration of the study in both the treated and the control experiment. After six weeks of growth in Aroclor treated soil, there was no sign of lethal phytoxicity to C. odorata, except at the 500 ppm treatments where chlorosis was observed towards the end of sixth week. Plant growth measured from the difference between the initial and final length of C. odorata and deduced in percentage as the percentage growth rate is as presented in Table 1.

Table 1: Percentage growth rate of C. odorata at different concentrations of Aroclor in soil.

| Treatments (ppm) | % growth rates | | |
	100µg/ml (cm)	200µg/ml (cm)	500µg/ml (cm)
T	45.89 ±0.13	40.01 ±1.12	39.41 ±0.24
C	NP	NP	NP

Values with the same * in the same column were not significant at 5% level according to Bonferoni test. T=Treatments, C=Control, NP=Not planted.

Mean percentage growth rate was highest at 100 ppm (45.9), least in 500 ppm (39.41). Meanwhile, the growth of C. odorata was found to be negatively correlated with increase in concentration of Aroclor in soil. Percentage growth rate at untreated control was slightly lower than that in 100 ppm treatments, this was not significant (p = 0.02) at p= 0.05. MLPP of C. odorata at any interval of time was observed to be influenced by the presence of Aroclor in its tissues. MLPP followed the same trend as seen in growth rate as well as increase in root length. At 100 ppm MLPP increased from initial 28 leaves to final 50 leaves, leaving a mean percentage increase of 78.6. Mean percentage increases at 200 and 500 ppm were 45.1 and 24.2 respectively (Table 2).

Table 2: Percentage change in mature leaves per plant in different concentrations of Aroclor.

Treatments (ppm)	% increase in mature leaves per plant		
	100 (mg/kg) soil	200 (mg/kg) soil	500 (mg/kg) soil
T	78.57 ±0.43*	45.16 ±1.04*	24.24 0.31*
C	NP	NP	NP

Values with the same * in the same column were significant at 5 % level according to Bonferoni test. T=Treatments, C=Control, NP=Not planted.

Mean Percentage change in MLPP at untreated control (C1) was 86.9; this was not significant from the 100 ppm treatments but significantly different from the 200 and 500 ppm treatments respectively. Root lengths of *C. odorata* at different concentrations of Aroclor 1254 and 1260 were equally synonymous with what was observed in growth rate and MLPP. Percentage change in root length was high at 100 ppm (78.3), but reduced considerably at 200 and 500 ppm (59.1 and 56.1) respectively Table 3.

Table 3: Percentage change in root length at different concentrations of Aroclor.

Treatments (ppm)	Percentage change in root length		
	100 ppm	200 ppm	500 ppm
T	78.28 ±0.33*	59.13 ±0.08	56.12 ±0.89
C	NP	NP	NP

Values with the same alphabets in superscript in the same column were significant at 5% level according to Bonferoni test. T= Aroclor 1254, C=Control, NP=Not planted

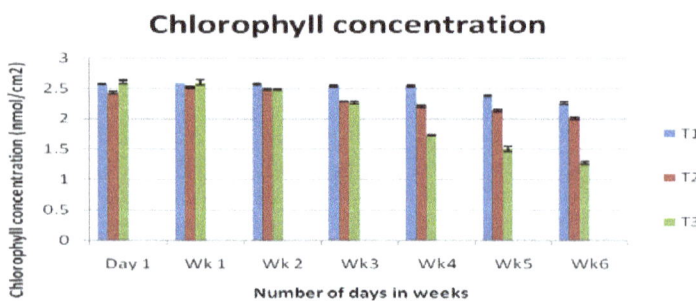

Figure 1: Leaf chlorophyll concentration showing variations in leaf color in different Aroclor concentrations over time (Error bars indicate standard error of the mean), T_1= T at 100 ppm, T_2= T at 200 ppm, T_3= T at 500 ppm

Table 4: Change in leaf chlorophyll concentration (Values are means of three replicates).

Treatments	Initial (Day 1) nmol/cm^2	Final (Week 6) nmol/cm^2	Percentage change in chlorophyll concentration (%)
T_{100}	2.57 ±0.02	2.26 ±0.06	12.06 ±0.37
T_{200}	2.43 ±0.14	2.01 ±0.07	17.28 ±0.16
T_{500}	2.60 ±0.11	1.28 ±0.19	50.76 ±0.25*
C	NP	NP	NP

Values with the same * in the same column were significant at 5% level according to Bonferoni test. T_{100}= T at 100 ppm, T_{200}= T at 200 ppm, T_{500}= T at 500 ppm, C= Control, NP=Not planted

Although the leaves of *C. odorata* at 100 and 200 ppm treatments showed no observable colour change throughout the duration of the experiment when compared with untreated control, chlorophyll measurement chart however indicated alternatively. At 100 ppm, mean initial chlorophyll measurement was 2.57 nmol/cm^2, it increased slightly to 2.59 nmol/cm^2 at the first week and reduced from the second week (2.57 nmol/cm^2) to the sixth week (2.26 nmol/cm^2) Figure 1. Differences existed between chlorophyll concentration of the initial and final measurements of the chlorophyll meter; this is depicted as the change in percentage leaf chlorophyll concentration Table 4. The mean percentage change in chlorophyll throughout the six weeks of experimentation was 12.06, 17.28, 50.76, in their respective treatments (T_{100}, T_{200}, T_{500}). However, pale green colour change was observed at the 500 ppm treatment towards the sixth week of experimentation as was shown by their high percentage change in chlorophyll concentration.

Phytoremediation of soil contaminants is attributed to amongst other factors to the ability of plants to grow rapidly in the contaminated soil so as to be able to exert the physiological processes involved in the translocation of the contaminants in solutes through the plants tissues. *Chromolaena odorata* demonstrated this characteristic by its ability to grow in the PCB-contaminated soil throughout the period of the experiment. In this experiment, PCB concentrations of 100 and 200 ppm treatments were not phytotoxic to *C. odorata* as the plant was able to complete the growth duration of the experiment in those treatments or that the plant was able to manage such effects. At 500 ppm PCB concentrations, *C. odorata* was slightly affected by phytotoxicity of the pollutant towards the sixth week of growth in the treated soil, although it completed the experimental period. PCB contamination between 0-260 ppm has been reported not to be phytotoxic to various plants tested for its phytoremediation ability, but higher concentration of PCB was seen to cause stress to the plants (Lee, 2013; Zeeb *et al.*, 2006; Ficko *et al.*, 2011). The response showed by *C. odorata* towards 500 ppm of PCB may have been the cause of the stress; it could also be attributed to other factors not measured.

3.1 Residual PCB Recovered from the Soil and Plant

Mean total PCB concentration in plants tissues were 6.4, 11.7 and 55.8 µg for 100, 200 and 500 ppm treatments respectively, while the control was zero. Final soil PCB concentrations were measured to be 2.06, 3.01, and 4.05 ppm for 100, 200, and 500 ppm treatment concentrations respectively. At same time, the residual PCB concentration at control was 15.76 ppm/kg of soil. Results of total PCB recovery in soil and plants tissues are presented in Table 5. PCB concentration factors also know as bioaccumulation factors (BAFs) were 0.022, 0.032, and 0.065 for the respective treatments. In author's previous unpublished study, a comparative study was set out to test the impending effects of volatilization and microorganisms in soil PCB remediation. The mean values observed were higher than the corresponding values obtained at uninhibited treatments. This shows that the inhibition of volatilization as well as microbial actions accounted for PCB retention by the soil. However, such effect was higher in former than in later, and the reduction was however, not significantly different from the reduction of the treated and planted. This is in agreement with Aslund *et al.*, (2008), which reported that the primary uptake pathway of PCB into plants should be root uptake and possible translocation and consistent with other studies on other POP uptake in plants (Mattina *et al.*, 2007). Other scholars have also reported on the influence of microbes on uptake of PCB by plants (Xu *et al.*, 2010; Anyasi 2012).

Table 5: PCB recovery results: Initial soil PCB concentration, Residual soil PCB concentration, total PCB concentration, percentage PCB absorbed, percentage change in PCB, and PCB concentration factor.

Treatments	Initial soil PCB conc. (ppm)	Final soil PCB conc. (ppm)	Total PCB in plants tissue (µg)	% PCB absorbed	% change in PCB	PCB concentration factor (PCB-RF)
T100	10.5 ±0.20*	2.06 ±0.04	6.40 ±0.12*	2.10 ±0.00	80.4 ±0.60*	0.022 ±0.00
C	9.4 ±0.02	8.50 ±0.11	NP	NP	9.6 ±0.30	NP
T200	19.8 ±0.00*	3.01 ±0.02	11.70 ±0.15*	0.03 ±0.00	84.8 ±0.06*	0.032 ±0.00
C	18.5 ±0.03	15.76 ±0.13	NP	NP	14.8 ±0.10	NP
T500	27.2 ±0.20	4.05 ±0.08	55.80 ±0.70*	11.98 ±0.12*	85.1 ±0.20*	0.065 ±0.00*
C	18.5 ±0.20*	5.76 ±0.00	NP	NP	14.8 ±0.00*	NP

Values with the same * in the same column were significant at 5 % level according to Bonferoni test. T= Aroclor 1254, C=Control, NP=Not planted

3.2 Total PCB Recovered from Plant

Absorption of PCB by *C. odorata* occurred mostly at the root tissues of the plant. For example at 100 ppm, the concentration of PCB in the root was found to be 0.22 µg/g of soil with a total root biomass of

7.25 g, resulting to a total root PCB of 1.6 µg. Although the stem and leaf biomass were above 10 g respectively which was higher than what was obtained in the root, the total PCB in both the stem and leaf was still not applicable because there was no PCB dictated in the stem and leaf tissues. Total PCB concentrations found in the tissues of *C. odorata*, ranges from 3.1 to 64.6 ppm, the value was seen to increase as the concentration of the treatment was increased. This is in agreement with the study of Pinsker, (2011), which reported that initial soil PCB has a great effect on the amount of PCB absorbed by plants, its translocation as well as the concentrations of the residual PCB in the soil at the end of a phytoremediation study. There was percentage reduction of PCB concentration from 80.4 to 86.6 within the six weeks of experiment. These values are appreciable when compared with the mean reduction of PCB per month of other plants species that were used for various PCB phytoremediation studies. The mean PCB reductions were reported to be in the range of 0.1-14.8 % (Dzantor *et al.*, 2000; Ficko *et al.*, 2010). Although total PCB concentration in most of the plant tissues of the Aroclor-treated experiments was not applicable, there was still reduction in the amount of PCB in soil at the end of the experiments. Such effect was also observed in the unplanted control samples (C2=14.8 %) and could be attributed to natural attenuation and perhaps other parameters not measured.

3.3 PCB Reduction in the Soil after Six Weeks of Growth with *C. Odorata*

Highest reduction in PCB concentration was found in soil with 500 ppm concentrations as observed in T_{200}, T_{500}, C4 and C3 where they was reduction of PCB of about 85.1, 84.8, 84.1 and 82.6 % respectively. Treatment with initial concentration of 100 ppm had percentage reduction of 80.4. However, the control sample without plant (C) only reduced by 14.8 ppm, this may have been aided by nature (Table 5). In this study, greater amount of PCB found in the plants tissue were concentrated in the roots, it could be as a result of the diffuse root system of the plant; an importance feature of any phytoremediation plant. Total root concentrations of PCB were reported to be in the range of 0.26-17.85 ppm (Result not shown). Increased root concentration of PCB leads to a synonymous increase in bioaccumulation factors (BAF). Bioaccumulation factor determines plants ability to accumulate and concentrate a greater quantity of PCBs than the

surrounding soil. This phenomenon is important as it provides one with the idea on how to measure the ability of plants to draw PCB towards the roots when it absorbs water and nutrients from the soil known as imbibition. The measurement of BAF in *C. odorata* was in range of 0.01 to 0.4 which is greater than what was observed with Alfalfa by Zeeb *et al.*, (2006), and some other studies (Low, 2009; Aslund *et al.*, 2007/8). From this, it can be explained that *C. odorata* was able to draw PCBs towards its root with its BAF value within the range of measured value of BAF of other PCB phytoremediation plants. This generates a new interest in the use of *C.odorata* in the phytoextraction of PCB.

4. Conclusions

This study has been able to demonstrate that *C. odorata* is able to reduce the concentrations of PCB at the 100, 200 and 500 ppm of Aroclor treated soil significantly than unplanted control. At the end of six weeks of experiment, there was sustenance of growth of the plant which causes phytotreatment of soil PCB from a PCB contaminated soil. The fact that *C. odorata* was able to survive this different contaminant regime of Aroclor, is an evident that the plant is a promising candidate for uptake of PCB from a PCB-contaminated soil and such effect could be enhanced with soil amendments (bioaugmentation) that aids microbial presence in the rhizosphere as well as growing the plants for a longer duration in the contaminated soil. The major weakness of the method that was used in this study was that the reduction of the PCB in the soil may have been aided by other factors but such was measured in the author's other unpublished study.

Acknowledgement

The South African National Research Foundation (NRF) is acknowledged for funding this project.

References

[1]   Andersson P, 2000. Physico-chemical characteristics and quantitative structure-activity relationships of PCBs. Department of Environmental chemistry, Umea University-Sweden. Pp. 1-10.

[2]   Anyasi R O, 2012. Bioremediation of polychlorinated biphenyls (PCBs)-contaminated soil by phytoremediation with Chromolaena odorata (L) RM King and Robinson. MSc Thesis, University of South Africa. Pp. 77-78.

[3]   Aslund M L W, Zeeb, B.A., Rutter, A. and Reimer, R. 2007. In situ phytoextraction of polychlorinated biphenyls (PCB)-contaminated soil. *Science of the Total Environment,* 374(1), 1-12.

[4]   Bhandary A, 2007. Remediation technologies for soil and groundwater. US Environmental Council. Science, pp. 17-23.

[5]   De Rouw A, 1991. The invasion of *Chromolaena odorata* (L.) King & Robinson (ex Eupatorium odoratum) and competition with a native flora in a rain forest zone, South-west Cote d'Ivoire. *Journal of Biogeography,* 18: 13-32.

[6]   Ficko S A, Rutter A, Zeeb B A, 2011. Phytoextraction and uptake patterns of weathered polychlorinated biphenyls-contaminated soils using three perennial weeds species. *Journal of Environmental Quality,* 40: 1870-1877.

[7]   Graham C, Ramsden J J, 2008. Introduction to global warming: Complexity and security. IOS Press, pp. 147-184.

[8]   Idris A, Ahmed M, 2003. **Treatment of polluted soil using bioremediation – A review.** Faculty of Chemical and Environmental Engineering, University of Putra, Malaysia, pp. 1-18.

[9]   Lee J H, 2013. An overview of phytoremediation as a potential promising technology for environmental pollution control. *Biotechnology and Bioprocess Engineering,* 18: 431-439

[10] Low J E, Whitefield Aslund M L, Rutter A, Zeeb B A, 2009. Effect of plant age on PCB accumulation by Cucurbita pepo ssp. pepo. Journal of Environmental Quality, 39: 245-250.

[11] Mattina M J I, Berger W A, Eitzer B D, 2007. Factors affecting the phytoaccumulation of weathered, soil-borne organic contaminants: analyses at the ex Planta and in Planta sides of the *plant root. Plant and Soil,* 291: 143–54.

[12] Mosaddegh M H, Jafarian A, Ghasemi A, Mosaddegh A, 2014. Phytoremediation of benzene, toluene, ethylbenzene and xylene contaminated air by D. deremensis and O. microdasys plants. *Journal of Environmental Health Science and Engineering,* 10: 12-39

[13] Pilon-Smith E, 2005. Phytoremediation. *Plant Biology,* 56: 15-39.

[14] Pinsker N I, 2011. Phytoremediation of PCB-contaminated soil: Effectiveness and regulatory policy. MSc. Thesis, Virginia Commonwealth Universisty. Pp. 56-73.

[15] Rodriguez J G, Lafuente A, 2002. A new advanced method for heterogeneous catalyzed dechlorination of polychlorinated biphenyls in hydrocarbon solvent. *Tetrahedron Letters,* 43: 9581–9583.

[16] Singh S, Thorat V, Kaushik C P, Raj K, D'Souza S F, 2009. Potential for *Chromolaena odorata* for phytoremediation of 137Cs from solution and low level nuclear waste. *Journal of Hazardous materials,* 162: 743-745.

[17] Tanhan P, Pokethitiyook P, Kruatrachue M, Chaiyarat R, Upatham S, 2011. Effects of soil amendments and EDTA on lead uptake by Chromolaena odorata: Greenhouse and Field Trial Experiments. *International Journal of Phytoremediation,* 13: 897-911.

[18] US EPA 1997. **Method 1613, Revision B, Tetra-**through octachlorinated dioxins and furans by isotope dilution HRGC/ HRMS, September 15, 1997, 40 CFR 136 (FR 48405), Washington, DC. Pp. 79.

[19] Xu L, Teng Y, Li Z, Norton J M, Luo Y, 2010. Enhanced removal of polychlorinated biphenyls from Alfalfa rhizosphere soil in a field study: The impact of a rhizobial innoculum. *Science of the Total Environment,* 408: 1007-1015.

[20] Zeeb B A, Amphlett J S, Rutter A, Reimer K J, 2006. Potential for phytoremediation of polychlorinated biphenyls-(PCB)-contaminated soil. *International Journal of Phytoremediation,* 8: 197-221.

7

Adsorption of Pb (II) Ions from Aqueous Solutions by Water Hyacinth (*Eichhornia Crassipes*): Equilibrium and Kinetic Studies

Davis Amboga Anzeze [1], **John Mmari Onyari,**[1] **Paul Mwanza Shiundu** [2]**, John W Gichuki** [3]

[1] Department of Chemistry, School of Physical Sciences, College of Biological and Physical Sciences, University of Nairobi,
P.O. Box 30197- 0100, Nairobi, Kenya
amboga.davis@yahoo.com; jonyari@uonbi.ac.ke

[2] Department of Chemical Sciences and Technology, The Technical University of Kenya ,Haile Selassie Avenue, P.O. Box 52428 - 00200, Nairobi – KENYA
pmshiundu@kenpoly.ac.ke

[3] Environmental Protection Department (EPD), Big Valley Rancheria Bond of Pono Indiana, 2726 Mission Rancheria road, Lakeport CA 95453, USA.
jgichuki@big-valley.net

*Abstract-*The ***Eichhornia crassipes*** *roots for the removal of Pb (II) ions from aqueous solutions has been investigated. The adsorption of Pb (II) ions was found to be affected by solution pH, contact time, adsorbent dosage, initial metal ion concentration, and temperature. The equilibrium was analysed using Langmuir and Freundlich isotherm models. The data was found to have a closer correlation with the Freundlich isotherm as evidenced by a higher correlation coefficient (R^2). The biosorption capacity for E. crasippes was found to be 16.350 mg g^{-1}. The Kinetics data was also subjected to pseudo-first-order and the pseudo-second-order kinetic models. The data could be explained better using the pseudo-second-order kinetic model.*

Keywords: Biosorption, Water hyacinth (*E. crassipes*), Heavy metal ions, Isotherms, Kinetics.

1. Introduction

Heavy metals in water have been a major preoccupation for researchers for many years due to their toxicity towards aquatic life, human beings and environment [1]. This is because they do not biodegrade unlike organic pollutants, making their presence in industrial effluents and drinking water a public health concern [2].

Despite their poisonous nature the heavy metals are used in various ways. Lead is an essential element in the manufacture of bullets, pipes, lead-acid accumulators and welding. Lead in the environment has negative impacts which include severe damage to the kidney, nervous system, reproductive system liver and brain [3, 4]. Therefore it must be removed from the water and waste waters in order to safeguard human and environmental health [5].

The conventional removal methods for lead such as chemical precipitation, ion exchange, electrochemical deposition solvent extraction are not only expensive but they also generate secondary sludge which at times is more difficult to treat [6]. The methods are also ineffective in removal of heavy metals at low concentrations of < 100 mg L^{-1} [7]. Due to the stringent environmental regulations on the minimum acceptable amounts of Pb (II) ions in drinking water, irrigation water and industrial effluents, there is need to come up with novel cost effective technologies for Pb (II) ions removal.

Previously several materials have been investigated for removal of heavy metals from water and waste waters. This includes heartwood powder of *Areca catechu* [8, 9, 10, 11]

The objective of this study was to investigate the application of *E. crasippes* an invasive weed species which has potential to successfully colonize, spread, and subsequently displace vegetation and disrupt ecosystems [12] as an adsorbent for Pb (II) ions removal from aqueous solutions.

The effects of several physico-chemical parameters such as pH, adsorbent dosage, contact time, initial metal ion concentration and temperature that affect adsorption were investigated. Equilibrium isotherm models and kinetic models were applied to the data obtained for a better understanding of the adsorption process.

2. Materials and Methods

1.1. Biomass Preparation

E. crassipes plants were harvested from Winam Gulf, Kisumu bay at Kisat and Hippo point of L. Victoria, Kenya. The collected biomass was washed several times with tap water to remove adhering dirt. The washed biomass was then cut into roots, shoots and stems and the parts dried separately for 2 week. The dried brown plant biomass were then transported to the University of Nairobi laboratories were they were further dried and later ground and sieved to various particle sizes(<75μm, >75< 300 μm, >300< 425 μm and > 425 μm). The material was washed again using distilled water, then dried in an oven for 48 hours at 70ºC then stored in plastic containers awaiting biosorption experiments. Preliminary studies indicated that though there was no big difference in the adsorption rates of the various parts the roots were better. Therefore roots were used in this adsorption experiments.

1.2. Chemicals

All chemicals used in the present work were of analytical grade. The stock solution of Pb^{2+} ions was prepared in 1.0 g L^{-1} concentration using 1.6066 g $Pb(NO_3)_2$ (Sigma Aldrich) then diluted to appropriate concentrations .The pH of the solutions was adjusted using 0.1 mol L^{-1} HCl and 0.05 mol L^{-1} NaOH solutions.

1.3. Analysis of Metal Ions

The concentration of Pb (II) ions in the biosorption media was determined using Atomic absorption Spectrophotometer (Varian Spectr AA), equipped with air acetylene burner. The hollow cathode lamp was operating at 8 mA. Analytical wavelength was set at 283.3 nm.

1.4. Biosorption Experiments

Biosorption experiments were conducted at room temperature (26 ºC) by agitating a given mass of biosorbent with 20 mL of metal ions solution of desired concentration in 100 ml polypropylene containers using an orbital shaker at a speed of 200 rpm for 20 min except for contact time experiments. The effect of solution pH on equilibrium biosorption of metal ions

was investigated under similar experimental conditions between 2.0 and 7.0. After the adsorbate has had the desired contact time of interaction with the adsorbent, the samples were filtered using Whatman no. 42 filter paper and the residual concentration analyzed using CTA- 2000 AAS. However experiments involving effect of contact time used filter paper no 2. For studies on effect of temperature the adsorption studies were carried out at 25, 30, 40, 50, 60 and 70 ºC.

The amount of biosorption (q) was calculated by using the equation below.

$$q = \frac{(Co-Ce)V}{m} \tag{1}$$

The biosorption efficiency, A %, of the metal ion was calculated from:

$$A \% = \frac{(Co - Ce)}{Co} \times 100 \tag{2}$$

Where C_o and C_e are the initial and final metal ion concentrations (mg L^{-1}) respectively. **V** is the volume of the solution (L) and **m** is the amount of biosorbent used (g).

3. Results and Discussion

3.1. Effect of pH on Metal Biosorption

Hydrogen ion concentration is one of the important factors that influence the adsorption behavior of metal ions in aqueous solutions. It affects the solubility of metal ions in solution, replaces some of the positive ions found in active sites and affects the degree of ionization of the adsorbate during the process of biosorption. This is because it affects solution chemistry and also the speciation of the metal ions. The effect of initial pH on biosorption of Pb (II) ions onto *E.crasippes* was evaluated in the pH range of 2.0 to 7.0. Studies in pH range above 7.0 were not attempted as there is precipitation of lead (II) hydroxides. From the Figure 1.

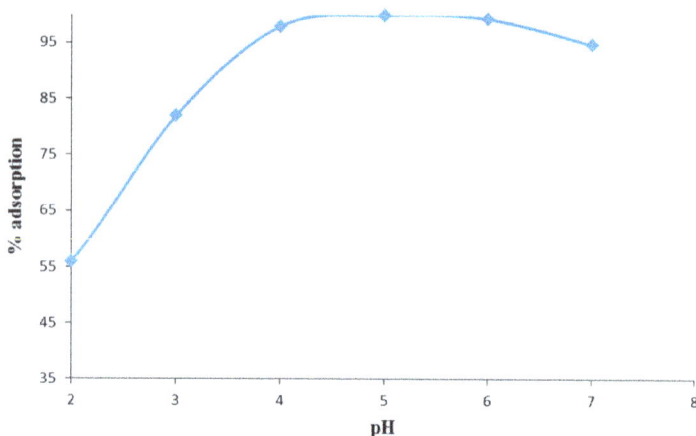

Figure 1. Effect of pH on % adsorption of Pb (II) ions.

From Figure 1 it could be seen that Pb (II) ions adsorption increased as the pH increased. At low pH values, protons occupy the biosorption sites on the biosorbent surface and therefore less Pb (II) ions can be adsorbed because of electrostatic repulsion between the metal cations and the protons occupying the binding sites.

When the pH was increased, the biosorbent surface became more negatively charged and the biosorption of the metal cations increased drastically until equilibrium was reached at pH 5.0 - 6.0. At pH of >6 .0 there is formation of hydroxylated complexes of the metal ions and these complexes compete with the metal cations for the adsorption sites hence a reduction in the effective metal cations removal. Therefore adsorption experiments at pH above this were not considered.

3.2. Effect of Adsorbent Dosage

The number of available binding sites and exchanging ions for the biosorption depends upon the amount of biosorbent in the biosorption system. This is attributed to the fact that it determines the number of binding sites available to remove the metal ions at a given concentration. The dosage also determines the adsorption capacity of the adsorbent with an increase in mass reducing the adsorption capacity as the mass increase from 0.125 g to 2.5 g per 20 mL of adsorbate. The effect of biomass dosage on adsorption of Pb (II) ions is indicated in Figure 2.

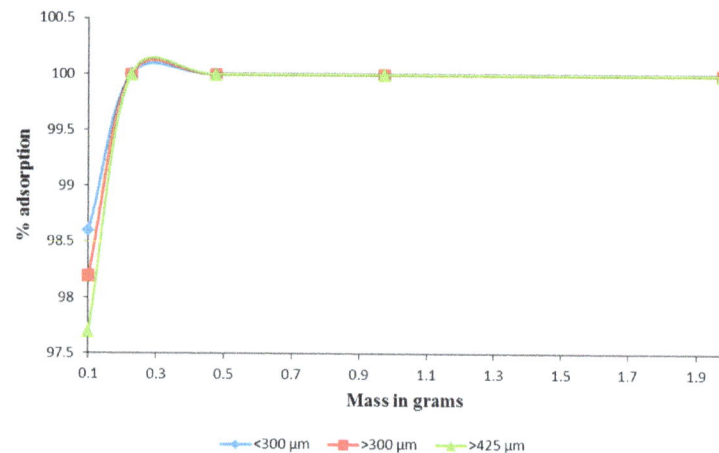

Figure 2. Effect of weight and particle size of E.crassipes on adsorption of 10 mg/L Pb (II) ions.

An increase in the percentage adsorption is attributed to an increase in the number of binding sites for the metal cations. Similar results were recorded in the literature for other adsorbents. However the mass could not be increased infinitely as at some point all the solution is sequestered leaving no residual solution for concentration determination.

3.3. Effect of Initial Metal Concentration.

The initial concentration remarkably affected the uptake of Pb (II) ions in solution. The efficiency of Pb (II) ions adsorption by *E.crasippes* at different initial concentrations (20-600 mg L^{-1}) was investigated as shown in Figure 3.

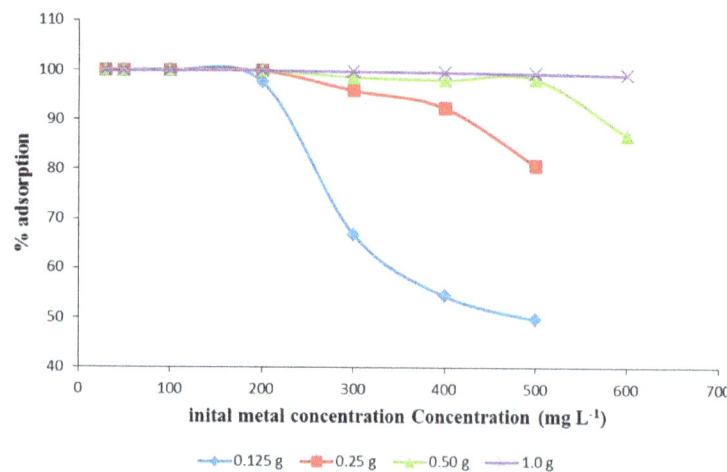

Figure 3. Effect of initial concentration of Pb (II) ions on adsorption.

At a lower concentration, the adsorption sites take up the available Pb (II) ions much quickly due to less competition among the metal ions for the available binding sites which are fixed in this case. However, as the concentration increases the competition for the limited binding sites sets in as the binding sites become saturated.

3.4. Effect of Contact Time

Contact time is an important parameter for any successful use of the biosorbents for practical purposes. Effect of contact time on adsorption of Pb (II) ions was investigated keeping the biomass in contact with the metal ion solution for different time periods between 0 to 60 minutes. It was noted that as adsorption proceeds, the sorbent reaches saturation state, at this point the sorbed solute tends to desorb back into solution (Figure 4).

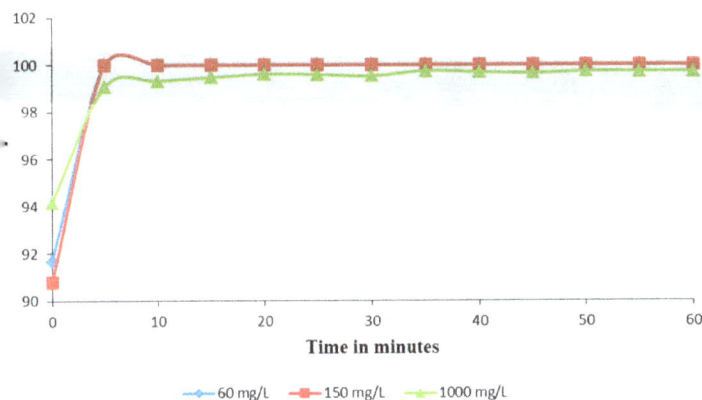

Figure 4: Effect of contact time on adsorption of Pb (II) ions

Eventually, the rate of adsorption and desorption are equal at equilibrium. When the system attains equilibrium, no further net adsorption occurs. The time taken to attain equilibrium is very important for process optimization. The rate of adsorption is very fast at first and over 95 % of total biosorption of Pb (II) ions occurs in the first 5 minutes and thereafter it proceeds at a slower rate and finally no further significant adsorption is noted beyond 20 minutes of contact time. The very fast adsorption makes the material suitable for continuous flow water treatment systems.

3.5. Effect of Temperature

Temperature of the medium affects the removal efficiency of pollutants in aqueous solutions. This is because a change in temperature in turn affects the solubility of pollutants and also the kinetic energy of the adsorbing ions. Therefore the effect of temperature on adsorption of Pb (II) ions was investigated.

The results obtained indicated that the % adsorption increases with increase in temperature up to 40 ℃, after that any increase in temperature is accompanied by a reduction in % adsorption. This can be attributed to the fact that with increase in temperature of the solution, the attractive forces between the biomass surface and Pb (II) ions are weakened thus decreasing the sorption efficiency. This could be due to increase in the tendency for the Pb (II) ions to escape from the solid phase of the biosorbent to the liquid phase with increase in temperature. When the kinetic energy of particles is increased the rate of desorption is faster than the adsorption rate hence decreased adsorption efficiency.

3.6. Adsorption kinetics

Kinetic study provides useful information about the mechanism of adsorption and subsequently investigation of the controlling mechanism of biosorption as either mass transfer or chemisorption. This helps in obtaining the optimum operating conditions for industrial-scale batch processes.

A good correlation of the kinetic data explains the biosorption mechanism of the metal ion on the solid phase. In order to evaluate the kinetic mechanism that controls the biosorption process, the pseudo-first-order models were applied for biosorption of Pb (II) ions on the biosorbent.

3.6.1. Pseudo-first-order Kinetics

The Lagergren pseudo-first-order rate model is represented by the equation [13]:

$$\log(q_e - q_t) = \log q_e - \frac{k_1}{2.303}t \qquad (3)$$

Where q_e and q_t are the amounts of metal adsorbed (mg g^{-1}) at equilibrium and at time t respectively, and k_1 is the rate constant of pseudo-first-order biosorption (min^{-1}). The q_e and rate constant were calculated from the slope and intercept of plot of log (q_e- q_t) against time t.

3.6.2. Pseudo-second-order Kinetics

The pseudo-second-order equation assumes that the rate limiting step might be due to chemical adsorption. According to this model metal cations can bind to two binding sites on the adsorbent surface. The equation can be expressed as [14]:

$$\frac{t}{q_t} = \frac{1}{k_2 q_e^2} + \frac{1}{q_e} t \qquad (4)$$

Where k_2 is the rate constant of the pseudo-second-order adsorption (g mg^{-1}min^{-1}). If the adsorption kinetics obeys the pseudo-second-order model, a linear plot of t/q_t versus t can be observed as shown in Figure 5.

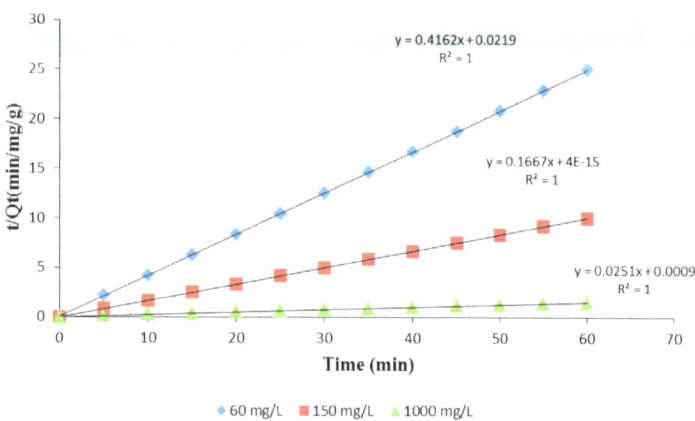

Figure 5. Pseudo-second-order plots for Pb (II) ions.

3.7. Biosorption Isotherms

For optimization of the biosorption process design , its imperative to obtain the appropriate correlation for the equilibrium data.Biosorption isotherms describe how adsorbate interacts with the biosorbent and the residual metal ions in solution during the surface biosorption. The isotherms also help in determination of adsorption capacityof the biosorbent for the metal ions. The data on Pb (II) biosorption was fitted with the Langmuir and Freundlich isotherms.

3.7.1. Langmuir Isotherm

The Langmuir isotherm assumes monolayer coverage of the adsorbate onto a homogeneous adsorbent surface and the biosorption of each cation onto the surface has equal activation energy. According to the model the number of adsorption sites are fixed

and once a site is filled no further adsorption will take place at that site. The surface becomes saturated. The Linear form of the Langmuir isotherm can be expressed as [15, 18, 19] :

$$\frac{C_e}{q_e} = \frac{1}{k_L q_{max}} + \frac{1}{q_{max}} C_e \qquad (5)$$

Where q_{max} is the maximum adsorption capacity of the adsorbent (mg g^{-1}), and k_L is the Langmuir biosorption constant (L mg $^{-1}$) related to energy of adsorption.Values of Langmuir parameters are calculated from the slope and intercept of linear plot of C_e/q_e versus C_e should be a straight line with a slope of $1/q_{max}$ and intercept of $1/q_{max} k_L$ when the biosorption follows Langmuir equation.The non-linear and linearized plots are represented in figure 6(a) and 6(b).

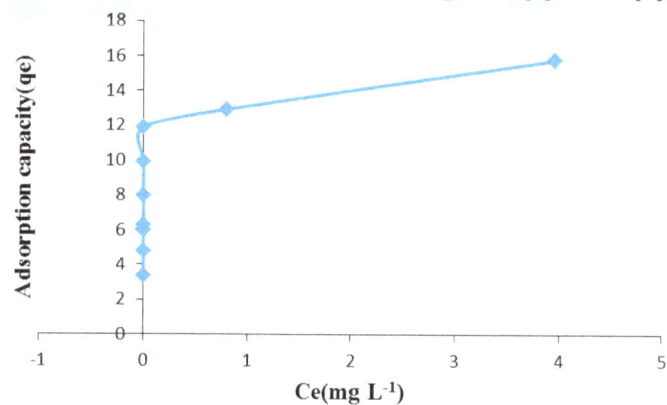

Figure 6a. Langmuir non linearized isotherm for adsorption of Pb (II) ions.

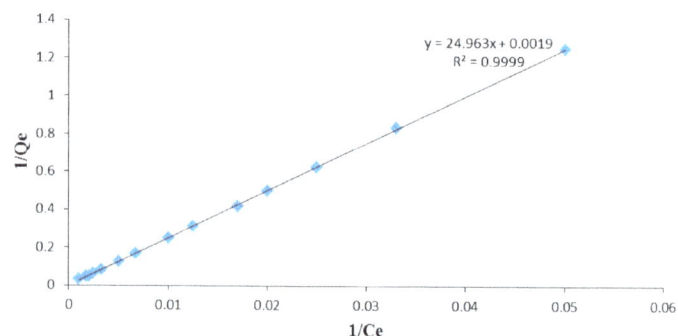

Figure 6b. Linearized Langmuir isotherm for adsorption of Pb (II) ions.

The essential characteristics of the Langmuir isotherm parameters can be used to predict the affinity between the adsorbent and adsorbate using separating factor or the dimensionless equilibrium parameter R_L, expressed as:

$$R_L = \frac{1}{1 + k_L C_o} \qquad (6)$$

Where k_L is the Langmuir constant and C_o is the initial concentration of metal ions. The separation factor determines the nature of adsorption. $R_L = 0$ means adsorption is irreversible, $0 < R_L < 1$ indicates favourable adsorption while $R_L = 1$ denotes unfavorable adsorption.

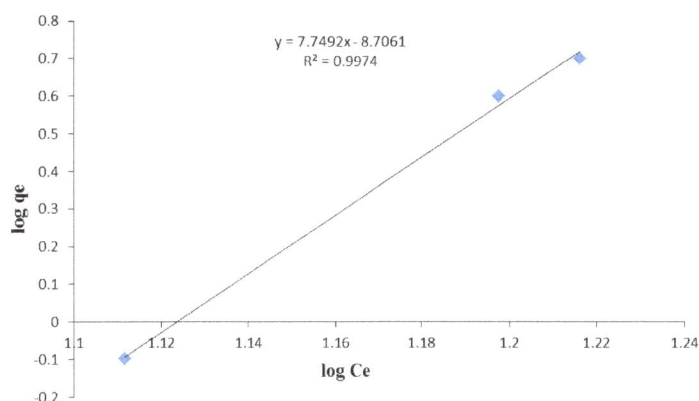

Figure 6c. Linearized Freundlich isotherm plot for the adsorption of Pb (II) onto E.crassipes.

3.7.2. Freundlich Isotherm

This is the earliest model for adsorption. The model envisages heterogeneous adsorption where there is interaction between adsorbed molecules on a surface. The Freundlich equation can be expressed as [15, 20]:

$$\log q_e = \log K_F + \frac{1}{n} \log Ce \qquad (7)$$

where K_F and $1/n$ are the Freundlich isotherm constants related to bonding energy and adsorption intensity respectively. n describes the heterogeneity of the adsorbent surface and its affinity for the adsorbate [16]. $\frac{1}{n}$ is the heterogeneity factor and n is a measure of deviation from linearity of adsorption, higher value of n (or a smaller value of $\frac{1}{n}$ indicates a stronger bond between the adsorbate and the adsorbent thus values of n larger than unity indicate a strong bond which implies favourable adsorption [17]. If n is less than 1, adsorption is a chemical process; if n is greater than 1 then adsorption is a physical process [11]. If the equation applies then a plot of log q_e versus log C_e will give a straight line of slope $\frac{1}{n}$ and intercept as K_F as shown in Figure 6(c).

4. Conclusion

This study demonstrates that *E.crassipes* is a promising adsorbent for the removal of Pb(II) ions from aqueous solutions. The adsorption process was affected by various physico-chemical parameters such as contact time, pH, initial concentration of the metal ions, shaking speed and temperature. The kinetic study revealed that the adsorption data obeyed the pseudo-second-order model better than the pseudo-first-order model given the higher correlation coefficient(R^2). It can therefore be concluded that *E.crasippes* is an effective alternative biomass for the removal of Pb(II) ions from wastewater because the material has a high adsorption capacity, naturally and abundantly available at a low cost. Therefore it can be used in the removal of heavy metal pollutants in water and waste water hence conserving the environment.

Acknowledgements

This work was surpported by grant through International Centre for Insect Physiology and Ecology (ICIPE) from the World Federation of Scientists (WFS).

References

[1]   Ibrahim, M.N.M., Wan Ngah, W.S., Norliyana, M.S., Dawood, W.R ., Rafatullah, M., Sulaiman, O., & Hashim, R.(2010). *A novel agricultural waste adsorbent for the removal of lead(II) ions from aqueous solutions.* Journal of Hazardous Materials. 182, 377-385.

[2]   Alslaibi, T. M., Abustan, I., Ahmad , A. , M. and Foul, A. A. (2013). *Cadmium removal from aqueous*

solutions microwaved olivestone activated carbon, Journal of Environmental Chemical Engineering 1; 589- 599

[3]   Gautam, R. K., Chattopadhayaya, M. C., Sharma, S.K. (2013). *Biosorption of heavy metals : Recent trends and challenges, in ; Charma, S, K. Sanghi (Eds), Waste water reuse and Management,* Springer, London pp 305-322.

[4]   Jeyakumar, S. R. P and Chandrasekaran, V. (2014). *Adsorption of lead (II) ions by activated carbons prepared from marine green algae: equilibrium and kinetics studies, Int J Ind Chem 5:10 DOI 10.1007/ s40090-014-0010-z.*

[5]   Amboga, D, A., Onyari, J. M., Shiundu, P. M &Gichu-ki, J. W. (2014). *Equilibrium and Kinetics studies for the biosorption of aqueous Cd(II) ions onto Eichhornia crassipes biomass.* IOSR Journal of applied Chemistry, vol. 7 Issue 1, ver. II, pp 29-37.

[6]   Barakat, M. A. (2011). *New trends in removing heavy metals from industrial wastewater, Arabian* Journal of Chemistry 4, 361–377.

[7]   Rangabhashiyam, S., Anu, N., Nandagopal, G., M. S. and Sevraju, N. (2014). *Relevance of isotherm models in biosorption of pollutants by agricultural byproducts; A review.* Jounal of Environmental Chemical Engineering 2: 398- 414.

[[8]   Chakravarty P, Sarma NS, Sarma H (2010) *Biosorption of cadmium(II) from aqueous solution using heartwood powder of Areca catechu.* Chemical Engineering Journal. 162:949–955.

[9]   Kelly-Vargas K, Cerro-Lopez M, Reyna-Tellez S, Bandala ER, Sanchez Salas JL (2012) *Biosorption of heavy metals in polluted water using different waste fruit cortex.* Phys Chem Earth 39:26–39

[10]   Abdel-Ghani NT, Hefny MM, El-Chaghaby GA (2008) *Removal of metal ions from synthetic wastewater by adsorption onto Eucalyptus camaldulenis tree leaves.* J Chil Chem Soc 53:1585–1587

[11]   Farhan, A. M., Al-Dujaili, H. A. and Awwad, M. A. (2013). *Equilibrium and kinetic studies of cadmium (II) and lead (II) ions biosorption onto Ficus carcia leaves,* International Journal of Industrial Chemistry 4: 24.

[12]   Holm, L. G., Plucknett, D. L., Pancho, J. V., & Herberger, H. P. (1977). *The world's worst weeds: Dis-tribution, and biology.* Honolulu: University Press of Hawaii.

[13]   Wan Ngah, W. S., & Hanafiah, M. A. K. M. (2008a). *Removal of heavy metal ions from wastewater by chemically modified plant wastes as adsorbents: A review.* Bioresource Technology 99: 3935–3948.

[14]   Ho Y. S, McKay G. (1999) *Pseudo-second order model for sorption processes.* Process Biochem 34:451–465

[15]   Foo, K.Y. and Hameed, B. H. (2010). *Insights into the modelling of adsorption isotherm system.* Chemical Engineering Journal, 156: 2-10.

[16]   Sari, A and Tuzen; M. (2009). *Equilibrium, Thermodynamic and Kinetic studies on aluminium biosorption from aqueous solution by brown algae (Padina pavonica),* J. Hazard. Mater, 171. 973-979.

[17]   Sarı, A. and Tuzen, M. (2007). *Biosorption of Pb2+ and Cd2+) from aqueous solution using green alga (Ulva lactuca) biomass,* J. Hazard. Mater,. 152 302–308.

[18]   Langmuir, I., (1918). *The adsorption of gases on plane surfaces of glass, mica and platinum,* J.Am. Chem. Soc 40, 1361-1403.

[19]   Meena, A.K. ,Mishra,G.K., Rai, P.K., Rajagopal, C., & Nagal,P.N. (2005). *Removal of heavy metal ions from aqueous solutions using carbon aerogel as an adsorbent.* Journal of Harzadous Materials, 122, 161-170.

[20]   Freundlich, H. M. F. (1906). *über die adsorption in läsungen.* Zeitschrift für Physikalische Chemie, 57, 385–470.

8

Evaluation of Management Scenarios for Controlling Eutrophication in a Shallow Tropical Urban Lake

Zikun Xing[1], Lloyd H. C. Chua[2], Jörg Imberger[3]
[1]School of Civil and Environmental Engineering, Nanyang Technological University
50 Nanyang Avenue, Singapore, 639798
zkxing@ntu.edu.sg
[2]School of Engineering, Deakin University
Geelong Waurn Ponds Campus, 75 Pigdons Road, Waurn Ponds VIC 3216, Australia
lloyd.chua@deakin.edu.au
[3]Centre for Water Research, University of Western Australia
M023, 35 Stirling Highway, Crawley 6009, Western Australia, Australia
jimberger@cwr.uwa.edu.au

Abstract – *Urban lakes are typically smaller, shallower, and more exposed to human activities than natural lakes. Although the effects of harmful algal blooms (HABs) associated with eutrophication in urban lakes has become a growing concern for water resources management and environmental protection, studies focussing on this topic in relation to urban lakes are rare and knowledge of the ecological dynamics and effective management strategies for controlling eutrophication in urban lakes is lacking. This study applied an integrated three-dimensional hydrodynamics-ecological model for a small shallow tropical urban lake in Singapore and evaluated various management scenarios to control eutrophication in the lake. It is found that in-lake treatment techniques including artificial destratification, sediment manipulation and algaecide addition are either ineffective or possess environmental concerns; while watershed management strategies including hydraulic flushing and inflow nutrients reduction are more effective and have posed less environmental concerns. In this study, inflow phosphorus reduction was found to be the best strategy after evaluating the advantages and drawbacks of the management strategies studied. Runoff from the watershed exerts significant influence on urban lakes and thus an integrated water resources management at the watershed level is critical for the control of eutrophication.*

Keywords: urban lakes, water quality modelling, ELCOM-CAEDYM, eutrophication, management scenarios.

1. Introduction

Compared to natural lakes, urban lakes are characterized by having a smaller size, shallower depth, and larger ratio of watershed area to lake surface area [1]. A large proportion of the urban lakes are man-made with the purposes of storing rainwater, water supply or recreational uses [2]. Urban lakes tend to be more impacted by human activities from receiving larger and poorer quality inflows from urban watersheds. Thus, urban lakes receive higher nutrient loadings and have more severe water quality issues such as eutrophication and reduced health [3, 4], compared to natural lakes. With excessive addition of nutrients, especially phosphorus, the water body can become eutrophic or even hypereutrophic with undesired effects such as harmful algal blooms which would pose serious environmental and management problems [5]. In recent decades, urbanization has led to significant increases of nutrient enrichment levels in urban lakes and as a result, there have been more occurrences of algal bloom events [6, 7]. While the water quality problems in urban lakes remains a challenge for watershed and lake management, there are comparatively fewer studies focusing on urban lakes and we have little knowledge of their ecological dynamics and management strategies to effectively control eutrophication [1, 3].

In this paper, we present a numerical modelling study for a small shallow urban lake located in a tropical region using an integrated hydrodynamics-ecological model. After calibrating the model, several management scenarios including watershed management and in-lake treatment were tested with the model. The results obtained and the advantages and disadvantages of the management scenarios studied are evaluated and discussed.

2. Methodology

Kranji Reservoir (1°25'N, 103°43'E) is a small and shallow, tropical urban reservoir located in Singapore (Figure 1A) formed by the impoundment of Kranji River in the 1970s. The reservoir experiences episodes of eutrophication and recurrent cyanobacteria blooms. The lake surface area is 3.0×10^6 m² and catchment area is 5.6×10^7 m². The average depth of Kranji Reservoir is 5 meters while the depth at the deepest location is about 20 meters. The reservoir receives base flow and storm runoff from its main tributaries and the direct precipitation on the reservoir surface.

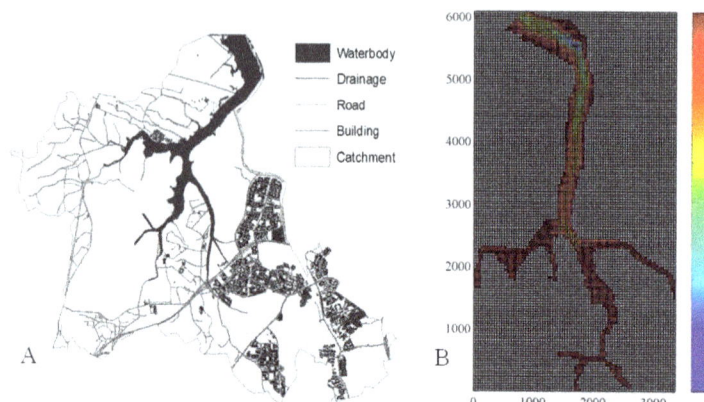

Figure 1. (A). Kranji Catchment and Reservoir and (B). Bathymetry of Kranji Reservoir used in ELCOM.

We apply an integrated three-dimensional hydrodynamics-ecological model to Kranji Reservoir, the Estuary Lake and Coastal Ocean Model-Computational Aquatic Ecosystem Dynamics Model (ELCOM-CAEDYM) developed by the Centre for Water Research, University of Western Australia [8, 9]. ELCOM-CAEDYM has been applied and evaluated widely in many different aquatic ecosystems around the

world, including lakes, reservoirs, estuaries and oceans. ELCOM solves the hydrostatic, Boussinesq, unsteady Reynolds-averaged Navier-Stokes (RANS) and scalar transport equations [8]. CAEDYM provides a dynamic coupling with ELCOM, particularly calculating ecological processes including oxygen dynamics, nutrient cycling, and phytoplankton dynamics [9]. For the ELCOM-CAEDYM model of Kranji Reservoir (Kranji model), we have assumed boundary condition of turbulent benthic boundary layer and used initial conditions from measured vertical water temperature and water quality profiles in the reservoir. The discretization of Kranji model consists of horizontal orthogonal grids with 40×40 m wide cells as shown in Figure 1B. The grid size of 40 m is chosen taking into account the bathymetry map resolution and the simulation speed of the model. For the vertical direction, we have used a total of 21 layers for the maximum depth of 22.5 meters allowing for water level fluctuations. The near surface vertical resolution is 0.5 meter and decreasing gradually towards the bottom of water column to improve computational accuracy in the surface layers.

In this study, we have run the ELCOM-CAEDYM simulations for 2005 under a total of five management scenarios, including tests for hydraulic flushing, phosphorus load reduction, artificial destratification, sediment manipulation and algaecide addition. The first two belong to watershed management strategies while the others are in-lake treatment techniques. The details of the settings and results of each management scenario are presented and evaluated in the next section.

3. Results

3.1. ELCOM-CAEDYM model for Kranji Reservoir

Calibration of the Kranji model was carried out by using a combination of biogeochemical parameters obtained from both literature review of other tropical lakes and field measurements and laboratory analysis of water and sediment samples collected from Kranji Reservoir. The model was calibrated by tweaking of important biogeochemical parameters to fit observed data. The Kranji model was calibrated and validated against a two-year record (2005 and 2006) of water temperature and water quality variables [10].

The simulated results are compared with field measurements for the water surface of Kranji Reservoir in 2005 (see Figure 2). The calibrated model was found to be capable of predicting total chlorophyll-a (TCHLA), total phosphorus (TP) concentrations, total organic carbon (TOC) and total nitrogen (TN) concentrations fairly well. As TCHLA and TP concentrations are the most critical water quality variables for eutrophic lakes, the model can be used as an effective prediction tool to evaluate the management scenarios for controlling eutrophication in Kranji Reservoir. However, due to the scope of this study, the present model has not quantified the uncertainties from model parameters and input data. We will investigate the sensitivity of model parameters and the uncertainties of the model results in further studies.

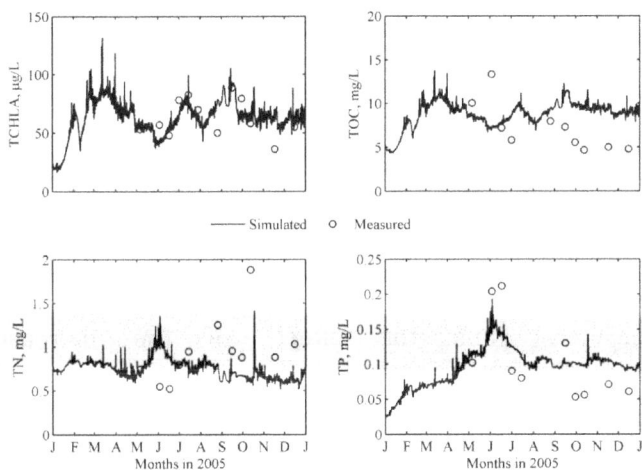

Figure 2. Comparison between measured and simulated water quality variables including total chlorophyll-a (TCHLA) concentration, total organic carbon (TOC) concentration, total nitrogen (TN) concentration and total phosphorus (TP) concentration at the water surface of Kranji Reservoir in 2005.

3.2. Water Quality in Kranji Reservoir

Three water quality variables measured 1-2 times per month over a 9 month period in 2005 and the TN:TP ratio computed based on these measurements are plotted in Figure 3. The TCHLA concentration was found to exceed the eutrophic level of 10 µg/L [11] for the months considered, with an annual mean concentration of 77 µg/L. Algal blooms occurred from June to September where the mean TCHLA concentration reached 116 µg/L during this period. The total nitrogen (TN) and total phosphorus (TP) concentrations in the reservoir both increased in late July and gradually decreased thereafter, indicating increased nutrients input from inflows during the former period. The TN:TP ratio was found to be higher than the Redfield Ratio [12] of 7.2 (by weight) for most of the time throughout the year, thus the phytoplankton

in Kranji Reservoir is considered to be phosphorus limiting. During the algal bloom period from June to September, the TN:TP ratio stays relatively low, indicating that the relative abundance of phosphorus combining with warmer weather during Jun to Sep tends to foster the algal blooms. In contrast, the TN:TP ratio increases significantly to well above 10 from October to December, coinciding with the notable decrease of TCHLA concentration. It suggests that the limited phosphorus combining with colder weather during Oct to Dec tends to suppress the algal blooms.

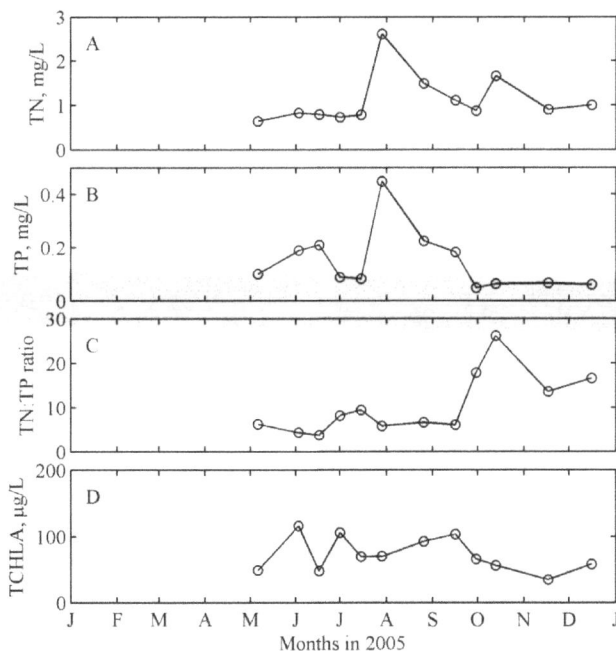

Figure 3. (A). Total nitrogen (TN) concentration, (B). Total phosphorus (TP) concentration, (C). TN: TP ratio and (D). Total chlorophyll-a (TCHLA) concentration measured at the surface of Kranji Reservoir in 2005.

3.3. Management Scenarios Evaluation

To control the eutrophication in Kranji Reservoir therefore, several management scenarios were evaluated. Of the watershed management strategies, hydraulic flushing is believed to be an effective technique in diluting the high-nutrient lake water and flushing out poorer quality water, thus reducing the phytoplankton population [13]. In the Kranji model, the inflow volumes from the tributaries were doubled. In reality, doubling of inflows could result from a diversion of inflows from nearby drainages. The model results show that a doubling of inflows can result in a reduction of 23.7% in the mean TCHLA concentration for 2005 (see Figure 4A). Though the reduction is significant, this method is not always feasible due to the high costs involved and availability of alternate sources of inflows.

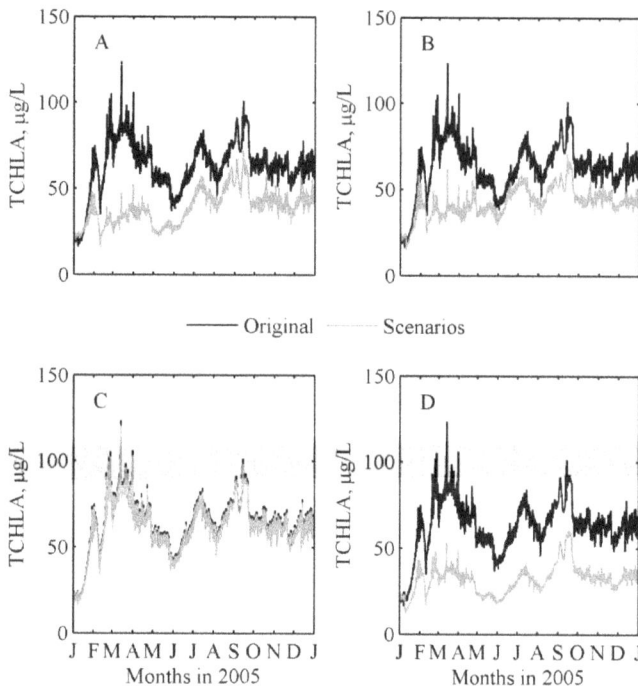

Figure 4. Simulations of TCHLA concentration under various management scenarios: (A) Hydraulic flushing, (B) Phosphorus reduction, (C) Sediment manipulation, (D) Chemical treatment.

Reduction or removal of nutrients loadings from the inflows is one of the most commonly used lake restoration techniques. It can be achieved by best management practices such as wetland filtration [14, 15] or direct control and treatment of the point source discharges [16]. Since it is observed that algae in Kranji Reservoir is phosphorus-limited, we decreased the phosphorus loadings in inflows by 50%. Such a reduction can be achieved through engineered systems or legislating against the use of phosphorus containing detergents. Our model results (Figure 4B) show that a 50% reduction in phosphorus loading can result in a 21.05% decrease in the annual mean TCHLA concentration. Inflow nutrient reduction is found to be an effective technique to control eutrophication. However, integrated water resources management in the watershed is required and techniques of filtration or treatment of the inflows need to be developed and implemented.

Aeration systems in lakes have been widely used to control eutrophication [17, 18]. Artificial destratification achieved by bubble-plume systems can add oxygen to the water column and improve mixing. In the Kranji model, simulations without bubblers and with bubbler operating at a rate of 0.20 m^3/s were carried out and compared. Figure 5 shows plots of the simulated vertical profiles of mean TCHLA concentration at the location of bubbler with and without bubbler operation. Slight reduction of TCHLA concentration was noticed, affecting only the top 2 meters of water column with no discernible differences in the deeper parts of the water column. As a whole, bubbler operation resulted in only a 0.01% decrease in TCHLA concentration. Thus artificial destratification has limited use in Kranji Reservoir, probably because it is a shallow and polymictic system. The reservoir undergoes diurnal cycles of stratification and destratification induced by daytime heating and night-time cooling. And unlike most temperate deeper lakes, there is no existence of seasonal thermocline in the reservoir. Thus the effects of the increased destratification contributed by the aerator on the reservoir water column are limited in comparison to diurnal destratification due to convective cooling.

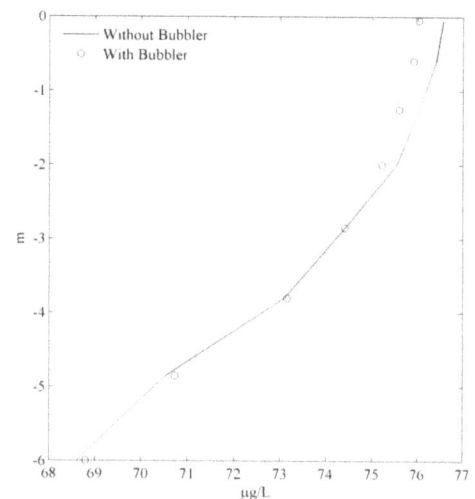

Figure 5. Simulated vertical profiles of mean TCHLA concentration at the location of bubbler in Kranji Reservoir under the conditions without bubbler and with bubbler operating.

Sediment manipulation in lakes is useful when internal nutrients sources from sediments account for a significant portion of the total nutrients for lake algal growth [19]. Typical techniques of sediment manipulation include sediment dredging and active barriers to reduce phosphorus release from lake sediments [20]. In the Kranji model, the effects of sediment manipulation have been tested by setting the initial nutrients concentration in the sediments to be zero, simulating the management scenario after a complete dredging of lake sediments. The results (Figure 4C) show that despite an extreme reduction in phosphorus release, only slight improvements in the lake water quality, a mere 2.6% reduction in TCHLA concentration, was obtained. Sediment manipulation is expensive and cannot solve the problem of water quality deterioration completely if there exist continuous excessive nutrients loadings from inflows.

Chemical treatments with the addition of algaecides such as copper sulphate, are considered to be an economical and effective technique to control cyanobacteria blooms [21]. In the Kranji model, the mortality rate of cyanobacteria was arbitrarily set to two times the calibrated value to simulate the effects of algaecide addition. The dosing of algaecide resulted in an average 36.17% decrease of TCHLA concentration (see Figure 4D). This represents the largest reduction in TCHLA among all the five management scenarios considered. Although the effect of controlling phytoplankton population is the most significant, there are potential drawbacks of this technique. The effects

are temporal and the undesired taste, odour and toxicity associated with the use of algaecides which may cause adverse changes to other lake processes and ecosystem stability [22] are main concerns.

4. Conclusion

The effects of the five management scenarios, in terms of reductions in the TCHLA concentration and their respective advantages and drawbacks are summarized in Table 1. It was found that hydraulic flushing, inflow nutrients reduction and algaecide addition had significant effects on the reduction of TCHLA concentration while artificial destratification and sediment manipulation had limited effects. Though algaecide addition is economical and effective, there are serious concerns in adopting this technique [22]. Compared to the latter three in-lake treatments, watershed management strategies were found to be the most effective with fewer drawbacks since inflows from the watershed can exert significant effects on the water quality of the lake. Of the management scenarios considered, phosphorus reduction was considered to be the best strategy. Integrated water resources management on the watershed level is critical towards preventing eutrophication.

Table 1. Effects of management scenarios on mean TCHLA concentration in Kranji Reservoir and the advantages and disadvantages of each management scenario

Scenario	TCHLA reduction	Advantages	Disadvantages
Hydraulic flushing	-23.68%	Significant reduction in TCHLA	Expensive; not always feasible, need integrated water resources management
Inflow nutrients reduction	-21.05%	Significant reduction in TCHLA	Need integrated water resources management and treatment of inflows
Artificial destratification	-0.01%	Slight improvements at the location of bubblers	Negligible effects for the whole reservoir
Sediment manipulation	-2.60%	Slight improvements in lake	Expensive; the reduction of THCLA is much lower than some other techniques
Algaecide addition	-36.17%	Significant reduction in TCHLA; economical and effective in controlling algal blooms in lake	Effects are temporal; taste, odour and toxins may still remain in water; may cause changes to other lake processes

Acknowledgements

This work is supported by the Singapore National Research Foundation under its Environment & Water Technologies Strategic Research Programme and administered by the Environment & Water Industry Programme Office (EWI), under project 1002-IRIS-09.

References

[1] T. Schueler and J. Simpson, "Why urban lakes are different," Watershed Protection Techniques, vol. 3, 2001, pp. 747-750.

[2] L. Naselli-Flores, "Urban Lakes: Ecosystems at Risk, Worthy of the Best Care," in Taal2007: The 12th World Lake Conference, 2008, pp. 1333-1337.

[3] S. Birch and J. McCaskie, "Shallow urban lakes: a challenge for lake management," Hydrobiologia, vol. 395-396, 1999, pp. 365-377.

[4] C. Walker, J. L. Lampard, A. Roiko, N. Tindale, A. Wiegand, and P. Duncan, "Community well-being as a critical component of urban lake ecosystem health," Urban Ecosystems, vol. 16, 2013, pp. 313-326.

[5] V. H. Smith, "Cultural eutrophication of inland, estuarine, and coastal waters," in Successes, limitations, and frontiers in ecosystem science, ed New York: Springer, 1998, pp. 7-49.

[6] J. Lv, H. Wu, and M. Chen, "Effects of nitrogen and phosphorus on phytoplankton composition and biomass in 15 subtropical, urban shallow lakes in Wuhan, China," Limnologica, vol. 41, 2011, pp. 48-56.

[7] Y. Zeng, "A risk assessment on the alga bloom in city-a case of the "six seas" urban lakes in Beijing," Procedia Environmental Sciences, vol. 2, 2010, pp. 1501-1509.

[8] B. R. Hodges, J. Imberger, A. Saggio, and K. B. Winters, "Modeling basin-scale internal waves in a stratified lake," Limnology and Oceanography, vol. 45, pp. 1603-1620, 2000.

[9] J. R. Romero, J. P. Antenucci, and J. Imberger, "One- and three-dimensional biogeochemical simulations of two differing reservoirs," Ecological Modelling, vol. 174, 2004, pp. 143-160.

[10] E. Y.-M. Lo, E. B. Shuy, K. Y. H. Gin, L. H. C. Chua, S. B. K. Tan, J. Imberger, J. Antenucci, M. Hipsey, and T. Brown "Water quality monitoring, modeling and management for Kranji catchment/reservoir system—Phases 1 and 2 Rep., Public Utilities Board, Singapore.," 2008.

[11] H. Klapper, Control of eutrophication in inland waters: Ellis Horwood Limited, 1991.

[12] A. C. Redfield, "The biological control of chemical factors in the environment," American scientist, vol. 46, 1958, pp. 205-221.

[13] D. L. Roelke and R. H. Pierce, "Effects of inflow on harmful algal blooms: some considerations," Journal of Plankton Research, vol. 33, 2011, pp. 205-209.

[14] F. Cui, Q. Zhou, Y. Wang, and Y. Q. Zhao, "Application of constructed wetland for urban lake water purification: trial of Xing-qing Lake in Xi'an city, China," Journal of Environmental Science and Health Part A, vol. 46, 2011, pp. 795-799.

[15] E. F. Lowe, L. E. Battoe, D. L. Stites, and M. F. Coveney, "Particulate phosphorus removal via wetland filtration: an examination of potential for hypertrophic lake restoration," Environmental management, vol. 16, 1992, pp. 67-74.

[16] S. W. Effler, S. M. O'Donnell, A. R. Prestigiacomo, D. A. Matthews, and M. T. Auer, "Retrospective Analyses of Inputs of Municipal Wastewater Effluent and Coupled Impacts on an Urban Lake," Water Environment Research, vol. 85, 2013, pp. 13-26.

[17] J. Grochowska, R. Brzozowska, and M. Łopata, "Durability of changes in phosphorus compounds in water of an urban lake after application of two reclamation methods," Water Science & Technology, vol. 68, 2013, pp. 234-239.

[18] D. Hanson and D. Austin, "Multiyear destratification study of an urban, temperate climate, eutrophic lake," Lake and Reservoir Management, vol. 28, 2012, pp. 107-119.

[19] W. F. James, J. W. Barko, H. L. Eakin, and P. W. Sorge, "Phosphorus budget and management strategies for an urban Wisconsin lake," Lake and Reservoir Management, vol. 18, 2002, pp. 149-163.

[20] B. Hart, S. Roberts, R. James, J. Taylor, D. Donnert, and R. Furrer, "Use of active barriers to reduce eutrophication problems in urban lakes," Water Science & Technology, vol. 47, pp. 157-163, 2003.

[21] H. Qian, S. Yu, Z. Sun, X. Xie, W. Liu, and Z. Fu, "Effects of copper sulfate, hydrogen peroxide and

N-phenyl-2-naphthylamine on oxidative stress and the expression of genes involved photosynthesis and microcystin disposition in Microcystis aeruginosa," Aquatic Toxicology, vol. 99, 2010, pp. 405-412.

[22] L. Song, T. L. Marsh, T. C. Voice, and D. T. Long, "Loss of seasonal variability in a lake resulting from copper sulfate algaecide treatment," Physics and Chemistry of the Earth, vol. 36, 2011, pp. 430-435.

9

A Comparison of Antimony in Natural Water with Leaching Concentration from Polyethylene Terephthalate (PET) Bottles

Mihyun Jo[1], Taeyuel Kim[1], Sirim Choi[1], Jongpil Jung[1], Hee-il Song[1], Hyunjin Lee[1], Gyoungsu Park[1], Jogyo Oh[1], Jai-young Lee[2]

[1]Gyeonggi Province Institute of Health and Environment, North branch, Drinking Water Inspection Team
1 Cheongsaro, Uijeongbu-si, Gyeonggi-do, Republic of Korea 11780
jomh@gg.go.kr
[2]University of Seoul, Environmental Engineering
163 Seoulsiripdaero, Dongdaemun-gu, Seoul, Republic of Korea 02504
leejy@uos.ac.kr

Abstract – Antimony (Sb) is one of the trace hazardous compounds in drinking water. We investigated Sb concentration on natural environment such as river, reservoir, groundwater and raw water for bottled water. The natural content of Sb in northern Gyeonggi province in South Korea showed range from 0.00~1.64 µg/L. The average of Sb in 47 brands of bottled water was 0.57 µg/L on market. As a results of leaching experiment, was leached from polyethylene terephthalate(PET) bottles under storage condition at 35, 45 and 60℃. Sb concentration was increased from 1.04 to 9.84 µg/L under 60℃ after 12weeks. UV-ray irradiation to bottled water not significantly induced antimony leaching for 14days.

Keywords: Antimony, Leaching, Polyethylene Terephthalate (PET), Bottled Water.

1. Introduction

Antimony (Sb) is one of the trace hazardous components in drinking water. Polyethylene terephthalate (PET) widely used to container for bottled water. Antimony trioxide (Sb_2O_3) is one of the most important catalysts widely used for solid-phased poly condensation of PET. It offers high catalytic activity, does not induce undesirable color, and has a low tendency to catalyse side reactions [1]. Polyethylene terephthalate (PET) has wide acceptance for use in direct contact with food, can be recycled and can be depolymerized to its monomer constituents [2]. In a recent study of Sb in bottled water in Europe and Canada, it was shown that the water become contaminated during storage because of Sb leaching from PET [3]. People tend to bottled water in cars for weeks or months. The temperature in car can reach to 75℃ at ambient temperature of 33℃ in summer [4]. It can be expected that storage at high temperature may enhance contaminant release into water from PET bottles [5]. The bottled water market is continuously growing amount up from 2.1 million ton of 2004 to 3.5 million ton of 2013 in South Korea [6]. Especially, our institute has concerned safety of bottled water for customer. Because there are 62% of manufacturing plants for bottled water located in Gyeonggi province in South Korea.

Recommended standard for tap water have regulated 20 µg/L of Sb since 1998 in Korea. Recently, recommended standard for bottled water regulated in 2014 that should be inspected within 15 µg/L. Europe Union, World Health Organization, United States and Japan also have drinking water standards for Sb at 5 µg/L, 20 µg/L, 6 µg/L and 2 µg/L, respectively[7,8,9,10]. Bottles made using PET typically contain 100~300 of mg/kg Sb in the plastic. In contrast to bottles, Sb is found at natural environment from rocks, groundwater and river. Concentrations of Sb in crustal rocks is about

0.3 mg/kg and pristine groundwater and surface water normally range from 0.1 to 0.2 μg/L[11]. The International Agency for Research on Cancer (IARC) was classified as possibly carcinogenic to humans; Group 2B. It can cause nausea, vomiting and diarrhea when exposed MCL in short periods. The exposure of long term elevated Sb can lead to increased blood cholesterol and decreased blood sugar [12].

The objective of this study was to investigate Sb concentration in natural environment such as source water from river or reservoir, tap water, natural springs, raw water for bottled water. Next, brands of bottled water on market were collected and analysed. Then, Sb content was compared between natural water and PET bottled water.

Finally, the Sb leaching experiments for PET bottled water were conducted according to storage duration, temperature and ultraviolet (UV) to know amount of migration Sb from PET bottles into drinking water.

2. Materials and Methods
2.1. Sb Concentration in Natural Environment

Natural Sb contents were investigated for source water from river or reservoir and tap water in 15 of water supply plants, groundwater from 50 of mineral springs located in northern Gyeonggi province in South Korea. The raw water for bottled water sampled from 54 of intake holes in 13 of bottled water manufacturer plants in northern Gyeonggi province. Sb concentration was analysed by inductively coupled plasma-mass (Bruker aurora, Germany).

2.2. Investigation of Sb Concentration in PET Bottled Water and Comparison with Raw Water for Bottled Water

To investigate leaching effect of Sb into water of PET bottles, 47 commercial brands of PET bottled water were collected in market include domestic manufactured products and imported products. The domestic products, imported water and deep sea water were 35, 7 and 7 bottles respectively. And then Sb concentration was compared with PET bottled water which was produced on that day and raw water for bottled water sampling in 54 intake holes.

2.3. Leaching Experiment in PET Bottled Water

The PET bottled water stored in 4, 21, 35, 45 and 60°C for 12 weeks used two bottles for each temperature. Sb concentration was analysed for every 2 weeks during that period. Polypropylene (PP) and glass bottles were used for control sample to compare to leaching amounts of Sb from PET bottles. Next, PET bottled water stored for 2, 6, 12, 24, 36, 48hrs, 3, 7, 9, 12, 14 days under ultraviolet (UV)-ray. Control samples were PP bottles and glass bottles. pH of bottled water at 6-8 poses no effect on Sb release [12]. Therefore, leaching experiment for pH influence was not conducted.

2.4. Leaching Rate of Sb under Ambient Condition

The Sb concentration for 50 brands of bottled water was analysed under room temperature and natural sunlight after six months. Because boxes of bottled water before reaching customers were mostly stored in uncontrolled storage, yard in country. PET bottled water can easily expose in sunlight and no constant temperature. That's why we conducted leaching experiment under ambient condition. Initial concentration of Sb was analysed in June of 2016 and then after six month checked later concentration.

2.5. Power Function Model on Temperature-and Time-Dependant for Sb Leaching

The results of leaching experiment for temperature used to calculate in power function model. The rate of change in Sb leaching was best fit by a power rather than by first- or second order reaction kinetics with respect to Sb concentration [12]. The equation of power function model was presented in Eq. 1 as:

$$C = a \, [\text{time}]^k \tag{1}$$

The a is the fitted initial Sb concentration a is the fitted initial Sb concentration (C, ppb) at time zero and k is a temperature-dependent power function constant. Time is in hours.

2.6. Estimation of Exposure Time to Reach Standard Concentration in PET Bottles

When PET bottles place in high temperature condition, it is needed to know how many times to reach standard concentration. Estimated exposure time was from Eq.2 [12] as:

$$\text{Time(days)} = ([\text{STD ppb}]/Sb_0)^{1/k} \tag{2}$$

The Sb_0 was initial concentration when PET bottles stored in 60°C. k was power function model exponent which was plotted by temperature and time dependent relation for Sb leaching. STD ppb was concentration for each nation such as South Korea (15ppb), Europe (5ppb), U.S. (6ppb) and Japan (2ppb).

3. Results

3.1. Natural Sb Concentration for Source and Tap Water in Water Supply Plants

Natural Sb concentration in source water from river or reservoir in fifteen water supply plants was 0.13 μg/L of average. The average of tap water also indicated same value at 0.13 μg/L of average. Although natural concentration was very low to not enough to compare variation of concentration, it means that treatment process to be tap water might be to not influence Sb concentration. In Korea, the recommended standard of Sb content for tap water is 20 μg/L. The average of tap water in northern Gyeonggi area was 0.65% compared to the standard. Figure 1 indicated natural Sb concentration in source water and tap water from fifteen water supply plant located in northern Gyeonggi area. HR, SW and GP plants were located in Gypeong. GY and IS plants were located in Goyang. Most of samples showed approximately 0.1 μg/L. Otherwise source water and tap water from MS plant located in Paju city had higher concentration than others. Sb in source water was 0.72 μg/L and tap water showed 0.84 μg/L. In previous study, typical concentration of dissolved Sb in unpolluted water is less than 1 μg/L[13].

Table 1. The water supply plants located in northern Gyeonggi province.

Region	Plant	Source	Capacity (1,000ton/day)
Gapyeong	HR	Groundwater	7
	SW	Groundwater	2
	GP	Groundwater	19
Goyang	GY	River	250
	IS	Reservoir	250
Guri	TP	River	30
Pocheon	GI	River	1.7
	ID	Groundwater	1.5
Yeoncheon	YC	River	33
Paju	MS	River	96
Dongducheon	DD	River	40
Namyangju	HD	River	55
	WB	Reservoir	215
	DS	River	450
Uijeongbu	GN	River	8

3.2. Natural Sb Concentration for Mineral Springs

The average of the drinking water from 50 mineral springs for each of 10 sites in 5 cities in northern Gyeonggi province was 0.02 μg/L that were 0.1 times lower than Sb average concentration of groundwater which was investigated by WHO. According to data from ministry of environment in Korea, Sb concentration for groundwater was 0.24 μg/L on average.

Figure 2 showed Sb concentration on average for mineral springs located in each city. These values were under 0.03 μg/L that was very low. There were 6 non-detected places among 50 mineral springs.

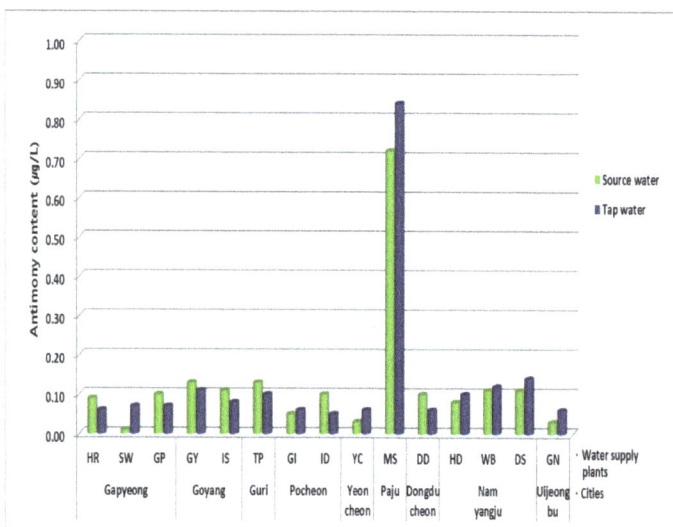

Figure 1. Antimony in source and tap water sampled in water supply plants located in northern Gyeonggi province.

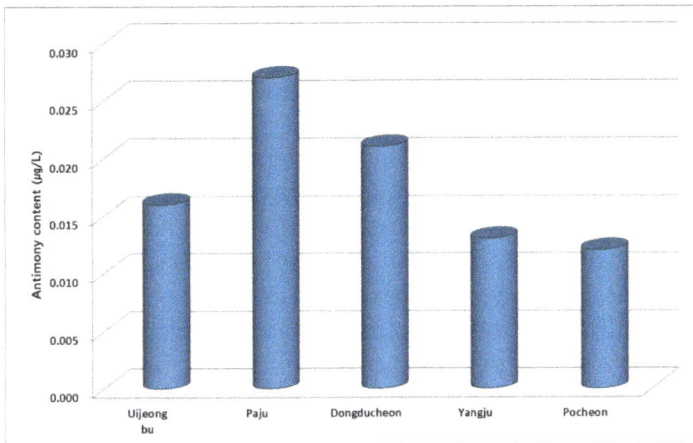

Figure 2. The average of antimony concentration in mineral springs for each city.

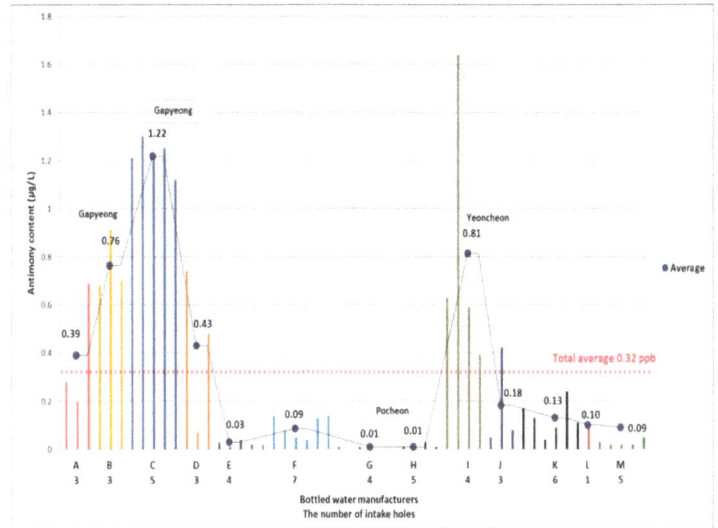

Figure 3. Antimony in mineral water from 54 intake holes in 13 bottled water manufacturing plants.

The Sb in source water from river-bed, surface water and reservoir was 0.13 µg/L. On the other hand, mineral springs showed 0.02 µg/L in northern Gyeonggi area. The Sb concentration in source water was 6.5 times higher than groundwater from mineral springs. Because the Sb might be to leached from soil and sediment with other metals in river.

3.3. Sb Concentration in Mineral Water for Bottled Water

We investigated mineral water from 54 intake holes of 13 bottled water manufacturing plants located in northern Gyeonggi province. Mineral water was draw aquifer up below approximately 200meter. Aquifer Detection rate of Sb was 90.7%. Average concentration in mineral water was 0.32 µg/L and maximum was 1.64 µg/L and there were 5 non-detected intake holes. Figure 3 has shown Sb concentration in mineral water sampled from 54intake holes. From A to M in graph mean each bottled water company and the numbers below these alphabet indicated how many intake holes had in that bottled water manufacturing plants.

The C company located in Gapyeong had highest concentration of Sb. Average concentration for 5 holes was 1.22 µg/L. And then I company located in **Yeoncheon showed 0.81 µg/L from 4 intake holes.** Futhermore the highest value was shown at 1.64 µg/L sampled from second intake hole in I company. The G and H companies had the lowest concentration at 0.01 µg/L. As we reported that source water from river and reservoir included 0.13 µg/L of Sb. And Sb from mineral springs was 0.02 µg/L. The mineral water for bottled water had higher contents than source water and mineral springs. Figure 4 indicated median concentration of Sb in mineral water for each area in northern Gyeonggi area. The 3 bottled water manufacture plants located in Gapyeong had high value at 0.80 µg/L. The 5 companies located in Pocheon had lowest value at 0.06 µg/L. Therefore we have known Sb concentration was affected by properties of local area. So we tried to investigate geologic map for Sb to find more specific proof but there were no specific geologic information for Sb in Korea yet.

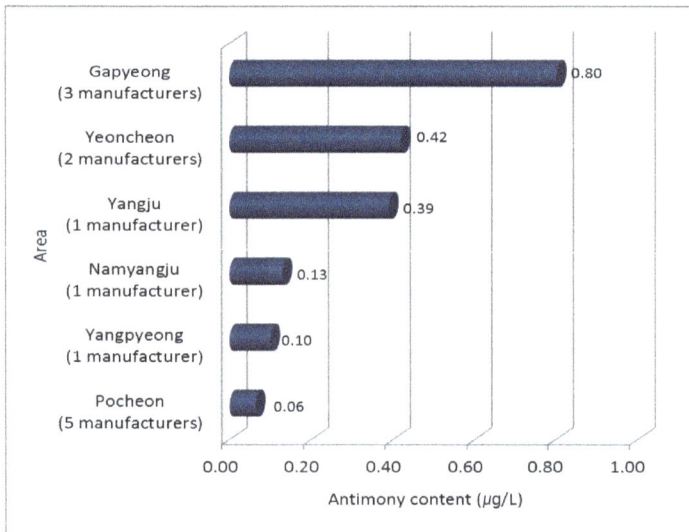

Figure 4. Median value of antimony for each area in northern Gyeonggi province.

3.4. Sb Concentration for PET Bottled Water on Market

We collected 47 brands of PET bottled water in market. 35 products was domestic bottled water. Imported products were 7 and deep sea water was 7 products. Detection rate of Sb was 100% in these products, while the detection rate in mineral water from manufacturing plants was 90.7% . Total average concentration of 47 samples was 0.57 μg/L on the other hand mineral water was 0.32 μg/L. In Korea, the monitoring standard for Sb is 15 μg/L. Therefore the mean concentration in PET bottled water was 0.04% compare to 15 μg/L of monitoring standard to consider safe to drink.

Table 2. Sb concentration in brands of PET bottled water collected in market (μg/L).

PET Bottled Water	Ave.	Max.	Min.	Total Ave.
Domestic (mineral water)	0.75	1.09	0.12	
Imported (mineral water)	0.55	1.08	0.19	0.57
Deep sea water	0.41	0.65	0.27	

Table 2 was shown that Sb concentration for 47 brands of PET bottled water collected in 5 markets. The PET bottled water was classified as three parts like domestic, imported mineral water and deep sea water. Because

bottled water was divided into mineral water and deep sea water by law in Korea. Domestic brands of PET bottled water contained 0.75 μg/L and one of the products had 1.09 μg/L of Sb. And minimum concentration was 0.12 μg/L. In case of imported PET bottled water was 0.55 μg/L of average. Maximum and minimum values in imported samples were similar with domestic products. We guessed that imported PET bottled water would contain more Sb concentration than domestic ones because of the long transport time. However, the average of Sb domestic products got higher value. Deep sea water of bottled water contained lower concentration than mineral water of bottled water.

3.5. Sb Leaching Experiment from PET Bottled Water According to Time and Temperature

Sb leaching experiment was conducted by storing PET bottled water at 4, 21, 35, 45 and 60 °C for 12 weeks to analyse every two weeks. After 2weeks, it observed that Sb was leached from PET bottles at 35, 45 and 60 °C. Figure 5 indicated leaching effects of Sb from PET bottles according to time and temperature.

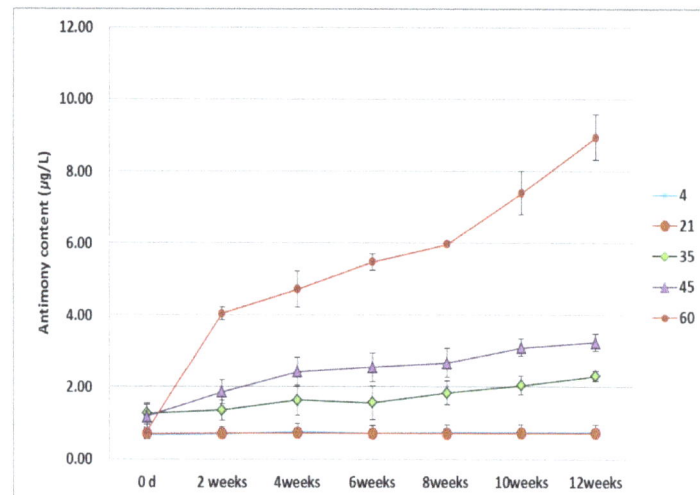

Figure 5. Sb concentration leached from PET bottles when stored at 4, 21, 35, 45, 60°C for 12 weeks.

There were no change for concentration at 4 and 21°C for 12weeks. However it was exposed that Sb was released from PET bottles after 2weeks at 35, 45 and 60°C. In case of storing 35°C bottles, initial concentration of Sb was 1.6 μg/L and time goes by final concentration of Sb indicated 2.50 μg/L which was 1.6 times compare to before 12weeks. At 45 °C, initial Sb in PET bottle water was 1.71 μg/L and then after 12weeks it contained 2.1times contents. Finally, there are significant releasing effects from PET bottled water at 60 °C for every two weeks. After 2weeks, it observed significant increasing concentration of Sb from 1.04 µg/L to 4.31 µg/L to be 4.1 times leaching effects. After 12weeks, Sb was observed at 9.84 µg/L in PET bottled water.

3.6. Sb Leaching Experiment to Irradiate UV-Ray for PET Bottled Water

PET bottled water was exposed by UV-ray for 2, 6, 12, 24, 36 and 48hrs, 3, 7, 9, 12 and 14days respectively. As a result, releasing of Sb started to irradiate UV-ray after 6hrs. Figure 6 showed variation of Sb concentration in PET bottled water that indicated releasing amounts. After 6hrs, 0.19 µg/L of Sb was leached from PET bottled water. There were no significant changes of amount for releasing Sb as time goes by. It was founded that the Sb variation was similar range from 0.1 to 0.2 µg/L until 14days. And mean of variation was 0.16 µg/L. Therefore, Duration for UV-ray irradiation was not contributed to releasing Sb from PET bottles.

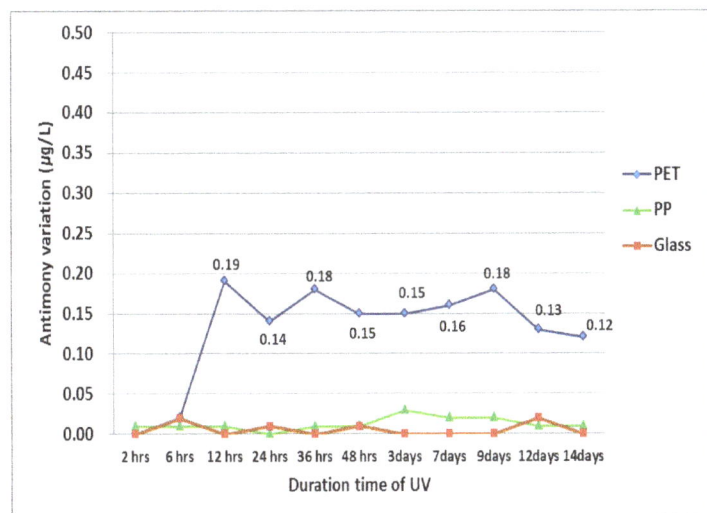

Figure 6. Antimony variation to expose UV-ray for PET, PP and Glass bottles.

3.7. Leaching Rate of Sb under Ambient Condition from PET Bottled Water

The results for Sb leaching experiment for temperature and UV-ray were mentioned above sections. Also, Sb leaching rate under ambient condition was observed from June of 2016 to December of 2016. The 50 brands of PET bottled water were store in room temperature for six months. Because most consumers would put PET bottled water in part of the kitchen or

terrace. Figure 7 showed the leaching rate of Sb. 43 PET bottled water of the 50 samples were observed for releasing Sb. Most samples indicated 20~60% of leaching compare to initial concentration. Maximum leaching rate was approximately 160%.

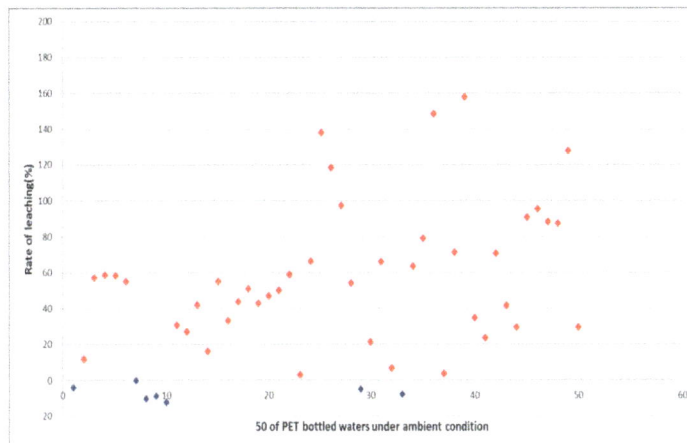

Figure 7. The Sb leaching rate of 50 brands of PET bottled water were stored in ambient condition after six month.

3.8. Sb Leaching Applied for Power Function Model on Temperature-And Time-Dependant

The rate of antimony leaching was correlated by a power function model [10]. So, we also calculated equation of Eq.1 using results of leaching experiment for Sb at 35, 45 and 60°C. Table 3 indicated value for k and R^2 for each temperature. All of R^2 were over 0.90 so we knew that this studies for leaching Sb also got well fitted with power function model.

Table 3. The results of power function model for Sb leaching experiment for exposure time.

Exposure time(°C)	a	k	R^2
35	1.60	0.0587	0.91
45	1.71	0.0978	0.97
60	1.04	0.2953	0.93

3.9. Estimation of Exposure Time to Reach Standard Concentration in PET Bottles

We calculated the expected exposure time to reach standard concentration for standards for each country in case of PET bottled water stored in 60°C. The Eq.2 was used for estimation and Sb_0 was 1.04 µg/L. The k value was 0.93 indicated in Table 3. In case of South Korea, the Sb concentration will reach to 15 µg/L of recommended standard after 350days. The 15.7 days were needed to reach 6 µg/L of standard in U.S. The 8.5 days of exposure time can exceed 5 µg/L of standard in

Europe. And, 9.2 hours were required to reach 2 µg/L of recommended standard in Japan.

4. Conclusions

In the northern Gyeonggi province of South Korea, natural concentrations of antimony was investigated in 13 of water supply plants for source water from river or reservoir and tap water were range from 0.01~0.84 µg/L and mean was 0.13 µg/L for both source water and tap water. 50 mineral springs for public showed 0.02 µg/L that were very low to not need to be worry. And mineral water for bottled water which contained 0.32 µg/L on average to be investigated from 54 intake holes in 13 bottled water manufacturing plants. As a result of 47 brands of PET bottled water, average concentration of Sb was 0.57 µg/L. and detection rate was 100%. Otherwise detection rate of mineral water was 90.7%. PET bottled water brands, the average value was 0.57 µg/L and detection rate was 100%. Otherwise detection rate of raw water to bottled water showed 90.7%. Sb in PET bottled water contained higher concentration than natural water such as river, mineral springs, mineral water for bottled water. As a result of leaching experiment for PET bottled water, releasing of Sb was revealed temperature-dependent. Sb concentration from PET bottled water in 35, 45 and 60°C started to increase after 2weeks. The leaching amount was rapidly increased till 9.84 µg/L after 12weeks in 60°C, though, it was less than recommended standard in Korea. However UV-ray irradiation to bottled water for 14days induced increasing antimony release into water very slightly that variation for Sb concentration was constant from 0.1 to 0.2 µg/L. Therefore we realized that it needed to store the PET bottled water in storage condition to maintain 4, 21°C. High temperature can induce leaching Sb from PET bottled water.

References

[1]   B. Duh, "Effect of antimony catalyst on solid-state polycondensation of poly (ethylene terephthalate)," *Polymer*, vol. 43, pp. 3147-3154, 2002.

[2]   W. A. MacDonald, "New advances in poly(ethylene terephthalate) polymerization and degradation," *Polymer International*, vol. 51, no. 10, pp. 923-930, 2002.

[3]   W. Shotyk, M. Krachler and B. Chen, "Contamination of canadian and european bottled waters with antimony leaching from PET containers," *Journal of Environmental Monitoring*, vol. 8, no. 2, pp. 288-292, 2006.

[4]   R. Manning and J. Ewing, "Temperature in Cars survey," pp. 1-21, 2009.

[5]   Y. Y. Fan, J. L. Zheng, J. H. Ren, J. Luo, X. Y. Cui and L. Q. Ma, "Effects of storage temperature and duration on release of antimony and bisphenol A from polyethylene terephthalate drinking water bottles of China," *Environmental Pollution*, vol. 192, pp. 113-120, 2014.

[6]   Ministry of Environment, *Environmental statistics yearbook*. R. K., 2015, pp. 125.

[7]   CEC, "Council directive relating to the quality of water intended for human consumption," 80/778/EEC, 1980.

[8]   WHO, *Guidelines for drinking-water*, 3rd ed., vol. 1, recommendations, 2006, pp. 305.

[9]   USEPA, "National primary drinking water standards," U.S., 2009.

[10]   W. Hiroshi, "Revision of drinking water standards in Japan," Japan, 2003, pp. 84.

[11]   H. J. M. Bowen, "Environmental chemistry of the elements," London, Academic Press, 1979.

[12]   P. Westerhoff, P. Prapaipong and A. Hillaireau, "Antimony leaching from polyethylene terephthalate(PET) plastic used for bottled drinking water," *Water Research*, vol. 42, pp. 551-556, 2008.

[13]   B. Nelson, C. Yu-Wei, I. Occurrence and F. Montserrat, "Antimony in the environment: a review focused on natural waters," *Earth-Science Reviews* 57, pp. 125-176, 2002.

10

Nano-Iron Carbide-Catalyzed Fischer-Tropsch Synthesis of Green Fuel: Surface Reaction Kinetics-controlled Regimes in a 3-φ Slurry-Continuous Stirred Tank Reactor

Nicolas Abatzoglou and Benoit Legras
Université de Sherbrooke, Department of Chemical and Biotechnological Engineering
2500 boul. Université, Sherbrooke, Quebec, Canada J1K 2R1
Nicolas.Abatzoglou@USherbrooke.ca

Abstract- *Liquid fuel derived from renewable resources is one of the technologies under development as part of "biorefining" platforms. Fisher-Tropsch synthesis (FTS) is a commercial technology producing alternative fuels from coal (CTL- Coal-To-Liquid) and natural gas (GTL; Gas-To-Liquid). FTS Biomass-To-Liquid (BTL), although not yet at the market level, is a continuously-growing field, and its successful commercialization depends on improving techno-economic sustainability.*

In a previous study by the authors' research team, a plasma-synthesized nano-iron carbide catalyst (PS-Nano-FeC) demonstrated direct relationship to the presence of iron carbide high-FTS and water-gas-shift (WGS) activity, with high-catalyst stability and regenerability in a 3-phase slurry, continuously-stirred tank reactor (S-CSTR). Although these results, along with a recently-published phenomenological model, are indicators of the industrial potential of this catalyst, transport phenomena, chemical mechanisms and their intrinsic kinetics are needed as reactor scale-up tools.

In the work reported here, the PS-Nano-FeC catalyst was tested in a S-CSTR with hexadecane as liquid carrier. We evaluated the optimal operating conditions for a surface-reaction, kinetics-controlled regime.

The results include: (1) Reactant conversion and product yields; (2) Fresh and used catalyst instrumental analyses; (3) A model considering all transfer and surface kinetics, accompanied by proof of the rate-controlling step.

Keywords: Green Fuels; Biorefinery; Fischer-Tropsch synthesis; Nanocatalyst; Slurry reactor; Kinetics; Plasma-spray

Nomenclature

a_b	bubble surface area (m²/m³ solution)
BTL	Biomass-To-Liquid
C_i	concentration of i species in liquid phase
CTL	Coal-To-Liquid
F_i	molar flow rate of component I at the exit
FID	Flame Ionization Detector
FTS	Fisher-Tropsch Synthesis
$G-L$	Gas-Liquid
GTL	Gas-To-Liquid
HC	Hydrocarbon
k	specific reaction rate (m³/g$_{cat}$.s)
k_b	mass-transfer coefficient for gas absorption
k_c	mass transfer coefficient to the catalyst (m/s)
$L-S$	Liquid-Solid
m	mass concentration of the catalyst (g$_{catalyst}$/m³ solution)
m_i^C	Carbon molar selectivity of product i
N_i^C	Number of carbon atom in HC i
PS-Nano-FeC	Plasma-Synthesized-Nano-Iron Carbide
r_b	diffusional resistance (external mass transfer)
P	Pressure
R_i	Generation rate of species i in (mass per time unit par mass or volume unit)
R_A	apparent reaction rate (mol/m3 solution.s)
$S-CSTR$	Slurry-Continuous-Stirred-Tank-Reactor
SEM	Scanning Electron Microscopy
T	Temperature

TGA   Thermo-Gravimetric Analysis
XRD   X-Ray Diffraction
WGS   Water-Gas-Shift

Greek Symbols

χ_{FT}^{CO} Molar fraction of CO consumed in FTS

η Catalyst effectiveness factor

1. Introduction

Escalating crude oil prices and environmental considerations are motivating great interest in shifting from fossil to biomass and waste resources as feedstock of transportation fuels. Biomass and waste gasification involves a combination of partial oxidation and steam reforming, leading to synthesis gas production with controlled hydrogen (H_2)/carbon monoxide (CO) ratio. The latter is defined by the needs of Fischer-Tropsch Synthesis BTL processes. FTS reactions are commonly simplified as a combination of Fischer-Tropsch FT (Eq. 1) and water-gas-shift (WGS) reactions (Eq. 2):

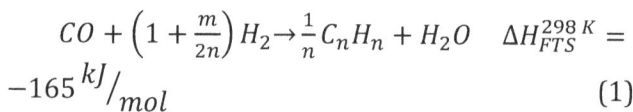

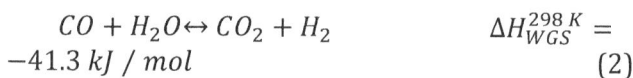

$$CO + \left(1 + \frac{m}{2n}\right)H_2 \rightarrow \frac{1}{n}C_nH_n + H_2O \quad \Delta H_{FTS}^{298\,K} = -165\,{}^{kJ}/_{mol} \tag{1}$$

$$CO + H_2O \leftrightarrow CO_2 + H_2 \quad\quad \Delta H_{WGS}^{298\,K} = -41.3\ kJ\ /\ mol \tag{2}$$

where n is the average number of carbon atoms and m is the average number of hydrogen atoms of hydrocarbon (HC) products. Water is a primary product of FTS reactions, and CO_2 is mainly produced by WGS reaction. Iron-based catalysts have high WGS rates [1]. Although WGS is not a desired reaction, its extent may be exploited to adjust poor H_2/CO reactant mixtures. Ruthenium and cobalt-based catalysts have negligible WGS activity, but the low cost and high stability of iron catalysts, especially when they are alkali-promoted, make the latter more attractive FTS catalysts. Most of these iron catalysts, used for many years industrially [2, 3], are prepared by precipitation techniques [4, 5, 6]. In a previous paper, Blanchard et al. [7] presented a novel catalyst preparation method employing a plasma-spray technique for the production of a nano-iron-carbide catalyst. This preparation method allows large batch production amounts and gives spherical core-shell iron carbide-graphitic carbon nanoparticles. The graphitic shell provides easy handling of the catalyst in ambient conditions without significant oxidation extent. The carbon-based shell needs preliminary *in situ* H_2 reduction to activate the catalyst prior to FTS.

A major aspect in the development of commercial FTS is the choice of reactor type [8]. The 2 most favourite systems for FTS are the multi-tubular trickle bed reactor and the slurry bubble column reactor. Their advantages and disadvantages have already been published [9]. The slurry reactor has the following main advantages over its competitors: 4 times lower differential pressure over the reactor; lower catalyst loading; excellent heat transfer characteristics resulting in stable and homogeneous reactor temperature; introduction of fresh catalyst is possible during runs; and lower investment capital. To scale-up such a FTS reactor, kinetics studies are necessary to identify the rate-controlling steps and range of operating conditions, allowing for surface-reaction kinetics control of the phenomenon. In the present work, the reactor tested was a 3-phase slurry, continuously-stirred tank reactor (S-CSTR) that is considered be an ideal, fully-mixed reactor, meaning that the following attributes/hypotheses were considered: perfect mixing (gradientless reactor, i.e., global uniformity of concentration and temperature throughout reactor volume). Conditions under which surface-reaction kinetics are the rate-limiting step were determined, and a 50-h FTS experiment was run under optimum conditions to examine catalyst stability over time-on-stream.

2. Materials and Experimental Methodology

2.1. Catalyst Preparation

Iron carbide nanoparticles were prepared according to a previously-reported plasma spray technique [7].

2.2. Catalyst Characterization

The catalyst produced by plasma was characterized prior to its reduction by X-ray diffraction (XRD) and scanning electronic microscopy (SEM). XRD analyses disclosed the presence of 2 iron carbide phases (Fe_5C_2 and Fe_3C) in a graphitic carbon matrix; some unconverted metallic iron was also identified (Figure 1). SEM analyses (Figure 2a) revealed unconverted iron spheres and 20- to 100-nm dispersed nanoparticles (Figure 2b).

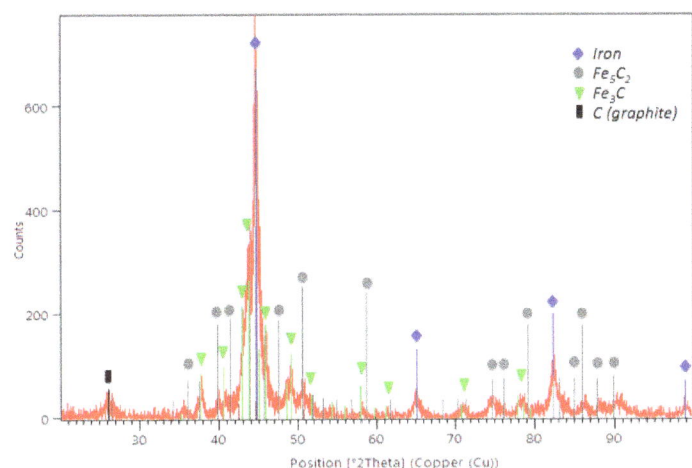

Figure 1. XRD analysis of nano-iron carbide catalyst prior to reduction.

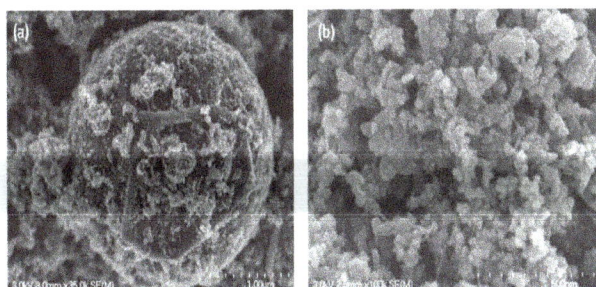

Figure 2a and 2b. SEM analyses of plasma catalyst prior to its reduction step.

The particle size distribution of plasma synthesis powder was assessed by laser diffraction with a Hydro 2000S Malvern Mastersizer Instrument. Measurement was undertaken with the nanopowder dispersed in dry ethanol. Number particle size distribution (Figure 3) confirmed that particle diameter was below 100 nm. The specific surface area of the catalyst was 67 m^2/g and was measured by N_2 physisorption after overnight degassing at 200°C, according to the Brunauer-Emmett-Teller method in a Micromeritics ASAP2020 apparatus. Thermogravimetric analysis (TGA) by a TA Instruments apparatus, coupled with a mass spectrometer to follow the CO_2 produced, was performed for elementary analysis of the synthesis catalyst. A 10°C/min heating rate was imposed between ambient temperature and 1,000°C under 50-ml/min airflow. Air oxidation of carbides, amorphous and graphitic carbons led to iron III oxide, Fe_2O_3, as the only solid end-product. 34% iron weight composition was ascertained in the plasma synthesis catalyst. For FTS reaction, gas space velocity was based on elemental iron mass calculations.

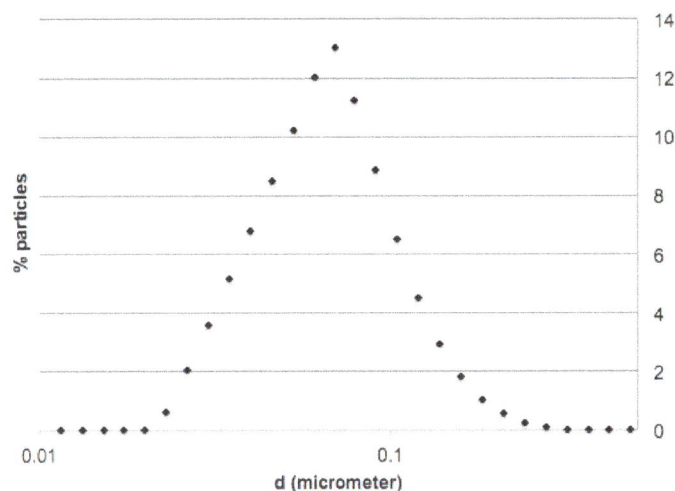

Figure 3. Nano-iron carbide particle size distribution.

3. FTS Reaction

3.1. Experimental Set-Up

The reaction apparatus is depicted in Figure 4. It includes an Engineer's Autoclave set-up designed for high pressure and temperature reactions. The reactor is of 600-ml capacity and is equipped with a magnetic stirrer engaging a turbine impeller in liquid phase operating up to 3,000 rpm. Inlet reactant gas mixture composition is controlled by a Brooks mass flow meter connected to a premixed gas tank whose composition is representative of reformed biosyngas derived from thermal gasification with pure oxygen (11% CO_2, 33% CO and 56% H_2), a pure hydrogen gas tank to adjust H_2/CO ratio and perform catalyst reduction pre-treatment, as well as an argon (Ar) gas tank to purge the system during stop and carrier liquid (hexadecane) feed. Inlet gas is fed into the reactor by a perforated, annealed pipe as sparger/diffuser at the reactor bottom. Outlet flow is heat-traced at 150°C to a high-pressure trap maintained at ambient temperature. Pressure is regulated by a micrometric needle valve. Expanded gas enters a 0°C trap and goes through a micrometric filter before exit gas flow monitoring by a BIOS Definer dry volumeter. Gas composition is analyzed continuously with a Varian micro-GC CP-4900 equipped with 2 columns, CP-Sil and CP-COX, coupled with a thermal conductivity detector. Periodic off-line gas analysis was performed by a Varian GC CP-3800 equipped with successive columns (Haysep T and Hayesep Q) coupled with a flame ionization detector (FID) to quantify C_2H_4 and C_2H_6 concentrations. Off-line analysis of the liquid product collected in the high-pressure trap was conducted with a Varian GC CP-3800 equipped with a capillary 0.25 mm x 0.5 µm x 100 m (CP7530) column coupled with a FID detector.

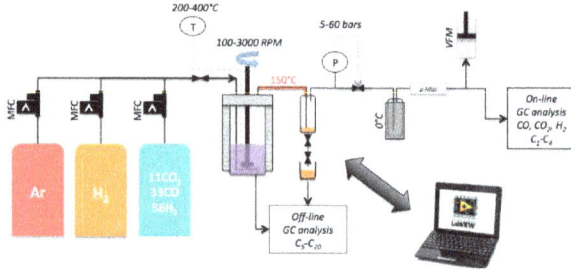

Figure 4. Experimental set-up.

Prior to testing, the FTS nano-iron-carbide catalyst was activated *in situ* at 400°C for 6 h under a pure H_2 flow rate of 6.7 ml/g_{cat}/min and 500 rpm stirring. Hexadecane served as inert liquid carrier after saturation with Ar to avoid oxygen feed. 200 ml was fed into the reactor by means of an Ar-pressurized tank.

3.2. Mass Balance Calculations

As Blanchard et al. (2010) described previously, this set-up does not allow direct full mass balance. However, conversion of the reactants can be calculated appropriately considering their absolute amount in the inlet and outlet streams, including gas hold-ups in the reactor and high-pressure condenser. With this protocol, CO and H_2 conversions were calculated, as was CO_2 production. To consolidate mass balance and access to carbon molar selectivity values, the following hypotheses were made: (a) the formation of oxygenated HC products was assumed to be negligible, and (b) all liquid HC products were considered to have an average atomic H/C ratio=2.2.

CO consumption kinetics to HC was expected to follow Eq. 3.

$$R_{CO} = -R_{FTS} - R_{WGS} \tag{3}$$

Carbon molar selectivity m_i^C of product i in FTS was calculated from experimental carbon mole fractions relative to CO consumption in FTS reactions (Eq. 4).

$$m_i^C = \frac{N_i^C F_i}{\chi_{FT}^{CO} F_{CO}} \tag{4}$$

where N_i^C represents carbon number in HC i, F_i is the oulet i molar flow rate, χ_{FT}^{CO} is the molar fraction of CO consumed in FTS reactions, and F_{CO} is the CO inlet flow rate.

Average 3% and maximum 5% mass balance errors were calculated for each run. Conversion and selectivity values were computed after reaching steady state and waiting for at least 3 gas residence times.

3.3. Kinetics Study

Prior to analyzing the influence of operating conditions on reactant conversions and selectivity, it is always necessary to check for mass and heat transfer limitations. Eq. 5 expresses the reaction rate as a function of all resistances in a 3-phase slurry reactor [10].

$$\frac{C_i}{R_A} = \frac{1}{k_b a_b} + \frac{1}{m}\left[\frac{1}{k_c a_c} + \frac{1}{k\eta}\right]$$

$$\frac{C_i}{R_A} = r_b + \frac{1}{m} r_{cr} \tag{5}$$

where C_i represents the concentration of i species in liquid phase (mol/m³ solution), R_A is the apparent reaction rate (mol/m³ solution.s), k_b is the mass-transfer coefficient for gas absorption (m/s), a_b is the bubble surface area (m²/m³ solution), m is mass concentration of the catalyst ($g_{catalyst}$/m³ solution), k_c is the mass transfer coefficient to the catalyst (m/s), a_c is the external surface area of particles (m²/g), η is the catalyst effectiveness factor, and k (m³/g_{cat}.s) is the specific reaction rate. Different resistances were extracted to determine the limiting step: r_b(s) expresses gas absorption resistance, and r_{cr}(s), diffusional resistance to (external mass transfer) and in (internal mass transfer) the catalyst. Considering the nanometric size of the catalyst particles, the effectiveness factor was close to 1, and internal mass-transfer resistance could be neglected. It was experimentally demonstrated [11] that intra-particle diffusion limitation was significant in case of FTS for particle diameters above 1 mm; below 63 µm, no intra-particle diffusion limitation was apparent and fully intrinsic kinetics were observed [12]. r_{cr} expression can be so reduced to Eq. 6.

$$r_{cr} \cong \frac{1}{k} \tag{6}$$

Stirring rate was the first parameter studied. Efficient stirring maintains homogeneous catalyst suspension and bulk liquid concentration and minimizes gas bubble size. Thus the resistances attributed to gas-liquid absorption and to diffusion between bulk liquid and external surface of the catalyst particles respectively are minimized. The conversion of gas reactants was monitored under conditions unfavourable to gas absorption (relatively low pressure and high temperature, e.g., 10 bars and 275°C, respectively) and relatively large catalyst load (16.5 g) requiring high stirring rates to reach homogeneous suspension and fast reaction kinetics. The stirring rate was varied between 250 and 2,500. Corresponding results are presented in Figure 5. Above 2,000 rpm, CO conversion reaches a plateau, indicating that higher stirring rates do not improve the reaction rate. The results reported in Figure 5 led us to choose the safe value of 2,500 rpm in subsequent runs aimed at minimizing gas absorption resistance.

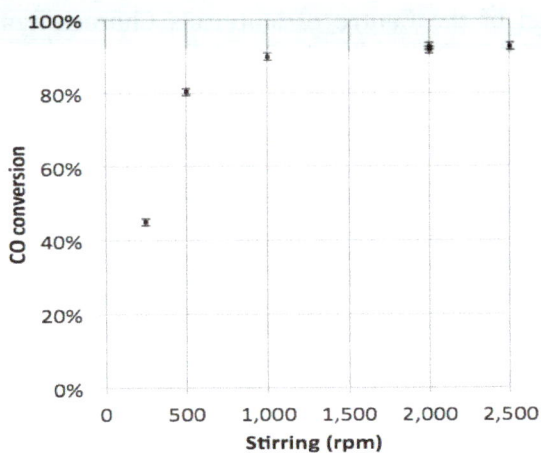

Figure 5. CO conversion at different stirring rates (P=10 bars, T=250°C, m_{cat}=16.5g, Q_{in}=400 cm³ (STP), $CO/CO_2/H_2$=33/11/56).

To ensure a surface-reaction-kinetics-limited regime, catalyst loading was varied to estimate gas absorption limitation. Three catalytic tests were undertaken with 4.5, 2.25 and 1.125 g catalyst loading in 200 ml hexadecane. CO conversion was monitored, and a $C_{CO}/(-R_{CO})=f(1/m)$ chart was built. CO concentration in hexadecane was estimated (13), and R_{CO} was the apparent CO consumption rate. Figure 6 shows that surface reaction kinetics + liquid-solid diffusion resistances were together overwhelmingly higher than gas absorption resistances. However, since the experiment was performed at 2,500 rpm, liquid-solid

diffusion resistance was not limiting, as see in Figure 5. Consequently, surface-reaction-kinetics resistance was definitely the rate-limiting step, and this procedure allowed the phenomenological evaluation of such resistance.

3.4. Sources of Experimental Errors and Preliminary Statistical Analysis

Reactor inlet flow is controlled with a mass flow meter. The measured precision is +/-3%. The BIOS Definer dry volumeter used to measure the exit gas flow has a theoretical measurement error of the order of 0.1% but a real precision of 1% has been considered. GC analysis precision depends upon the products; except propane, exhibiting a fairly high error of 15%, all other gases measured have errors in the range of 1-2.5%. The reproducibility tests have shown a maximum error of 3% on reactants conversion as well as WGS and FTS selectivity calculations. Regarding the liquid FTS products, the errors related to each particular molecular weight component is a function of the latter. The basic hypothesis used is that all peaks have the same conversion factor. More details can be found in Abatzoglou et al. [14] and Blanchard [15].

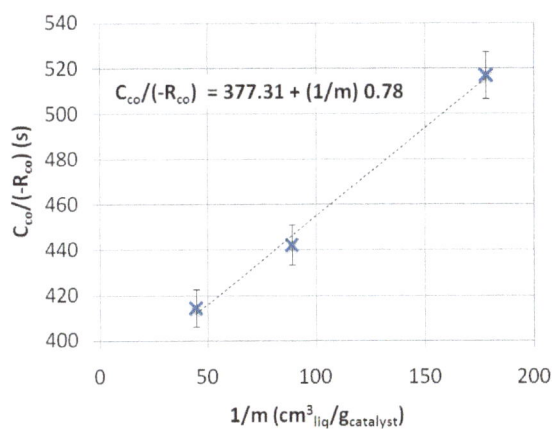

Figure 6. Effects of catalyst amount on CO consumption time (P=10.3 bars, T=275°C, 200 ml hexadecane, Q_{in}=200 cm³ (STP), $CO/CO_2/H_2$=33/11/56).

4. Catalyst Efficiency and Stability

Proof of catalyst efficiency and stability over time-on-stream was the last contribution of our paper. Figure 7 reports the results obtained under conditions described in the legend.

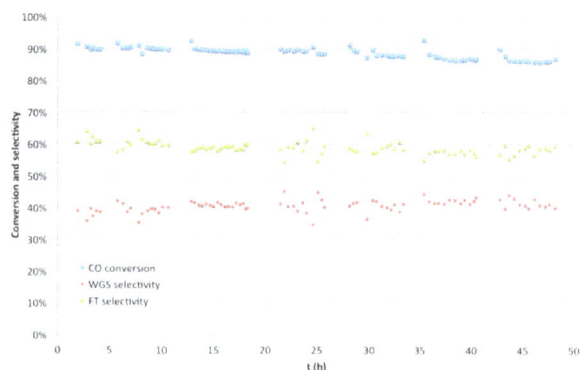

Figure 7. CO conversion and FT/ WGS selectivity during a 50-h semi-continuous run
(P=22 bars, T=250°C, $m_{nano-FeC}$=4.5 g, GHSV (ml/h.g_{iron})=8 400, $CO/CO_2/H_2$=33/11/56).

This experiment took place in the set-up depicted in Figure 4. The term semi-continuous is used because a part of the product, the heavier components, **accumulates in the reactor during the 50h duration run.** Although the catalyst evolves towards more oxidized forms, as shown in previous publication of our group [16], this experiment shows a remarkable stability in terms of reactants (CO) conversion and FTS selectivity.

5. Conclusion

Our work has produced new data on a novel plasma-produced FTS nano-iron carbide catalyst. This new knowledge is critical for eventual scale-up of the technology.

(1) Although, slurry-phase reactors are characterized by the advantage of higher heat transfer rates, thus rendering this process control easier and more efficient, mass transfer phenomena are complex and their control is rather difficult. In the case of the nanocatalysts used in this work, gaseous reactants are transferred (solubilized) in the liquid carrier+products phase (G-L transport), then diffuse at the surface of the catalysts (L-S transport) and are adsorbed by its active sites. Since these nanocatalysts have no internal porosity other than that of the surrounding carbon matrix, internal mass transfer, which is usually the slowest mass diffusion step, is not an issue. This work has determined stirring conditions, which ensure optimal G-L mass transfer rates in the S-CSTR used. These optimal conditions mean that the overall phenomena are not mass-diffusion controlled.

(2) As a consequence of the above, under rather unfavourable mass transfer conditions, the results clearly show that the nanometric characteristics of this catalyst allow surface-reaction-kinetics to be the rate-controlling phenomenon of FTS reactions.

(3) The catalyst has demonstrated excellent efficiency (more than 90% CO conversion) and stability in a 50-h run under optimal CO conversion conditions. Moreover, FTS and WGS selectivity, defined and measured in [14, 15], have been equal stable over time-on-steam for this 50h run.

A parametric analysis, using this catalyst as well other Fe and Co-based plasma-derived formulations is the next step of the running work.

Acknowledgements

The authors are indebted to the Natural Sciences and Engineering Research Council (NSERC) of Canada (Collaborative Research and Development, and Discovery grants program) and Enerkem Inc. for funding related to this project. The scientific input of Jasmin Blanchard and Prof. Nadi Braidy as well as the technical contributions of Henri Gauvin, Jacques Gagné and the personnel of the Centre of Materials Characterization (CCM) of the Université de Sherbrooke are gratefully acknowledged.

References

[1] G. P. Van Der Laan and A. A. C. M. Beenackers, "Intrinsic Kinetics of Gas-Solid Fischer-Tropsch and Water Gas Shift Reactions over a Precipitated Iron Catalyst," *Appl. Catalysis A: General*, vol. 193, no. 1-2, pp. 39-53, Feb. 2000.

[2] H. Schulz, "Short History and Present Trends of Fischer-Tropsch Synthesis," *Appl. Catalysis A: General*, vol. 186, no. 1-2, pp. 3-12, Oct. 1999.

[3] M. E. Dry, "The Fischer-Tropsch Process: 1950-2000," *Catalysis Today*, vol. 71, no. 3-4, pp. 227-241, Jan. 2002.

[4] D. B. Bukur, L. Nowicki, R. K. Manne, and X. S. Lang, "Activation Studies with a Precipitated Iron Catalyst for Fischer-Tropsch Synthesis: II. Reaction Studies," *J. of Catalysis*, vol. 155, no. 2, pp. 366-375, Sep. 1995.

[5] M. D. Shroff, D. S. Kalakkad, K. E. Coulter, S. D. Kohler, M. S. Harrington, N. B. Jackson, A. G. Sault, and, A. K. Datye, "Activation of Precipitated Iron Fischer-Tropsch Synthesis Catalysts," *J. of Catalysis*, vol. 156, no. 2, pp. 185-207, Oct. 1995.

[6] R. J. O'Brien, L. Xu, R. L. Spicer, and B. H. Davis, "Activation Study of Precipitated Iron Fischer-

Tropsch Catalysts," *Energy & Fuels*, vol. 10, no. 4, pp. 921-926, Jul. 1996.

[7] J. Blanchard, N. Abatzoglou, R. Eslahpazir-Esfandabadi, and F. Gitzhofer, "Fischer-Tropsch Synthesis in a Slurry Reactor Using a Nanoiron Carbide Catalyst Produced by a Plasma Spray Technique," *Ind. and Eng. Chemistry Res.*, vol. 49, no. 15, pp. 6948-6955, May 2010.

[8] S. T. Sie and R. Krishna, "Fundamentals and Selection of Advanced Fischer-Tropsch Reactors," *Appl. Catalysis A: General*, vol. 186, no. 1-2, pp. 55-70, Oct. 1999.

[9] R. Krishna and S. T. Sie, "Design and Scale-Up of the Fischer–Tropsch Bubble Column Slurry Reactor," *Fuel Process. Technol.*, vol. 64, no. 1-3, pp. 73-105, May 2000.

[10] H. S. Fogler, *Elements of Chemical Reaction Engineering*, 4th edition, New Jersey, USA: Prentice-Hall Int. Inc., 2006.

[11] M. F. M. Post, A. C. V. Hoog, J. K. Minderhoud, and, S. T. Sie, "Diffusion Limitations in Fischer-Tropsch Catalysts," *AIChE J.*, vol. 35, no. 7, pp. 1107-1114, Jul. 1989.

[12] Zimmerman, W. H. and Bukur, D. B., "Reaction Kinetics over Iron Catalysts Used for Fischer-Tropsch Synthesis," *Can. J. of Chemical Eng.*, vol. 68, no. 2, pp. 292-301, Apr. 1990.

[13] B. B. Breman, A. C. C. M. Beenackers, and, E. Oesterholt, "A Kinetic Model for the Methanol-Higher Alcohol Synthesis from CO/CO2/H2 Over Cu/ZnO-based Catalysts Including Simultaneous Formation of Methyl Esters and Hydrocarbons," *Chemical Eng. Sci.*, vol. 49, no. 24, pp. 4409-4428, Jan. 1994.

[14] N. Abatzoglou, J. Blanchard, and F. Gitzhofer, "Nano-Iron Carbide-Catalyzed Fischer-Tropsch Synthesis in Slurry Rectors: New Developments," Presented at the 20th European Union Biomass Conference and Exhibition, UK, 2012.

[15] J. Blanchard, "Fischer-Tropsch Synthesis in a Slurry Reactor Using a Nano-iron Carbide Catalyst," PhD Thesis, Dept. of Chemical Eng. And Biotechnological Eng., Université de Sherbrooke, Sherbrooke, Quebec, Canada, 2014.

[16] N. Braidy, C. Andrei, J. Blanchard, and N. Abatzoglou, "From Nanoparticles to Process: An Aberration-Corrected TEM Study of Fischer-Tropsch Catalysts at Various Steps of the Process," *Adv. Mat. Res.*, vol. 324, no. 197, pp.197-200, Aug. 2011.

Guidelines for Preliminary Design of Funnel-and-Gate Reactive Barriers

Benoît Courcelles, Ph.D.
Polytechnique Montréal, Department of Civil, Geological and Mining Engineering
CP 6079, Succ. Centre-Ville, Montréal, Qc, H3C 3A7, Canada
benoit.courcelles@polymtl.ca

Abstract - *Permeable Reactive Barriers represent an innovative remediation technique of contaminated aquifers. Three geometric configurations are encountered in the literature: a continuous wall, a funnel-and-gate system, and a caisson configuration. The present paper is focused on the design of the second and third geometric configurations and presents an analytical solution of the flow in a Permeable Reactive Barrier based on the Schwarz-Christoffel transformation. This analytical solution is coupled to residence time calculations to define a methodology of design taking into account the most important parameters on the design of a PRB: cut-off width, slenderness of the reactive cell, and hydraulic conductivity. Finally, the study provides a guidance diagram for the design of funnel-and-gate or caisson configurations, as well as a case study.*

Keywords: Permeable Reactive Barrier; Contamination; Groundwater; Analytical study; Design.

Nomenclature

$\alpha_1, \alpha_2, \alpha_3, \alpha_4$	External angles in the Schwarz-Christoffel transformation
γ	Complex number in the Schwarz-Christoffel transformation
ϕ	Velocity potential
$\phi_u^w - \phi_d^w$	Difference of velocity potential between two wells in an infinite domain [m]
$\phi_d^v - \phi_u^v$	Difference of velocity potential between two points in a uniform velocity field [m]
ψ	Stream function
D	Thickness of the aquifer [m]
f	Schwarz-Christoffel transformation
g	Inverse of the Schwarz-Christoffel transformation
h_u-h_d	Hydraulic head loss in a filter [m]
k_{filter}	**Hydraulic conductivity of a filter [m/s]**
L_{filter}	Length of a filter [m]
q	Flow rate per meter of depth of aquifer [m²/s]
Q	Total flow rate in a filtering gate [m³/s]
R	Half-length of the cut-off wall [m]
R_d	Radius of a drainage element [m]
S_{filter}	Cross-surface of a filter [m²]
T	Residence time [s]
V_0	Initial velocity [m/s]
$\chi_A, \chi_B, \chi_C, \chi_D$	Real constants in the Schwarz-Christoffel transformation
χ_S	Abscissa of the stagnation point (z-diagram) [m]
y_p	Half-width of the capture zone entering a well (z-diagram) [m]
Y_p	Half-width of the capture zone entering a PRB (z'-diagram) [m]
Y_S	Ordinate of the stagnation point (z'-diagram) [m]

1. Introduction

Permeable reactive barriers (PRBs) constitute a passive remediation technique for the treatment of polluted groundwater [1]. Their principle relies on the exploitation of hydraulic gradients to treat the groundwater in a reactive media able to degrade, adsorb or precipitate the pollutants.

Three main geometric configurations are available in the literature: (a) a continuous wall (CW) composed of reactive trenches or injection wells [1]; (b) a funnel-and-gate configuration (F&G) composed of two impermeable walls that direct the contaminated plume towards a filtering gate [2]; and (c) a caisson configuration (CC) similar to the previous one, but in which the flow in the filtering gate is in the upward direction [3].

As regards the implementation of each configuration, continuous walls represent the common type of PRB [1]. A design methodology is dedicated to this configuration and relies on the residence time of pollutants in the reactive media [4, 5]. On the contrary, only few practical tools are available in the literature for the design of funnel-and-gate PRBs and they are particularly focused on the hydraulic behaviour of PRBs [6-8]. However, the design of such PRBs relies on three technical aspects: (a) the reactive media must be appropriate to the pollutants, (b) the filters' size must be large enough to ensure a sufficient residence time [9-11], and (c) the reactive material must have a sufficient hydraulic conductivity to prevent any bypass of the system. The first aspect is a key issue in the design of PRBs and a particular attention has to be dedicated to the selection of the reactive or sorbent material when installing any PRB system [12]. The present paper will consider that the reactive media is selected adequately and the compatibility with the contaminant is not an issue.

Another key aspect for the design of the technology is an adequate site characterization [13]. Hydrologic characteristics of groundwater flows represent a challenge for the design of PRBs [14]. Assuming that the reactive material is adequately selected from laboratory tests, the two interdependent parameters for the design of PRBs are the residence time and the hydraulic capture width. Residence time refers to the contact time between the contaminated groundwater and the reactive media within the barrier. It ensures that the reactive barrier is large enough to meet regulatory requirements. Hydraulic capture width refers to the maximal width of the contaminated groundwater that can enter a filtering gate or go through a continuous wall.

Numerical modeling constitutes the most popular option for the design of PRBs and commercial software products such as MODFLOW [15] and FLONET [16] are extensively used to evaluate the effect of PRBs on regional flows. Nevertheless, this approach is cost and time consuming. Moreover, the comparison of alternative methods or geometries needs the preparation of different numerical models, as performed by Hudak [17]. To face this problem, several authors have considered an analytic approach for preliminary design and optimization of PRBs. Thus, Craig and al. chose the Analytic Element Method (AEM) to represent a continuous wall in a homogenous aquifer [18]. Their model was based on an elliptical inhomogeneity placed in a uniform flow as a representation of a PRB. This approach constituted a first step in analytical modeling and provided useful tools for preliminary designs. To expand on the approximation of an elliptic geometry, Klammler and Hatfield investigated another approach based on conformal mapping and obtained solutions for flow fields around a rectangular continuous wall [19]. This approach has been later extended to funnel-and-gate and drain-and-gate configurations [6,7] and the authors analysed the solutions for flow fields regarding widths and shapes of the capture zones under different scenarios. These anterior works are particularly useful for preliminary design of F&G, but they are mainly focused on hydraulic aspects of the design. As the residence time in reactive filters is an essential element, the originality of the present paper consists in considering the residence time as an additional criterion for preliminary design of permeable reactive barriers.

2. Schwarz-Christoffel theorem

The Schwarz-Christoffel theorem states that the interior of a closed polygon may be mapped into the upper half of a plane [20]. This transformation is illustrated in Fig. 1.a and b, where the function f transforms the real axis in the z-diagram into a polygon in the z'-diagram. The inverse transformation is represented by the function g. Eq. 1 represents the Schwarz-Christoffel transformation.

$$\frac{dz'}{dz} = \frac{d}{dz}[f(z)] = \frac{\gamma}{(x_A-z)^{\alpha_1/\pi} \cdot (x_B-z)^{\alpha_2/\pi} \cdot (x_C-z)^{\alpha_3/\pi} \cdot (x_D-z)^{\alpha_4/\pi} \cdots} \quad (1)$$

where γ is a complex number in the z-diagram; f is a complex function; $x_A, x_B, x_C, x_D,...$ are real constants in ascending order of magnitude; $\alpha_1, \alpha_2, \alpha_3, \alpha_4, ...$ are external angles of the polygon. Considering $\alpha_1=/\pi 2$; $\alpha_2=-\pi$ and $\alpha_3=\pi/2$, Eq. 1 becomes Eq. 2 and transforms the real axis (Fig. 1.c) into a polygon with two apexes at infinity (Fig. 1.d). In particular, this transformation states that a flow around a cut-off wall perpendicular to a uniform flow can be deduced from a uniform flow without any PRB. Indeed, the x-axis can be considered as a no-flow boundary condition for a uniform flow parallel to this axis in the z-diagram (Fig. 1.c). In the z'-diagram

(Fig. 1.d), this no-flow boundary is transformed into the negative part of the x'-axis, two segments [AB] and [BC] along the y'-axis, and the positive part of the x'-axis. Considering [AB] and [BC] as the upstream and downstream sides of an half cut-off wall, the basic set of boundary conditions in the z-diagram is mapped over a more complex geometry in the z'-diagram. As the flow in the lower half of the z'-diagram (respectively z-diagram) can be deduced from the flow in the upper half by symmetry, the analytical solution will be established for Y>0 (respectively y>0).

$$\frac{dz\prime}{dz} = \frac{d}{dz}[f(z)] = \frac{\gamma \cdot z}{\sqrt{z^2 - R^2}} \qquad (2)$$

where R is a real positive number equal to the half-length of the cut-off wall. After integration (Eq. 3) and introduction of two set of images, f (z=0) = iR and f (z=R) = 0, the analytical expressions of the function f and its inverse function g are respectively given in Eq. 4 and Eq. 5.

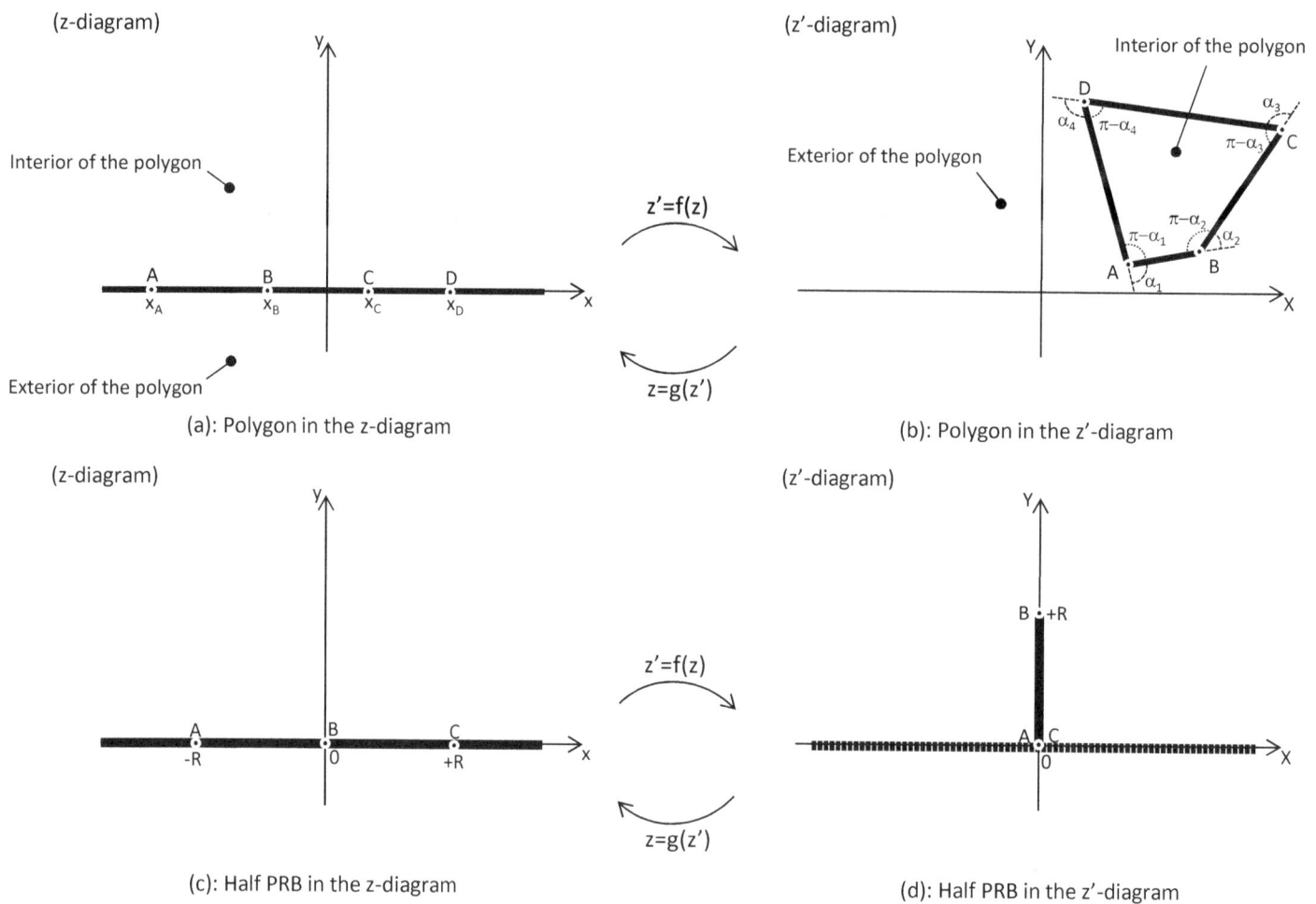

(a): Polygon in the z-diagram

(b): Polygon in the z'-diagram

(c): Half PRB in the z-diagram

(d): Half PRB in the z'-diagram

Figure 1. Schwarz-Christoffel's principle.

$$z' = f(z) = \gamma \int \frac{z}{\sqrt{z^2 - R^2}} \cdot dz + B \qquad (3)$$

$$z' = f(z) = \sqrt{z^2 - R^2} \qquad (4)$$

$$z = g(z') = \sqrt{z'^2 + R^2} \qquad (5)$$

Introducing the real and imaginary part of complex numbers z = x+iy and z' = X+iY, Eq. 5 can be rewritten as:

$$(x + iy)^2 = (X + iY)^2 + R^2 \qquad (6)$$

Development and separation of the real and imaginary parts leads to the following system of equations where the origins in the z and z'-diagrams are not considered in the domain (apex of the polygon in the z'-diagram):

$$\begin{cases} x^2 - y^2 = X^2 - Y^2 + R^2 \\ xy = XY \end{cases} \tag{7}$$

This system of equations is easily solved and leads to a fourth order polynomial whose roots are given in Eq. 8. Considering that y is strictly positive in the upper half of the z-diagram, the expression of y is given in Eq. 9.

$$x = \delta_X \cdot \sqrt{\frac{X^2 - Y^2 + R^2 + \sqrt{(X^2 - Y^2 + R^2)^2 + 4X^2Y^2}}{2}} \tag{8}$$

$$y = \delta_X \cdot \frac{XY}{\sqrt{\frac{X^2 - Y^2 + R^2 + \sqrt{(X^2 - Y^2 + R^2)^2 + 4X^2Y^2}}{2}}} \tag{9}$$

where δ_X represent the sign of X. The previous equations are particularly interesting because they represent a flow around a cut-off wall. In the next section, the geometry will be complicated by permitting a flow across the cut-off wall at the origin. This additional boundary condition represents the filtering gate (reactor)

2.1. Complex potential Ω around a PRB

In the present model, the widths of the cut-off wall and the reactive zone are neglected in comparison to the dimensions of the regional groundwater flow. Moreover, the filtering gate is represented by a sink and a source located respectively upstream and downstream of the cut-off wall. This assumption is particularly interesting to represent the flow in the z'-diagram as a function of a simple geometry in the z-diagram. Indeed, if we consider a sink located at point A and another at point C in the z-diagram (see Fig. 1.c), the image in the z'-diagram becomes a funnel-and-gate PRB.

Considering a steady, uniform and irrotational flow in the z-diagram, Eq. 10 represents the velocity potential ϕ and the stream function Ψ.

$$\begin{cases} \phi = -V_0 \cdot x \\ \psi = -V_0 \cdot y \end{cases} \tag{10}$$

where V_0 is a constant velocity parallel to the x-axis in the z-diagram. Considering a sink and a source of equal strength q respectively located at $(-R,0)$ and $(+R,0)$ in an initially static aquifer, the velocity potential ϕ and the stream function ϕ are given in Eq. 11 [21].

$$\begin{cases} \phi = -\frac{q}{4\pi} \cdot ln\left(\frac{(x-R)^2 + y^2}{(x+R)^2 + y^2}\right) \\ \psi = -\frac{q}{2\pi} \cdot \left(tan^{-1}\left(\frac{y}{x-R}\right) - tan^{-1}\left(\frac{y}{x+R}\right)\right) \end{cases} \tag{11}$$

where q is the flow rate per meter of depth of the aquifer (m^2/s). According to the superposition principle, the addition of Eq. 10 and Eq. 11 gives the velocity potential and the stream function of a uniform flow influenced by a sink and a source. Replacing x and y by Eq. 8 and 9, we obtain the velocity potential and stream function around a PRB. This potential function has been introduced into Winplot-2d and the equipotential lines are illustrated on Fig. 2 for a cut-off wall of 160 m coupled with a sink and a source of 12 m^2/d. This figure also illustrates the capture zone and the by-pass of the PRB (groundwater laterally away from the x'-axis, outside of the envelope curve).

As regards the hydraulic design of a PRB, the width of the plume caught by the filtering gate constitutes the most important information to be known. To determine this width, we studied the envelope curve in the z-diagram and considered its transformation by the function f. Looking at the stagnation points in the z-diagram, three configurations can be observed depending on the magnitude of the flow rate q. These configurations are: (a) two stagnation points on the x-axis (Fig. 3a), (b) a stagnation point at the origin, or (c) two stagnation points on the y-axis (Fig. 3b). In the last case, some flow occurs from the source to the sink (recirculation). Considering that the sink and the source are connected through the filtering gate, this configuration cannot be encountered on the field and the water always goes from the sink to the source via the reactor. In the z-diagram, the abscissas of the two stagnation points (x_S in Fig. 3.a.) are given in Eq.12 and we can easily demonstrate that their images are located on the y'-axis of the z'-diagram. Their ordinates Y_S is given in Eq. 13. As Y_s cannot be higher than R (half-length of the cut-off wall), the maximum flow rate q that can enter the filtering gate is $\pi \cdot R \cdot V_0$.

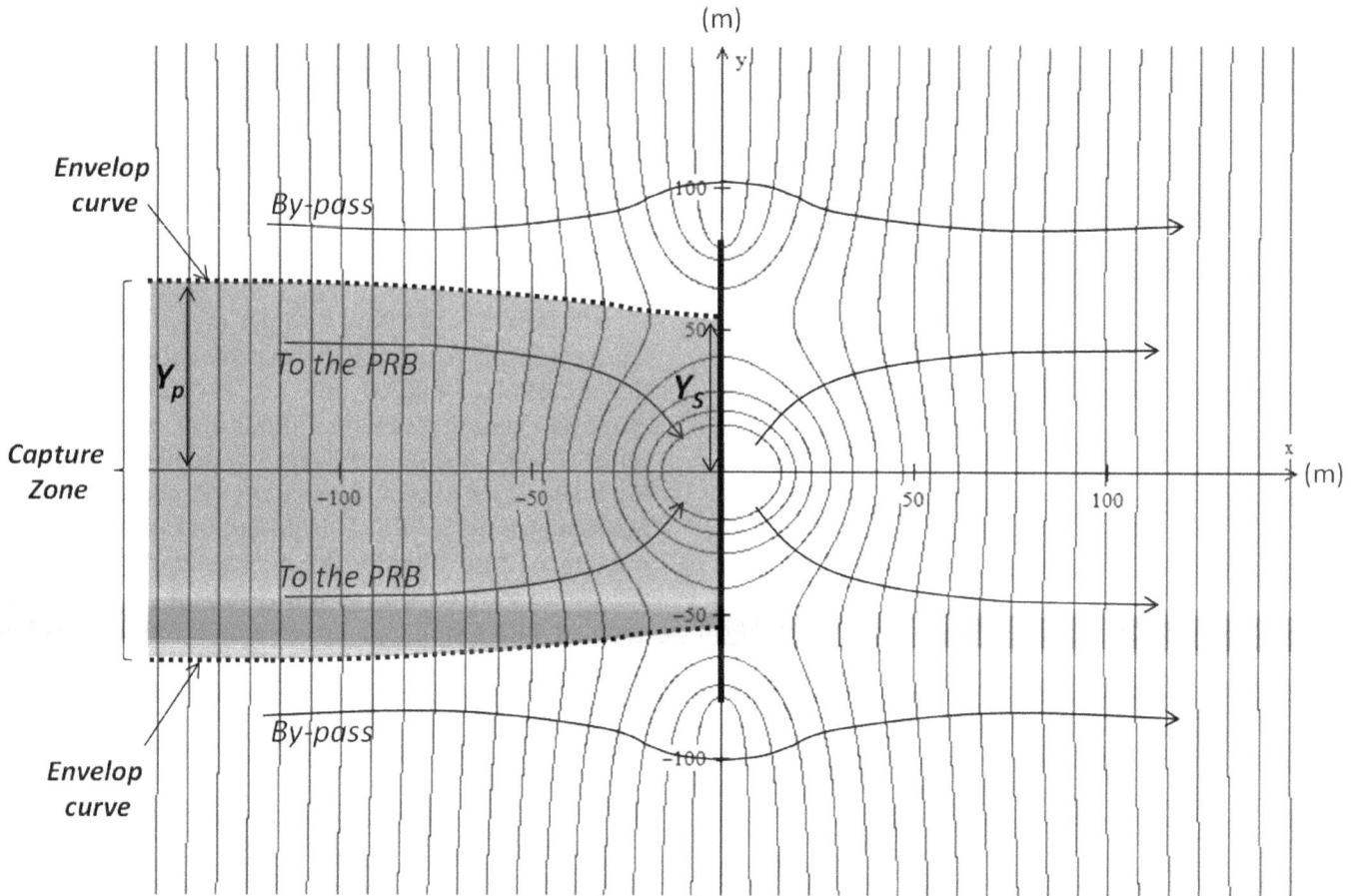

Figure 2. Equipotential around a Permeable Reactive Barrier (bold) and sketch of flow lines (arrows).

$$x_S = \pm R \cdot \sqrt{1 - \frac{q}{\pi \cdot R \cdot V_0}} \qquad (12)$$

$$Y_S = \pm\sqrt{R^2 - x_S^2} \qquad (13)$$

Moreover, the formulation of the stream line towards the stagnation point in the z-diagram is given in Eq. 14 [21].

$$\psi_0 = -\frac{q}{2\pi} \cdot \left(tan^{-1}\left(\frac{y}{x-R}\right) - tan^{-1}\left(\frac{y}{x+R}\right) \right) - V_0 \cdot y = -\frac{q}{2} \qquad (14)$$

Considering the limit of the stream function when $x \to -\infty$ (or $X \to -\infty$), we obtain the half-width of the capture zone entering the reactor (see Eq. 15 and Y_p in Fig. 4.b.).

$$Y_p = y_p = -\frac{\psi_0}{V_0} \qquad (15)$$

Combining Eq.14 and 15, we can demonstrate that the half width of the capture zone is proportional to q and inversely proportional to the initial velocity of the flow V_0 (see Eq. 16).

$$Y_p = \frac{q}{2 \cdot V_0} \qquad (16)$$

To illustrate the displacement of the envelope curve as a function of the flow rate q, different values of q from 0.1 to 1 m²/h are introduced in Eq. 14 and the solutions are provided on Fig. 4. For every single x∈ ┤-∞;χ_S], the corresponding ordinate y is calculated thanks to the Generalized Reduced Gradient method (GRG) due to the non-linearity of Eq. 14. The image of the envelope curve has then been calculated in the z'-diagram. The capture width Y_p is smaller than the ordinate Y_s of the stagnation point for small flow rates, whereas Y_p is greater than Y_s for high flow rates. Nevertheless, this illustration cannot yet be considered as a real flow net around a PRB as the flow rate in the reactor q is not a boundary condition, but is imposed by the geometry of the reactor and its related hydraulic head losses. As a consequence, the model is

extended in the next section by considering the geometry of the reactor and its impact on the flow rate.

(z-diagram)

(z-diagram)

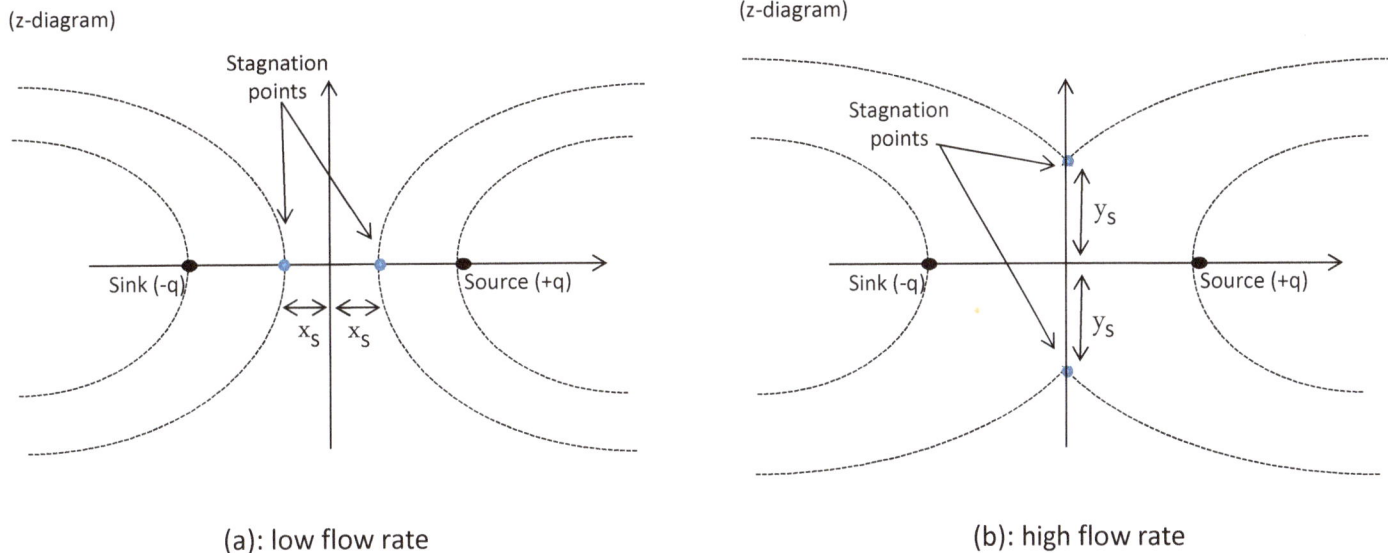

(a): low flow rate

(b): high flow rate

Figure 3. Uniform flow past a sink and a source of equal strength.

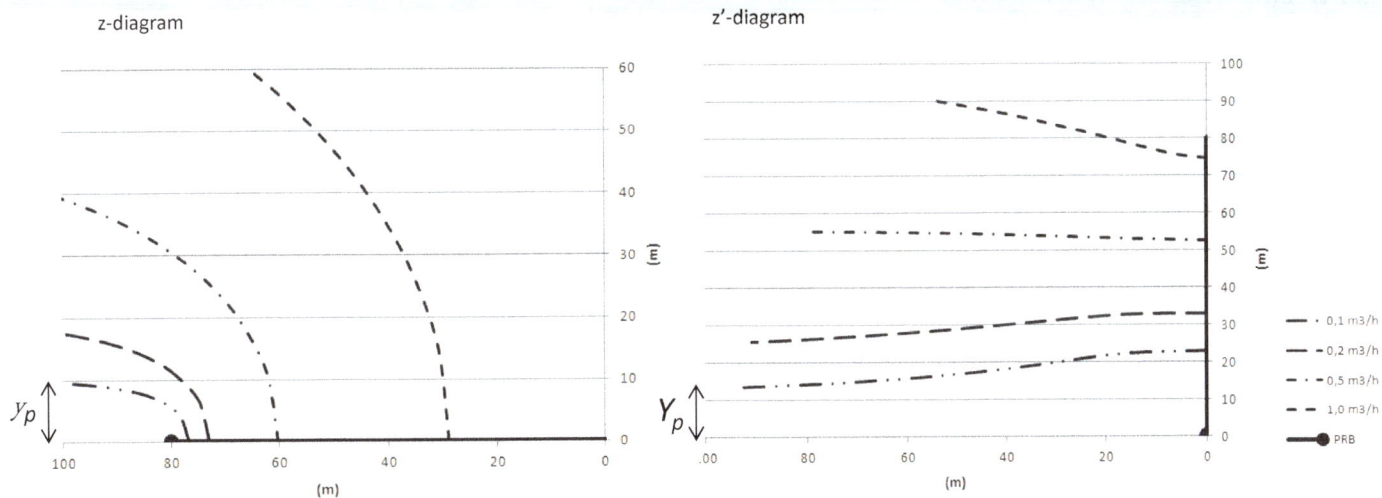

Figure 4. Envelope curves in the z and z'-diagrams.

2.2. Flow rate in a filtering gate

The previous representation of the filtering gate means that the velocity potentials are respectively set to $-\infty$ and $+\infty$ at the sink and the source. This assumption represents a bias because the velocity potential must be a continuous function across the reactor and the values at the entrance and the exit must be finite. To be more realistic, we considered the radius r_w of two wells representing the sink and the source in the z-diagram. Thus, the velocity potentials upstream (ϕ_u^w) and

downstream (ϕ_d^w) related to these two wells are defined in Eq. 17 and 18 and their difference is given in Eq. 19.

$$\phi_u^w = -\frac{q}{4\pi} \cdot ln\left(\frac{(2R+r_w)^2}{r_w^2}\right) \tag{17}$$

$$\phi_d^w = -\frac{q}{4\pi} \cdot ln\left(\frac{r_w^2}{(2R+r_w)^2}\right) \tag{18}$$

$$\phi_d^w - \phi_u^w = -\frac{q}{\pi} \cdot ln\left(\frac{r_w}{2R+r_w}\right) \tag{19}$$

This difference is combined to $\phi_d^v - \phi_u^v$ (velocity potential induced by a uniform velocity field in the z-diagram) to obtain the overall velocity field presented in Eq. 21.

$$\phi_d^v - \phi_u^v = -2 \cdot V_0 \cdot R \qquad (20)$$

$$\phi_d - \phi_u = \phi_d^v - \phi_u^v + \phi_d^w - \phi_u^w = -2 \cdot V_0 \cdot R - \frac{q}{\pi} \cdot \ln\left(\frac{r_w}{2R+r_w}\right) \qquad (21)$$

As r_w represents the radius of the gravel pack around the wells in the z-diagram, we can easily demonstrate that the image of the gravel pack around the sink (respectively the source) is a half-gravel pack located upstream (respectively downstream) of the cut-off wall in the z'-diagram. The radius of gravel packs in the z'-diagram (R_d) can be deduced from the function g (see Eq. 22).

$$R_d = \sqrt{(r_w + R)^2 - R^2} \qquad (22)$$

$$r_w = -R + \sqrt{R^2 + R_d^2} \qquad (23)$$

Otherwise, Darcy's law in a reactive filter states that the flow rate Q entering a filtering gate is proportional to its hydraulic conductivity, its surface and the hydraulic gradient. Considering that the head losses are generated by the porous media and negligible for all pipes or draining trenches that can be implemented at the entrance and the exit of the PRB, the flow rate in the reactive filter is provided in eq. 24.

$$Q = q \cdot D = k_{filter} \cdot \frac{h_u - h_d}{L_{filter}} \cdot S_{filter} \qquad (24)$$

where k_{filter} [m/s] represents the hydraulic conductivity of the filter, S_{filter} [m^2] and L_{filter} [m] respectively represent its cross-surface and length, $h_u - h_d$ [m] represent the hydraulic head loss between the entrance and the exit of the filter, Q is the flow rate in the filtering gate [m^3/s], q is the flow rate per meter of depth of the aquifer [m^2/s], and D is the thickness of the aquifer [m].

Considering that $\phi = k_{soil} \cdot h + cste$ in the z-diagram, the combination of Eq. 21, 23 and Eq. 24 leads to the flow rate in a PRB (Eq. 25).

$$q = \frac{2 \cdot V_0 \cdot R \cdot \frac{k_{filter}}{k_{soil}}}{D \cdot \frac{L_{filter}}{S_{filter}} + \frac{1}{\pi} \frac{k_{filter}}{k_{soil}} \cdot \ln\left(\frac{\sqrt{R^2 + R_d^2} - R}{\sqrt{R^2 + R_d^2} + R}\right)} \qquad (25)$$

This equation represents the flow rate per meter of aquifer that can enter a filtering gate. For design purpose, the geometry of the filter (S_{filter} and L_{filter}), its hydraulic conductivity (k_{filter}), the radius of the drainage elements (R_d) and the width of cut-off walls (R) have to be selected according to the site conditions: q (minimum flow rate to be treated with respect to the width of the plume), V_0 (Darcy velocity of the groundwater on site) and k_{soil} (permeability of the aquifer).

2.3 Residence time as a function of the flow rate

The first part of this paper was dedicated to the hydraulic aspect, which does not constitute the sole parameter for the design of a PRB. Indeed, the void volume of a filtering gate must be large enough to ensure a sufficient residence time. As a consequence, the design of a PRB must involve hydraulic and chemical considerations to prevent (a) any by-pass of the system, and (b) the break-through of the filter. The residence time T in a filter is deduced from the total flow rate Q and the porosity n of the reactive media, as mentioned in Eq. 26.

$$T = \frac{n \cdot S_{filter} \cdot L_{filter}}{Q} \qquad (26)$$

Isovalues of residence time can thus be plotted in the (S_{filter}, L_{filter}) plane, as presented on Fig. 5 for the following parameters: V_0=40m/yr, R=80 m, n=0.4, R_d= 1.25 m, and k_{filter}/k_{soil} =10. All couples of surface and length above a hyperbola generate a residence time greater than the corresponding isovalue and satisfy the chemical criterion. Fig. 5. Also contains a straight line corresponding to a specified flow rate q_s according to Eq. 25. All points under this line represent a higher flow rate than the specification (q_s). According to this figure, two conditions have to be satisfied simultaneously for the selection of a filter: (a) the residence time must be higher than the target, that is to say that the coordinates (S_{filter}, L_{filter}) must be above a selected hyperbola, and (b) the flow rate must be higher than q_s, that is to say that the coordinates (S_{filter}, L_{filter}) must be under the straight line. Combining the two previous conditions, a region of optimal geometry can be plotted as illustrated in grey on Fig. 5.

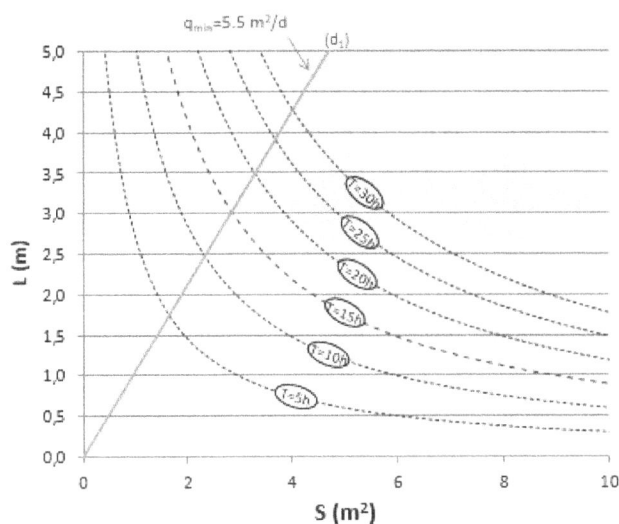

Figure 5. Design guidance diagram.

3. Practical application

Fig. 5 is a useful tool for the design of a PRB but it has to be redrawn when modifying the boundary conditions. To prevent this redrawing and to facilitate the design made by practitioners, Eq. 25 and 26 have been rewritten in terms of dimensionless variables.

$$\begin{cases} \dfrac{S}{D^2} = [A_1]^{1/2} \cdot \left[\dfrac{1}{A_2}\right]^{1/2} \cdot \left[\dfrac{1}{A_3} - \dfrac{1}{\pi} \cdot ln\left(\dfrac{\sqrt{1+A_4}-1}{\sqrt{1+A_4}+1}\right)\right]^{-1/2} \\ \dfrac{L}{D} = A \cdot \dfrac{D^2}{S} \end{cases} \quad (27)$$

where the dimensionless parameters are:

$$A_1 = \frac{q \cdot T}{n \cdot D^2} \qquad A_2 = \frac{k_{filter}}{k_{soil}} \qquad A_3 = \frac{2 \cdot V_0 \cdot R}{q} = \frac{Y_p}{R} \qquad A_4 = \frac{R_d}{R} \quad (28)$$

The system of Eq. 27 has been implemented in a **guidance diagram** provided on Fig. 6. This figure aims to design a PRB for all sets of boundary conditions and has to be used as follow:

i.   Considering the width of the plume Y_p, select a cut-off wall length that meets $R \geq {}^2\!/_\pi \cdot Y_p$,

ii. On Fig. 6.a, plot a vertical line at the corresponding $A_3 = Y_p/R$ and mark the intersection with the curve corresponding to the radius of the draining element ($A_4 = R_d/R$),

iii. Plot an horizontal line from this point and mark the intersection with the curve corresponding to the ratio $A_2 = k_{filter}/k_{soil}$ (See Fig. 6.b),

iv. Plot a vertical line from this point and note the value of α corresponding to the intersection with

the abscissa. α represents the slope of the upper boundary of the optimal geometry region in the (S_{filter}, L_{filter}) plan (see Fig. 6.c).

v.   Calculate the ratio $A_1 = q.T/n.D^2$ and select the corresponding hyperbola on Fig. 6.c. This hyperbola constitutes the lowest boundary of the optimal geometry region in the (S_{filter}, L_{filter}) plan.

vi. Select a filter section and length in the optimal region.

Considering the shape of the curves on Fig. 6.a., the design diagram demonstrates that a cut-off wall longer than $10\,Y_p$ (i.e. $Y_p/R > 0.2$) significantly increases the hydraulic efficiency of the system. With this assumption, a ratio k_{filter}/k_{soil} higher than 10 prevents any hydraulic issue. Indeed, the optimal region becomes mainly influenced by the residence time (slope of the upper boundary higher than 100).

In summary, the guidance diagram is useful for designers to choose the primary dimensions of a new PRB. Nevertheless, any designer has to keep in mind that the approach proposed here relies on the following implicit assumption: the width of the permeable reactive zone is extremely small relative to the total width. Hence, the reactive zone is essentially treated as a point feature and the head distribution in the vicinity of a more realistically dimensioned PRB would vary noticeably from heads generated from the model proposed here. Moreover, the initial hydraulic gradient is supposed uniform throughout the model, which can vary from the reality. As a consequence of these hypothesis, the model should be used for primary design only and to prevent a lot of trials and errors in the modelling.

4. Case study

For illustration purposes, a case study has been performed on a specific site characterized by the following elements:

i.   a width of plume Y_p equal to 10 m,

ii. an aquifer with a thickness of 3 m, and an hydraulic conductivity of 3.10^{-5} m/s,

iii. a gradient of 0.004, which leads to a flow rate to be treated of 113.5 m³/y (40 m³/y per meter of depth).

The contamination is composed of heavy metals (copper and zinc) and the treatment is based on the precipitation of metal hydroxides performed through an increase of pH. The targeted pH depends on the valence of the metals: a pH of about 10 allows the precipitation of hydroxides from divalent metal ions (Zn, Mn, Cu, Pb, Ni, Co and Cd), while a pH of around 6-7 is adapted to trivalent ions (Fe, Al, Cr). In our case, magnesium oxide has been selected as the pH at the equilibrium with pure magnesium oxide is around 10, which is particularly interesting for the precipitation of bivalent ions.

Laboratory tests leaded to a minimum residence time of 120 h and the reactive media is characterized by a permeability of 18.10^{-5} m/s and a porosity of 0.3.

(a)

(b)

(c)

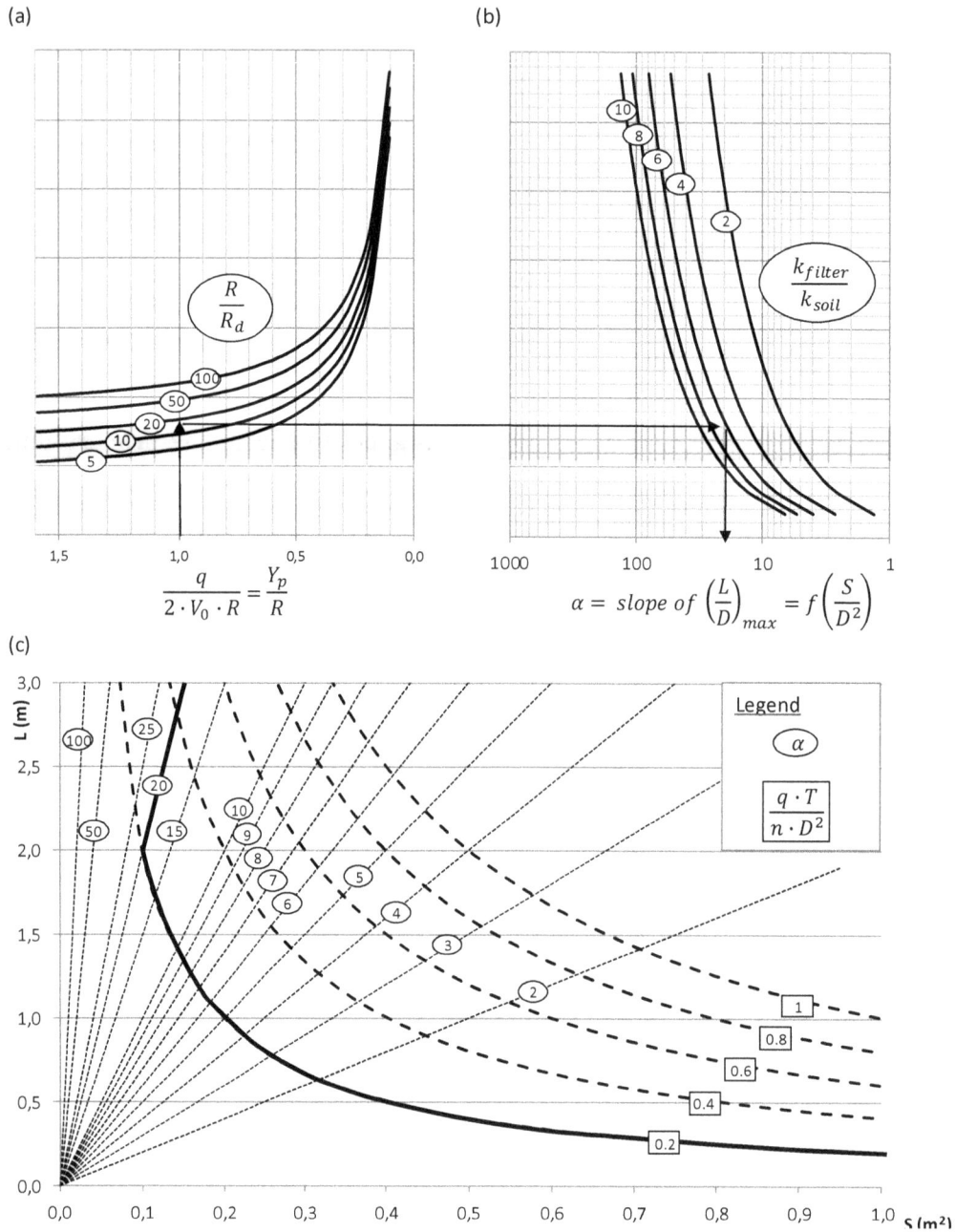

Figure 6. Guidance diagram for the design of funnel-and-gate permeable reactive barriers.

These characteristics lead to dimensionless parameters A_1 and A_2 respectively equal to 0.2 and 6. To evaluate the parameters A_3 and A_4, a length of 20 m has been chosen for the cut-off wall ($2R$). A higher value could be considered, but cut-off walls are expensive and

practitioners try to minimize their length. A lower value could also be considered, but the safety factor decreases as R approaches $2/\pi \cdot Y_p$. Considering a typical draining trench with a radius of 1 m, the construction on the guidance diagram leads to an optimal region above the black bold lines on Fig.6.c. (α = 20 and A_1 = 0.2).

The final dimension of the filter is then determined by practical consideration. For instance, shortest filters will be preferred to facilitate their manipulation. Limiting the length of the filter to 1 m, the minimum surface to warranty a sufficient residence time would be of 0.2 m^2, hence a radius of 0.5 m.

5. Conclusion

Based on the Schwarz-Christoffel transformation, we developed an analytical solution of the flow rate in a PRB. This study demonstrated that the cut-off width has the most important impact on the capture zone and it leads to a design methodology for the reactive cell. According to this methodology, the section and length of a reactive cell are selected to ensure (a) a minimum flow rate, and (b) a minimum residence time in the porous media. The first condition is essential to capture the entire plume, while the second is mandatory to treat efficiently the contaminated groundwater. To meet these conditions, the design of a funnel-and-gate PRB comprises two steps: (a) a minimum cut-off width is selected to ensure a sufficient capture width, and (b) the filter's dimensions are selected on figure taking into account hydraulic and chemical constraints (residence time).

References

[1] D. W. Blowes, C. J. Ptacek, et al., "Passive Remediation of Groundwater Using in Situ Treatment Curtains," *Geoenvironment 2000: Characterization, Containment, Remediation, and Performance in Environmental Geotechnics,* vol. 1-2, no. 46, pp. 1588-1607, 1995.

[2] D. C. McMurtry and R. O. Elton, "New Approach to in-Situ Treatment of Contaminated Groundwaters," *Environmental Progress,* vol. 4, no. 3, pp. 168-170, 1985.

[3] S. D. Warner, C. L. Yamane, et al., "Considerations for Monitoring Permeable Ground-water Treatment Walls," *J. of Environmental Eng.,* vol 124, no, 6, pp. 524-529, 1998.

[4] R. M. Powell, D. W. Blowes, et al., "Permeable Reactive Barrier Technologies for Contaminant Remediation," *NASA,* Washington, 1998.

[5] A. R. Gavaskar, "Design and Construction Techniques for Permeable Reactive Barriers," *J. of Hazardous Materials,* vol. 68, no. 1-2, pp. 41-71, 1999.

[6] H. Klammler and K. Hatfield, "Analytical Solutions for the Flow Fields Near Funnel-and-Gate Reactive Barriers with Hydraulic Losses," *Water Resources Research,* vol. 45, no. 2, Feb. 2009.

[7] H. Klammler, K. Hatfield, et al., "Analytical Solutions for Flow Fields near Drain-and-Gate Reactive Barriers," *Ground Water,* vol. 48, no. 3, pp. 427-437, 2010.

[8] H. Klammler, K. Hatfield, et al., "Capture Flows of Funnel-and-Gate Reactive Barriers Without Gravel Packs," *Advances in Fluid Mechanics,* vol. 69, pp. 319-330, 2010.

[9] R. Rumer and J. Eds. Mitchell, "Permeable Reactive Barriers: Assessment of Barrier Containment Technologies: A Comprehensive Treatment for Environmental Remediation Application," *Int. Containment Technol. Workshop,* Baltimore, Maryland, 1995.

[10] K. D. Warren, R. G. Arnold, et al., "Kinetics and Mechanism of Reductive Dehalogenation of Carbon Tetrachloride Using Zero-Valence Metals," *J. of Hazardous Materials,* vol. 41, no, 2–3, pp. 217-227, 1995.

[11] S. F. O'Hannesin and R. W. Gillham, "Long-Term Performance of an in Situ "Iron Wall" for Remediation of VOCs," *Ground Water,* vol. 36, no. 1, pp. 164-170, 1998.

[12] I. Kozyatnyk, et al., "Evaluation of Barrier Materials for Removing Pollutants from Groundwater Rich in Natural Organic Matter," *Water Sci. and Technol.,* vol. 70, no. 1, pp. 32-39, 2014.

[13] F. Obiri-Nyarko, S.J. Grajales-Mesa, and G. Malina, "An Overview of Permeable Reactive Barriers for in Situ Sustainable Groundwater Remediation," *Chemosphere,* vol. 111, pp. 243-259, 2014.

[14] S. Liu, X. Li, et al., "Hydraulics Analysis for Groundwater Flow Through Permeable Reactive Barriers," *Environmental Modelling and Assessment,* vol. 16, no. 6, pp. 591-598, 2011.

[15] M. McDonald and A. Harbaugh, "A Modular Three-dimensional Finite-Difference Ground-Water Flow Model," in *Techniques of Water-Resources Investigations Reports,* book 6, ch. A1, US Geological Survey, 1988.

[16] N. Guiger, J. Molson, et al., "Flonet v.1.02: Two-dimensional Steady-state Flownet Generator," Waterloo Centre for Groundwater Res., Univ. of Waterloo and Waterloo Hydrogeologic Software, Waterloo, Ontario, 1991.

[17] P.F. Hudak, "Comparison of Permeable Reactive Barrier, Funnel and Gate, Nonpumped Wells, and Low-Capacity Wells for Groundwater Remediation," *J. of Environmental Sci. and Health*, vol. 49, no. 10, pp. 1171-1175, 2014.

[18] J. R. Craig, A. J. Rabideau, and R. Suribhatla, "Analytical Expressions for the Hydraulic Design of Continuous Permeable Reactive Barriers," *Advances in Water Resources*, vol. 29, no. 1, pp. 99-111, 2006.

[19] H. Klammler and K. Hatfield, "Analytical Solutions for Flow Fields Near Continuous Wall Reactive Barriers," *J. of Contaminant Hydrology*, vol. 98, no. 1-2, pp. 1-14, 2008.

[20] V. L. Streeter, *"Handbook of Fluid Dynamics,"* New York, McGraw-Hill, 1948.

[21] H. Chanson, *"Applied Hydrodynamics: An Introduction to Ideal and Real Fluid Flows,"* CRC Press, 2009.

12

Investigation of Watershed-scale Surface Water Quality under Changing Land Use Conditions: A Case Study in Northern British Columbia

Gopal Chandra Saha, Jianbing Li*, Siddhartho Shekhar Paul, Ronald W Thring
Environmental Engineering Program, University of Northern British Columbia,
3333 University Way, Prince George, British Columbia, Canada V2N 4Z9
saha@unbc.ca; jianbing.li@unbc.ca; pauls@unbc.ca; thring@unbc.ca

Abstract – *The effects of land use change on surface water quality in an intensively used watershed in northern British Columbia were investigated. The water of the Kiskatinaw River watershed (KRW) was analyzed for its total organic carbon (TOC) concentration in 2004-2005 and 2010-2011, and the variation of water quality within sampling sites over time was examined. The results showed that TOC concentrations in 2010-2011 were lower than those in 2004-2005 due to decreased degradation of terrestrial vegetation, agricultural activities, and the loss of wetlands, as well as the strict regulation of industrial discharge from 2004-2005 to 2010-2011 in KRW. It was found that TOC concentrations were high in the agriculturally intensive sites. In addition to land use, the data indicated that more rainfall led to higher TOC concentrations at the nearby stream/river during spring runoff and summer due to more surface runoff and erosion. The land use change effects were more dominant on TOC concentration variation than stream flow.*

Keywords: Kiskatinaw river watershed, land use, water quality.

1. Introduction

Surface water quality is directly linked to the land cover of a watershed [1]. It can be degraded by changes in land cover due to various anthropogenic activities, resulting in threats to the aquatic ecosystem and posing serious challenges to drinking water supply authority [2,3]. For sustainable land use and resource management, it is crucial to understand how land use patterns affect surface water quality [4]. For example, Tong and Chen [5] found a positive correlation between the area of agricultural and urban lands and the loading of nitrogen and phosphorus within different watersheds in Ohio from 1988 to 1995. Smith et al. [6] found an increase of copper and zinc concentrations in a Canadian watershed from 1993-1994 to 2003-2004 due to an increase of agricultural activities. Lin et al. [7] found that the phosphorus loading in Lake Allatoona Watershed in USA increased by 17.5% from 1992 to 2001 due to a 20% decrease of forest area, a 225% increase of urban area, and a 50% increase of pasture area, respectively. Broussard and Turner [8] observed a significant increase of nitrate-nitrogen (NN) concentrations in 56 USA watersheds from 1900 to 2002 due to an increase in agricultural area. Huang et al. [9] found a negative correlation between the area of forest and grassland and the concentrations of total nitrogen (TN) and total phosphorus (TP) in Chaohu Lake Basin, China from 2000 to 2008, and they also observed a negative correlation between built up area and dissolved oxygen (DO) concentration in that basin.

The change in land cover and land management practices in a watershed have been recognized as major factors affecting its hydrological system which may then cause variation in its water quality [5]. The Peace Region in northern British Columbia in Canada is a diversified area where different types of land use activities (e.g., timber harvesting, agricultural, oil and gas, wildlife, and recreational) occur. However, very few research works have been reported which investigate the interaction between land use/land cover change and watershed-scale surface water quality variation in this region. The objective of this study is to fill this gap. The Kiskatinaw River Watershed (KRW) was used as a case study. The river water of KRW was analyzed for its total organic carbon (TOC) concentration in 2010-2011, and the water quality during this time period was then compared with that observed in 2004-2005, while land use changes from 2004-2005 to 2010-2011 were detected by using remote sensing analysis and Arc GIS.

2. Study Area

The Kiskatinaw River Watershed (KRW) is situated in northeastern British Columbia, Canada (Fig. 1), covering an area of about 2882 km². The City of Dawson Creek provides drinking water to its inhabitants and industrial users by collecting and treating water from the Kiskatinaw River. The KRW is an intensively used watershed as a result of the large and increasing scale of timber harvesting, oil and gas exploration/production and urbanization in recent years. It is a rain-dominated hydrological system, with peak flow occurring from late June to early July. It receives an average annual precipitation of 499 mm which consists of 320 mm of rain and 179 mm of snow. The average annual river flow rate of the Kiskatinaw River is 10 m³/s, but the flow drops to 0.052 m³/s in January [10]. In this study, TOC was selected as a study water quality parameter because it plays an important role in stream chemistry [11] by complexing metals and nutrients [12], affecting pH and alkalinity [13, 14] and acting as a substrate for microbial production [15] . It contains all forms of organic carbon, including petroleum hydrocarbons and natural organic matter [16]. Excessive TOC in source water could lead to the formation of carcinogenic by-product such as trihalomethane (THMs) following chlorination treatment [17, 18].

3. Methods
3. 1. Water Quality Sampling

The British Columbia (BC) Ministry of Environment conducted a water quality study within KRW in 2004-2005 at four sampling locations (i.e., Arras, Brassey, East Confluence, and West Confluence) (Fig. 1) [19]. In the present study, water quality samples were collected

in 2010-2011 at the same locations to measure the total organic carbon (TOC) concentrations. The Arras, East Confluence and West Confluence sites were sampled on seven occasions, and the Brassey site was sampled on six occasions. On two occasions, duplicate samples were collected at the same site and time in order to evaluate sample collection consistency [20]. The samples were collected using acid-washed 120-mL plastic bottles and stored in an ice-packed cooler, and were then sent to a certified laboratory for analysis. The TOC was measured using a Shimadzu (TOC-500) total organic carbon analyzer. The sampling and testing techniques used in 2010-2011 were consistent with those used in 2004-2005, and the variation of water quality between the two time periods was then examined.

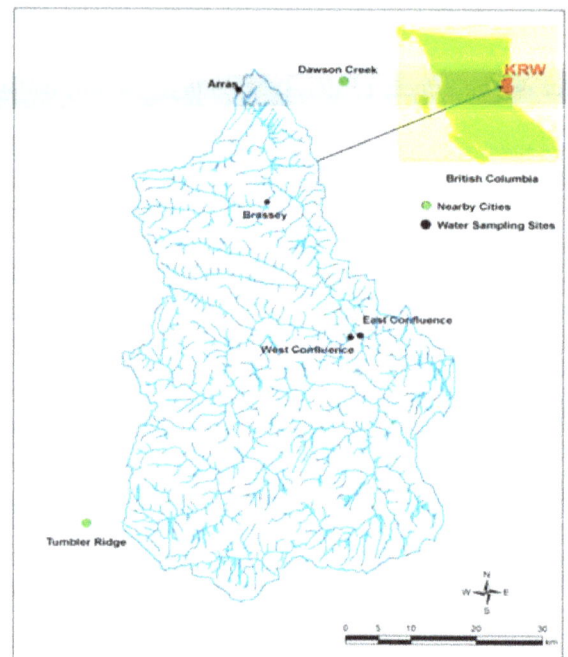

Figure 1. Overview of the KRW and the water quality sampling locations.

3.2. Hydrological Data Collection and Land Use Maps

The precipitation data of KRW and stream flow data at Arras site (i.e., outlet of KRW) during 2004-2005 and 2010-2011 were collected from Noel weather station and Farmington Water Survey Canada station, respectively. The land use maps for 1999 and 2010 in the KRW were generated using remote sensing analysis and Arc GIS based on collected Landsat satellite image of that corresponding year. The details of land use map generation can be found in Saha et al. [21]. The large-scale shale gas exploration/production activities began in 2005 [22]. As a result, the land use maps of 2010 and 1999 were used to represent land use conditions in 2010-2011 and 2004-2005, respectively.

4. Results and Discussion

4. 1. Land Use Change Analysis

Fig. 2 presents the land use map for 2010-2011, and it was found that forests account for a major portion of the KRW. The land use change between 2004-2005 and 2010-2011 is shown in Table 1. It is seen that the major land use change was from forest clear cut and wetland. The "forest clear cut" area includes the forest cut block areas which were cleared by industry (e.g., oil and gas) or for other purposes. This land use class comprises most of the gas development infrastructure, including drilling pads. Forest clear cut area increased by about 268%, while wetland area decreased by about 97%. The rapid change in forest clear cut was primarily due to large scale oil/gas exploration/production, while the rapid change in wetland area may be due to the shift of vegetation and oil/gas exploration/production in the study area. It was also found that forest and built up area (e.g., road, house, industrial infrastructures) increased by about 4% and 106% from 2004-2005 to 2010-2011, respectively. The agricultural (e.g., cropland and pasture) and open water (e.g. river, small channels) areas decreased by 57% and 17%, respectively.

Table 1. Land use changes from 2004-2005 to 2010-2011 in KRW.

Land use condition	Area (km²) in 2004/05	Area (km²) in 2010/11	Change (km²)	Change (%)
Forest	2431	2531	100	4
Forest clear cut	66	243	177	268
Agriculture	99	43	-56	-57
Built up area	16	33	17	106
Water	29	24	-5	-17
Wetland	241	8	-233	-97
Total	2882	2882		

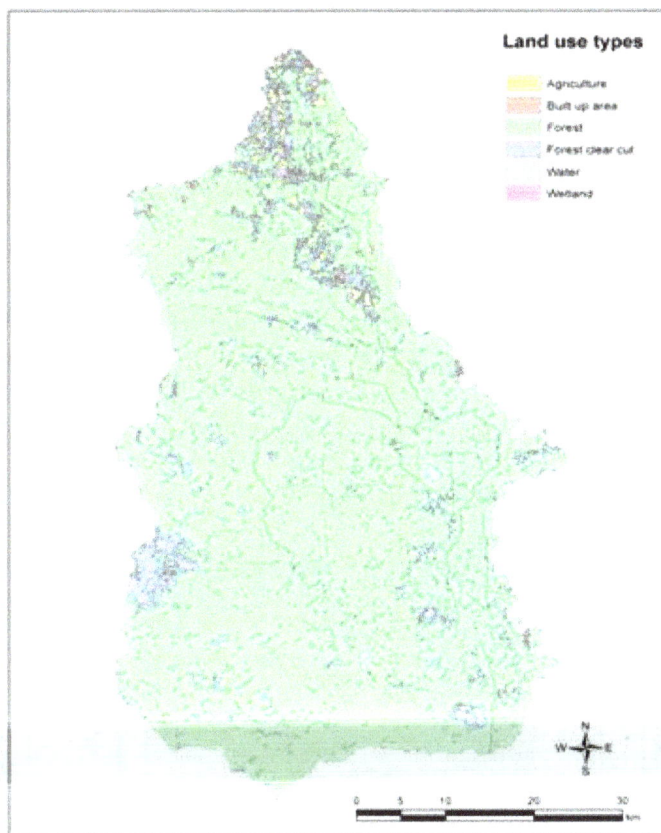

Figure 2. Land use map of KRW in 2010-2011.

4. 2. Precipitation and Stream Flow Trends

Fig. 3 shows the precipitation and stream flow in KRW from September 2010 to August 2011 and from September 2004 to August 2005. The annual precipitation in 2010-2011 and 2004-2005 was found to be 467 and 507 mm, respectively. The mean annual stream flow in 2010-2011 and 2004-2005 were 25.91 and 20.3 m³/s, respectively. It is to be noted that in June and July 2011, flooding occurred in KRW.

4. 3. Variation of Total Organic Carbon Concentration

The comparison of TOC concentrations at four sites between 2010-2011 and 2004-2005 is shown in Fig. 4. TOC concentrations at the Arras site (Fig. 4a) in 2010-2011 ranged from 4.3 to 13.5 mg/L, while in 2004-2005 they ranged from 7.3 to 17.7 mg/L. The median was 5.7 and 10 mg/L in 2010-2011 and 2004-2005, respectively. This illustrates a significant decrease of TOC concentrations at this site. The primary sources of TOC are natural sources (degradation of terrestrial vegetation, wetlands) as well as agricultural, industrial and residential waste discharges [19]. When terrestrial vegetation degrades, organic carbon can enter the soil. The surface runoff and erosion could then carry the organic carbon enriched soil into a nearby river/stream during snow melt and rainfall events. In general, TOC concentration is much higher in forested soils than agricultural soils [23, 24]. However, soil erosion rates from the agricultural area are higher than from the forest [25]. Therefore, a decrease in agricultural area could result in a decrease of soil carbon erosion [26] and thus a reduced TOC concentration in the nearby river/stream. In addition, several studies found that wetlands are the major contributor of TOC to streams and lakes [11, 27-30]. TOC enters wetland via groundwater and surface runoff from TOC enriched surrounding upland soils [31].

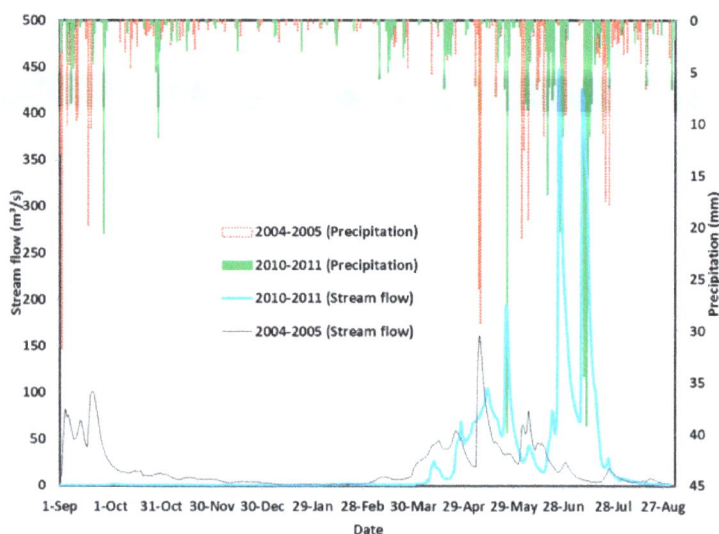

Figure 3. Precipitation and stream flow trends of KRW in 2010-2011 and 2004-2005.

The decrease of TOC concentrations at the Arras site may be due to the decrease of the degradation of terrestrial vegetation, agricultural activities, the loss of wetlands in the KRW, and the increase of forest clear cut and built up area, as well as the strict regulations imposed on industrial discharges between these time periods. In 2007, British Columbia Ministry of Environment introduced the Oil and Gas Waste Regulation (OGWR) under the Environmental Management Act of British Columbia [32], which reduced the discharge of hydrocarbons from the oil/gas industry to the environment and thus resulted in lower TOC concentration in the study area. From 2004-2005 to 2010-2011, wetland and agricultural activities in the study area decreased by 97% (233 km^2) and by 57%

(56 km^2), respectively. On the other hand, forest, forest cleat cut and built up area in the study area increased by 4% (100 km^2), 268% (177 km^2), and 106% (17 km^2), respectively. Although forest area, which contains soils with highest TOC concentrations than agricultural soils, increased by 100 km^2, TOC concentrations did not increase in the study area in 2010-2011. Due to the decrease of wetland, which is the main contributor of TOC to streams, by 233 km^2 and agricultural area by 56 km^2, TOC concentrations decreased in the study area in 2010-2011. These land use changes support the decrease of TOC concentrations in the KRW in 2010-2011. These results also correspond to the findings in other studies of changing TOC concentrations due to land use changes [30, 33-36].

It was also found that TOC concentrations at the Arras site were high in both 2004-2005 and 2010-2011 during high stream flow period, such as the high TOC concentrations observed during September 2004 and June 2011 when heavy rainfall events occurred. This indicates that more rainfall led to more TOC concentrations at the nearby stream/river during spring runoff and summer due to more surface runoff and erosion [25, 37]. In addition, the increase of forest clear cut (177 km^2) and built up area (17 km^2) intensified soil erosion rates by increasing surface runoff during heavy rainfall [38]. There was more precipitation in 2004-2005 than in 2010-2011, leading to higher TOC concentrations in 2004-2005 during spring runoff and summer. However, the higher mean annual stream flow in 2010-2011 than that in 2004-2005 occurred mainly due to flooding during June-July 2011(Fig. 3). Therefore, in August 2011 just following the flooding, TOC concentrations were higher than that in 2004-2005.

Fig. 4b shows that TOC concentrations at the East Confluence site in 2010-2011 ranged from 1.9 to 14.7 mg/L (with median of 5.1 mg/L), while in 2004-2005 they ranged from 9.5 to 18.2 mg/L (with median of 12.15 mg/L). The trend of TOC concentration variations at this site is almost similar to that of Arras site. TOC concentrations at the West Confluence site (Fig. 4c) in 2010-2011 ranged from 2 to 13.9 mg/L (with median of 5.9 mg/L), while in 2004-2005 they ranged from 5 to 14.3 mg/L (with median of 9 mg/L). TOC concentrations at the Brassey site (Fig. 4d) in 2010-2011 ranged from 11 to 20.2 mg/L (with median of 13.5 mg/L), while in 2004-2005 they ranged from 12 to 25.4 mg/L (with median of 17 mg/L). However, high TOC concentrations at the East and West Confluences, and

Brassey sites were not observed in August just following the flooding, indicating that land use change effects were more dominant on TOC concentration variation.

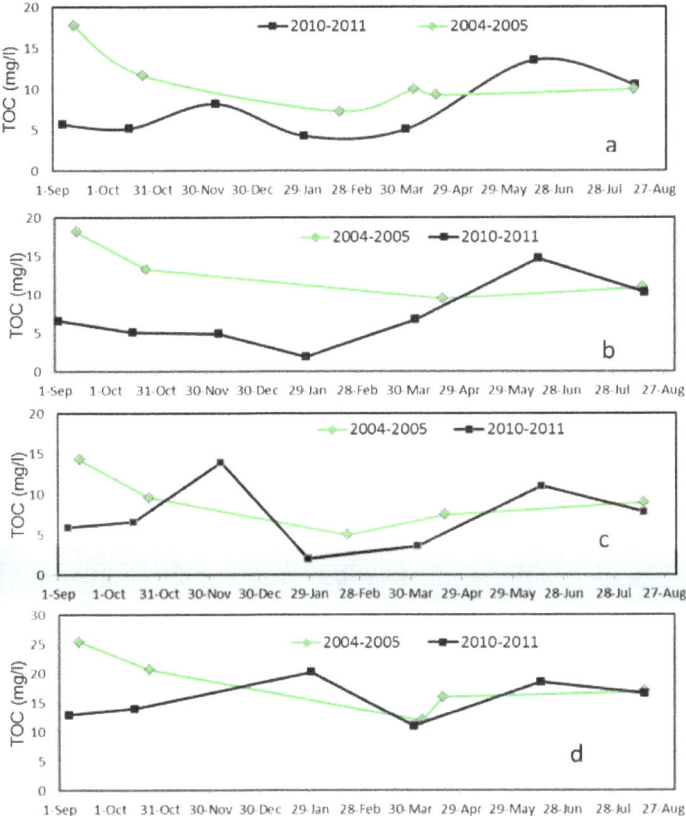

Figure 4. Comparison of TOC concentrations between 2010-2011 and 2004-2005 at (a) Arras, (b) East Confluence, (c) West Confluence, and (d) Brassey.

Fig. 5 shows the spatial variation of TOC concentrations in 2010-2011 in the KRW, and it can be found that the TOC concentrations were higher at the Brassey site compared to other sites. This may be caused by more agricultural activities and more degradation of terrestrial vegetation due to urban development (e.g., roads, roads crossing) in the Brassey area. In addition, soil erosion rates from the agricultural area are higher than from the forest [18], and urban development in the Brassey area intensified the soil erosion rates, which resulted in higher TOC concentration. Correll et al. [18] also found higher TOC concentration in agricultural watershed than forested watershed. TOC concentrations at the Arras site (i.e., outlet of the watershed) were also higher than that at the West Confluence site (upstream) because of discharge from Brassey with high TOC concentration to

Arras.

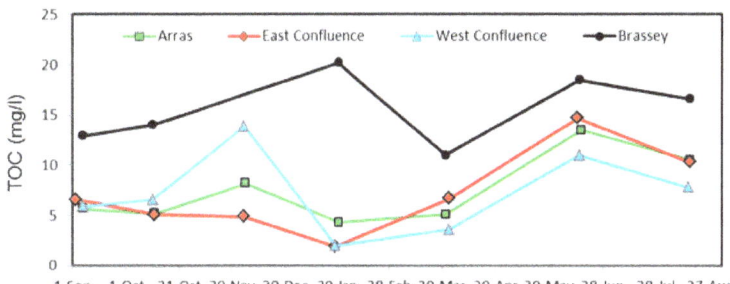

Figure 5. Spatial variations of TOC concentrations in the KRW in 2010-2011.

5. Conclusion

This study investigated the effects of land use change on surface water quality in the Kiskatinaw River watershed in northern British Columbia. The river water was sampled at four locations in 2010-2011 for analyzing total organic carbon (TOC) concentrations. The water quality results were then compared with those observed in 2004-2005 by the BC Ministry of Environment. Land use changes from 2004-2005 to 2010-2011 were detected by using remote sensing analysis and Arc GIS. The results illustrated that land use change played a primary role in the reduction of TOC concentration in the river water. The results also indicated that TOC concentrations were high at the agriculturally intensive sites than other types of land uses due to higher soil erosion rates. In addition to land use change, rainfall played a secondary role in TOC concentrations in the nearby stream/river during spring runoff and summer due to more surface runoff and erosion. Therefore, the results obtained from this study will be helpful for making better land use management plan in watershed to control TOC concentration.

Acknowledgements

This study was funded by Peace River Regional District, the City of Dawson Creek, Geoscience BC, EnCana, and BP Canada. The authors would like to thank Faye Hirshfield, Peter Caputa, Reg Whiten, and Chelton van Geloven for their help and support.

References

[1] B.J. Amiri, A. Nakane "Modeling the linkage between river water quality and landscape metrics in the Chugoku District of Japan", Water Resource Management, 23 (5), 2009, 931-956.

[2]   L. Sliva, D.D. Williams "Buffer zone versus whole catchment approaches to studying land use impact on river water quality", Water Research, 35(14), 2001, 3462-3472.

[3]   E. Nagoye, J. Machiwa "The influence of land use patterns in the Ruvu River watershed on water quality in the river system" Physics and Chemistry of the Earth, Parts A,B,C, 29(15-18), 2004, 1161-1166.

[4]   Y.P. Khare, C.J. Martinez, G.S. Toor "Water quality and land use changes in the Alafia and Hillsborough River watersheds, Florida, USA" JAWRA Journal of the American Water Resources Association, 48(6), 2012, 1276-1293.

[5]   S.T.Y. Tong, W. Chen "Modeling the relationship between land use and surface water quality", Journal of Environmental Management, 66(4), 2002, 377-393.

[6]   I.M. Smith, K.J. Hall, L.M. Lavkulich, H. Schreier "Trace metal concentrations in an intensive agricultural watershed in British Columbia, Canada" JAWRA Journal of the American Water Resources Association, 43(6), 2007, 1455-1467.

[7]   Z. Lin, D.E. Radcliffe, L.M. Risse, J.J. Romeis, C.R. Jackson "Modeling phosphorus in the Lake Allatoona watershed using SWAT: II. Effect of land use change", Journal of Environmental Duality, 38(1), 2009, 121-129.

[8]   W. Broussard, R.E. Turner "A century of changing land-use and water-quality relationships in the continental US" Frontiers in Ecology and the Environment, 7(6), 2009, 302-307.

[9]   J. Huang, J. Zhan, H. Yan, F. Wu, X. Deng "Evaluation of the impacts of land use on water quality: a case study in the Chaohu Lake basin", the Scientific World Journal, 2013, 7 pages.

[10]   Dobson Engineering Ltd. and Urban Systems Ltd. Kiskatinaw River Watershed Management Plan; File 0714.0046.01, 2003, [Online] Available at: http://www.dawsoncreek.ca/wordpress/wp-content/uploads/2011/08/KiskatinawWMP2003.pdf [Accessed on December 19, 2012].

[11]   P.J. Dillon, L.A. Molot "Effect of landscape form on export of dissolved organic carbon, iron, and phosphorus from forested stream catchments" Water Resource Research, 33(11), 1997, 2591-2600.

[12]   E. Tipping "Modeling the competition between alkaline-earth cations and trace-metal species for binding by humic substances" Environmental Science & Technology, 27, 1993, 520–529.

[13]   D.F. Brakke, A. Henriksen, S.A. Norton "The relative importance of acidity sources for humic lakes in Norway" Nature, 329, 1987, 432–434.

[14]   H. Laudon, O. Westling, S. Lofgren, K. Bishop "Modeling preindustrial ANC and pH during the spring flood in northern Sweden" Biogeochemistry, 54, 2001, 171–195.

[15]   M. Jansson, A.K. Bergstrom, P. Blomqvist, S. Drakare "Allochthonous organic carbon and phytoplankton/bacterioplankton production relationships in lakes" Ecology, 81, 2000, 3250-3255.

[16]   C.G. Schreier, W.J. Walker, J. Burns, R. Wilkenfeld "Total organic carbon as a screening method for petroleum hydrocarbons" Chemosphere, 39(3), 1999, 503-510.

[17]   British Columbia Ministry of Environment (BCME). Ambient water quality criteria for organic carbon in British Columbia. 1998, [Online] Available at: http://www.env.gov.bc.ca/wat/wq/BCguidelines/orgcarbon/index.html. [Accessed on January 12, 2014].

[18]   United States Environmental Protection Agency (USEPA). Basic information about disinfection byproducts in drinking water: Total Trihalomethanes, Haloacetic Acids, Bromate and Chlorite, 2003. [Online] Available at: http://water.epa.gov/drink/contaminants/basicinformation/disinfectionbyproducts.cfm [Accessed on January 19, 2014].

[19]   British Columbia Ministry of Environment (BCME). Water quality source identification in the Kiskatinaw Watershed near Dawson Creek, B.C. Interim report, 2006.

[20]   British Columbia Ministry of Water, Land and Air Protection (BCMWLAP). British Columbia Field Sampling Manual, 2003. [Online] Available at: http://www.env.gov.bc.ca/epd/wamr/labsys/field_man_03.html [Accessed on December 19, 2012].

[21]   G.C. Saha, S.S. Paul, J. Li, F. Hirshfield, J. Sui "Investigation of land-use change and groundwater-surface water interaction in the Kiskatinaw River Watershed, British Columbia (parts of NTS 093P/01, /02, /07-/10)", in Geoscience BC Summary of Activities 2012, Geoscience BC, Report 2013-1, 2013, p.139-148.

[22]   British Columbia Ministry of Energy and Mines (BCMEM). "The status of exploration and

development activities in the Montney Play region of northern BC". [Online] Available at: http://www.offshore-oil-andgas.gov.bc.ca/OG/oilandgas/petroleumgeology/UnconventionalGas/Documents/C%20Adams.pdf [Accessed on December 19, 2012]

[23] B.H. Ellert, E.G. Gregorich "Storage of carbon, nitrogen and phosphorus in cultivated and adjacent forested soils of Ontario" Soil Sciences, 161, 1996, 587–60.

[24] H.P. Apezteguia, R.C. Izaurralde, R. Sereno "Simulation study of soil organic matter dynamics as affected by land use and agricultural practices in semiarid Cordoba, Argentina" Soil Tillage Research, 102, 2009, 101-108.

[25] D.L. Correll, T.E. Jordan, D.E. Weller "Effects of precipitation, air temperature, and land use on organic carbon discharges from Rhode River Watersheds" Water Air & Soil Pollution, 128, 2001, 139-159.

[26] C. Boix-Fayos, J. de Vente, J. Albaladejo, M. Matinez-Mena "Soil carbon erosion and stock as affected by land use changes at the catchment scale in Mediterranean ecosystems" Agriculture, Ecosystems & Environment, 133, 2009, 75-85.

[27] H.F. Hemond "Wetlands as the source of dissolved organic carbon to surface waters" In: Organic acids in aquatic ecosystems ed. by E. M. Perdue and E. T. Gjessing, John Whiley & Sons, 1990.

[28] D. Hope, M.F. Billett, M.S. Cresser "Exports of organic carbon in two river systems in NE Scotland" Journal of Hydrology, 193, 1997, 61–82.

[29] H. Laudon, S. Kohler, I. Buffam "Seasonal TOC export from seven boreal catchments in northern Sweden" Aquatic Sciences, 66, 2004, 223-230.

[30] J. Shih, R.B. Alexander, R.A. Smith, E.W. Boyer, G.E. Schwarz, S. Chung "An initial SPARROW model of land use and in-stream controls on total organic carbon in streams of the conterminous United States" U.S. Geological Survey Open-File Report 2010–1276, 2010, 22 p.

[31] J.L. Richardson, J.L. Arndt, J.A. Montgomery "Hydrology of wetland and related soils" in JL Richardson & MJ Vepraskas (eds), Wetland Soils, Lewis Publishers, Boca Raton, 2001.

[32] British Columbia Ministry of Environment (BCME) Oil and gas waste regulation – users guide, 2007. [Online] Available at: www.env.gov.bc.ca/epd/industrial/regs/oil_gas/pdf/ogwr_guide.pdf. [Accessed on January 24, 2014].

[33] L. Arvola, A. Ralke, P. Kortelainen, M. Jarvinen "The effect of climate and land use on TOC concentrations and loads in Finnish rivers" Boreal Environmental Research, 9, 2004, 381-387.

[34] S. Bhat, J.M. Jacobs, K. Hatfield, J. Prenger "Relationships between stream water chemistry and military land use in forested watersheds in Fort Benning, Georgia" Ecological Indicators, 6, 2006, 458-466.

[35] D. Enters, A. Lucke, B. Zolitschka "Effects of land-use change on deposition and composition of organic matter in Frickenhauser See, northern Bavaria, Germany" Science of the Total Environment, 369, 2006, 178-187.

[36] P. Bragee, F. Mazier, P. Rosen, D. Fredh, A. Brostrom, W. Graneli, D. Hammarlund "Forcing mechanisms behind variations in total organic carbon (TOC) concentration of lake waters during the past eight centuries – palaeolimnological evidencs from southern Sweden" Biogeosci Discuss 10, 2013, 19969-20003.

[37] S.J. Kohler, I. Buffam, H. Laudon, K.H. Bishop "Climate's control of intra-annual and interannual variability of total organic carbon concentrations and flux in two contrasting boreal landscape elements" Journal of Geophysical Research 113(G3), 2008,

[38] N. Dov "Man, a Geomorphological Agent: An Introduction to Anthropic Geomorphology" Springer, pp. 121–122, 1983.

13

The Three Rs of Remediation: A Comparative Analysis of the Risk, Real Estate, and Regulatory Drivers Affecting Remediation Technology Applications

Afamia Elnakat, Ph.D.
Environmental Science & Engineering, The University of Texas at San Antonio
One UTSA Circle, San Antonio, Texas 78249
afamia.elnakat@utsa.edu

Abstract- Time and cost often compete in guiding engineers during the design phase of remediation plans. While time and cost are primary focuses for most private industry clients, regulators, and citizens; choosing the correct technology to address the fate and transport of contaminants does not necessarily align to time and cost allotments. To ensure the success of a remediation plan in addressing the removal of contaminants, it is important to focus on the incentives behind the remediation effort. This review provides a comparative analysis of the three drivers of remediation: Risk, Regulation, and Real Estate. Based on the driving R, the selected remediation engineering technology can be more appropriate in serving the ecosystem needs and sociopolitical implications. While time and cost will remain major drivers, this comparative study demonstrates that the success of contaminant removal is equally dependent on the tools and technologies applied to the geospatial and geopolitical environment.

Keywords: Remediation Engineering Drivers, Algorithm for Remediation Technologies, Contaminant Removal, Remediation Engineering Design.

Capsule: This review examines the 3Rs of remediation drivers: Risk, Regulation, and Real Estate, and cross-links these drivers to appropriate technologies in a simple to use algorithm.

1. Introduction

In 2004, the United States Environmental Protection Agency estimated that for the years 2004-2033 there will be approximately 294,000 hazardous waste sites to remediate and the cleanup costs will amount to nearly \$209 Billion United States Dollars [1]. In response to this market trend, federal, state, and local government and private industry will tackle the anticipated demand by seeking 1) technologies, 2) skilled professionals, and 3) smarter and more cost/time effective solutions to existing and up and coming complex environmental pollution [1]. Investors on the other hand, will pursue innovative technologies and promising real estate that has a potential for a lucrative technical and financial future. In alignment with market needs, universities continue to adjust their curricula in environmental science and engineering to prepare future professionals for this complex challenge where no two remediation sites are alike.

As electronic applications dominate our daily lives, the field of remediation engineering remains dependent on geomechanical design processes set in design blueprints and spreadsheets. The complexity of this field from assessing the classification of contaminants, to determining the parameters for lab testing, to investigating soil types and elevations, and calculating water flow and pump curves, necessitates introducing easy to use matrixes for sound and quick decision-making.

As electronic algorithms become more functional through the smart application of data utilization and automated technologies, the matrix environmental professionals utilize to design remediation plans is outdated in comparison. Even though a smart app that resolves remediation strategies may be futuristic, the objective of this paper is to delineate the 3Rs that drive remediation projects. Based on the driving R: Risk, Regulation or Real Estate – the selected remediation engineering technology can be more appropriate in serving the ecosystem needs and sociopolitical implications. While time and cost will remain major drivers, this comparative study demonstrates that the success of contaminant removal is equally dependent on the tools and technologies applied to the geospatial and geopolitical environment. Using a matrix, that acknowledges which specific technologies will better perform based on the driving R, will facilitate the design of remediation plans and assure the proper management of the contaminant resulting in the long-term reduction of health costs, ecosystem impacts, and federal funds.

2. Background

In describing the complexity of the environmental field of remediation, Suthan S. Suthersan said it best:

"Scientists and engineers practicing remediation engineering have to learn the nuances of investigative techniques, data collection, and treatment technologies. This education includes a new understanding of the physical and chemical behavior of the contaminant, the geologic and hydrogeologic impacts on the fate and transport of these contaminants, the human and environmental risks associated with contaminations, and the selection of appropriate technologies to provide maximum mass transfer and destruction of the contaminants" [2].

In determining the protocol for this study it became apparent that one of the main objectives is to produce a graphical, easy to use, algorithm to guide decision makers and engineers in quickly selecting some of the available technologies based on the driving factor for the project in question. In the same way electronic algorithms learn and adapt to the consumer choices and behaviors, an engineering algorithm can provide the field with a tool that can be applied to the various project needs and parameters of interest. The key is to simplify the study tool to provide a first step approach to the plethora of available tools and matrixes in remediation technologies. The tool provides a unique approach by focusing on one of the 3Rs as a project driver: Risk, Remediation, or Real Estate versus time and cost.

Before creating the graphical algorithm to serve as a quick and easy reference and in order to comparatively cross-examine technologies to project drivers, the following two targets will be addressed:

- First: Define and select remediation technologies that are appropriate for various scenarios and are well established in the field. This will be conducted by investigating the literature and industry experts to produce a list of ten proven and commonly utilized technologies.
- Second: Provide a comparative analysis of the 3R drivers as they relate to priorities and success factors.

3. Study Targets: Remediation Techniques and Comparative Analysis of 3Rs

3. 1. Remediation Techniques

The Federal Remediation Technologies Roundtable (FRTR) produced a remediation reference guide in a cooperative effort by various federal agencies to include the United States (U.S.) Department of Defense, the U.S. Army, the U.S. Department of Energy, the U.S. Geological Survey, the U.S. Environmental Protection Agency and more. The reference guide also included a Treatment Technologies Screening Matrix [3]. The matrix is composed of approximately 60 treatment technologies grouped into water versus soil clean up applications. The matrix also provides a cost and time performance measure. To date, the FRTR matrix and reference guide remain as one of the most comprehensive and easily accessible guides to environmental engineers working in the remediation industry.

This study will utilize technologies that are aligned with the FRTR database. To better focus this paper, the ten technologies selected for this study are dominated by in situ treatment options with the exception of "Pump & Treat" and "Dig & Haul" options. In situ remediation uses on site methods to treat the contaminant on location in efforts to save money, time, transportation cost, liability, and disposal fees [4]. Moreover, the ten selected remediation technologies are well established and referenced in textbooks used to educate future engineers such as CRC Press's textbook titled Remediation Engineering Design Concepts [2]. The ten selected technologies are listed and described as follows:

3. 1. 1. Pump & Treat

Used primarily for groundwater decontamination applications. Contaminated groundwater is pumped to the surface. The contaminant is then removed from the groundwater by aboveground treatment methods, such as filtration, and the clean water is re-injected or released to either the sewer or a surface water body.

Another option after pumping the contaminated water is to dispose the contaminated water as a stored hazardous substance. Pump and Treat is common in treating dissolved chemicals in water. In addition, the pumping process contains the contaminant plume and prevents it from migrating to other water resources to include drinking water wells and wetlands [2], [5].

3. 1. 2. Dig & Haul

Dig and haul is an ex situ approach to soil remediation techniques that utilizes construction equipment and heavy machinery. Dig and haul focuses on excavating contaminated soils and disposing of them as hazardous material waste in appropriate landfills. Timely excavation of the contaminated soils can play a significant role in preventing the spread of the contaminant through the water table and wind dispersion [5].

3. 1. 3. Air Sparging

Used separately or integrated in conjunction with other techniques, air sparging is the pumping of air to volatize volatile organic compounds (VOCs) and semivolatile organic compounds sorbed to soil and in the groundwater. The addition of air specifically beneath the water table expedites the movement of volatiles through the path of least resistance towards the surface. Air compressors are usually utilized through injection wells to produce sufficient air pressure to mobilize contaminants [3], [5]. The addition of oxygen for air sparging has the added benefit of creating an aerobic environment that could assist with biodegradation levels.

3. 1. 4. Vacuum Enhanced Recovery

Also known as dual phase extraction, Vacuum Enhanced Recovery removes through negative pressure (high vacuum system) both dissolved and free phase non-aqueous phase liquid (NAPL) contamination, and vapors in groundwater and the vadose zone. While energy intensive, this remediation technique is effective and unique for extracting both liquid and vapors at the same time. Vacuum Enhanced Recovery is also referred to in the literature as vacuum enhanced extraction, and bioslurping [2], [3].

3. 1. 5. Reactive Walls/Zones

An in situ approach to mass removal and treatment of contaminants in large areas where other mechanical technologies may be cost prohibitive and

troublesome due to generating hazardous waste for disposal. The approach of reactive walls/zones also known as "permeable reactive barriers" [5] is to utilize impermeable and permeable barriers and gates downgradient of the contaminant plume or a series of injection wells acting as a curtain to immobilize and/or transform the contaminant to non-harmful byproducts. This technique is also referred to in the literature as "funnel and gate systems" or "treatment walls" [2]. Decontamination usually occurs through chemical reagents addition, oxygen infusion, and/or filtration of the substance underground following a set hydraulic pattern.

3. 1. 6. Stabilization & Solidification

Stabilization and Solidification processes can range in implementation depending on whether the contaminant is encapsulated or bound onto a solid to prevent the contaminant migration and facilitate the excavation of the waste product [5]. The primary benefit of this technology is that solidification prevents contaminant leaching from surfaces and soils into the groundwater, surface water bodies, and stormwater runoff and drains. Specifically, the solidification process binds contaminants with reagents changing their physical properties -especially hardness, while stabilization refers to a chemical reaction that reduces waste leachability [6].

3. 1. 7. Soil Vapor Extraction

Soil Vapor Extraction is used to remediate soils from volatile and semi volatile organic compounds. Also known as soil venting and vacuum extraction, Soil Vapor Extraction utilizes in situ technologies to remove contaminant vapors through extraction wells with the use of blowers or vacuum pumps. The extracted vapors or "off-gasses" are then treated onsite with various above ground treatment methods such as thermal or catalytic oxidation or condensation. The Soil Vapor Extraction method is applicable to the vadose zone since extracting water will damage the system and will require the Vacuum Enhanced Recovery method instead. The mechanical set up of Soil Vapor Extraction is easily integrated to other remediation technologies such as Air Sparging [2], [5].

3. 1. 8. Bioventing

Bioventing refers to the use of low airflow to stimulate aerobic biodegradation of contaminants and mobilization of volatile compounds through the soil.

This technique is also compared to soil venting in the literature [3]. Bioventing not only reduces vapor treatment costs, but also can consequently remediate semi volatile organic compounds that are not directly volatized [2].

3. 1. 9. Bioremediation

Under both aerobic and anaerobic conditions, bioremediation is the degradation of organic contaminants in soil and water through microbial metabolism. Bioremediation can be enhanced through the addition of oxygen, hydrogen peroxide, nitrate, and nutrients. When acceleration of the naturally occurring biodegradation process is limited, bioaugmentation is utilized by adding exogenous microorganisms on site. The microorganisms, including bacteria and fungi, metabolize contaminated compounds to innocuous mineral products [2], [3].

3. 1. 10. Phytoremediation

Phytoremediation is the use of photosynthetic plants to uptake contaminants in the soil or water. Phytoremediation is especially effective for removing inorganic compounds, metals, pesticides, and explosives when the contaminant is at low levels tolerable by the plants. Phytoremediation provides the added benefit of soil stabilization -known as phytostabilization- by preventing wind and water runoff and dispersion of contaminants [5]. Phytoextraction is the process in which the plant extracts the contaminants above the soil surface into the plant shoot and leaves while contaminant degradation in the rhizosphere can also occur in the root zone of the plants with the symbiotic relationship augmenting the process by "incorporating bacterial, fungal, insect, and even mammalian genes into the plant genome" [2] to provide an opportunity for biodegradation at the root system.

3. 2. Comparative Analysis of the 3 R Drivers

3. 2. 1. Risk Drivers

When it comes to risk drivers, many stakeholders are involved in scrutinizing the remediation process of a contaminated site. These stakeholder entities range from state and federal agencies, environmental consulting engineers, to local citizen action and advocacy groups [7]. Yet, public pressure remains one of the strongest components in how risk is perceived and how regulatory standards react. Preferably, clients and companies will avoid environmental risk by taking risk precautions [8]. Unfortunately that is not always feasible based on cost and liability limitations. For example, depending on cost, a company might not be willing to invest large sums of money to avoid negligible risks perceived by the public. Contrarily, the public, investors, or companies may refuse to invest any money to reduce actual demonstrated risk due to associated costs or inconvenience [9]. This example is best demonstrated by the use of cell phones while driving in absence of local laws. Another validation are the survey results indicating that in the absence of state laws requiring passenger restraints/seat belts, only 15% of American drivers routinely use them [9].

As public perception is critical in the success of risk driven remediation projects, it is important to assess risk not only from a regulatory standpoint but also at a societal level. Society at large influences not only the level of perceived risk to human health but also the risk to ecosystem impacts, and the risk of mobilization, biomagnification, cross-contamination and chemical persistence. Meanwhile, regulatory drivers depend on established exposure limits by federal, state and local agencies and entities such as the Occupational Safety and Health Administration. In determining the regulatory exposure limits many risk assessment standards are considered for determining dosages. One value that is frequently used is "the No Observable (Adverse) Effect Level (NOEL). The NOEL includes effects, such as minor weight loss, that are not considered to be adverse. These values are applicable only to that species in which the test was conducted. Extrapolation to other species will require dosage adjustment" [9]. Regulatory standards are generally designed to stay within thresholds in which no adverse effect will occur. Especially when it comes to predicting cancer risk, many guidelines offer a NOEL of a 1/1,000,000 risk level as an acceptable risk [9]. Risk management thus, depends on evaluating various response alternatives, both regulatory and non regulatory [8].

Since risk assessment supports setting regulatory thresholds, and risk assessments are based on non-adverse and acceptable levels of risk as more stringent thresholds are cost prohibitive, it is important to then prioritize risk drivers as follows:

- Ability to "stop the bleeding": the technology that most rapidly immobilizes the contaminant to prevent migration and cross-contamination to include reducing short-term and long-term exposure limits through ingestion, inhalation and dermal absorptions.

- Cost-effectiveness: high energy intensive and mechanically dependent applications can be cost prohibitive as many of these facilities become a federal liability in the absence of private investors or responsible corporations.
- Public Acceptance of technology: it is important that the technologies utilized onsite or offsite not be intimidating to the public or ecosystem considering noise levels, byproducts and off gases (smell), risk of explosion, and accidents during operation.

3. 2. 2. Regulatory Drivers

In determining remediation mitigation plans, "both industry and government have their respective roles to play. Ever-increasing parts of that role, and the keys to success, are sound cost control practices and techniques" [7]. Due to the complex nature of these remediation plans, many of the regulation drivers place emphasis on schedule compliance rather than cost. There is a lack of guidance with regards to cost estimating (e.g., standardized cost databases and cost guides) to aid regulators in the process of identifying cost-effective and sound technology decisions and approaches. Nevertheless, corporations are becoming more aware that their environmental image and footprint is directly affecting their bottom line. The public attention on environmental impacts is significant and will continue to surge with increasing levels of public awareness [8]. Environmental compliance costs cannot only consider violation fees. There are costs related to complying with federal, state, and local requirements to include permitting costs, technology modification costs, changing source materials, environmental auditing, and creating and maintaining an environmental management system [8].

Therefore when prioritizing regulatory drivers, the following factors remain important in order of criticality, from most to less, as follows:

- Compliance of technique: prioritizing environmentally friendly low-impact and low-byproduct methods of decontamination, especially those of high public acceptance.
- Initial time to set up and implement cleanup technology: to achieve lower risk of mobilization of contaminant and halt the noncompliance regulatory fees.

- Cost-effectiveness: since as indicated by Richard A. Selg (1993), cost control practices are a key area for project success.

3. 2. 3. Real Estate Drivers

Real Estate drivers are compelled by a distinctive set of values that are primarily focused on profit, time and cost efficiencies, land value, property potential, and historical and social significance of the real estate. Private investors and land developers usually control the ownership of Real Estate driven remediation projects. Even though the developers will comply with regulatory requirements, the priorities rest in the efficiency of the cleanup versus the techniques utilized. Nevertheless, investors are now more aware of environmental concerns. This awareness is driven by the company's earnings, net worth, cash flow, acquisition potential, divestiture, and financing strategies that are impacted by environmental obligations and liabilities [8].

Remediation projects in the real estate industry are featured by specific characteristics. The language utilized to enhance and encourage property advancement in the future includes words such as "adaptive reuse" and "reconstruction" versus remediation. These words are selected carefully to entice and incentivize the future customer. Future customers include shoppers and recreational users that will invest in purchasing services and property in the residential, commercial, or retail real estate sectors. Examples of real estate developments on remediated sites include golf courses, commercial office space buildings, shopping centers, historical landmarks and more.

Richard Selg indicates that the "most common approach to eradicating cost growth is to revisit the basis of the estimate and to ensure that the current estimate accounts adequately for known scope as well as uncertainties surrounding the accomplishment of the current scope of work" [7]. In real estate driven remediation projects time is money setting the priorities as follows:

- Time Effectiveness: Efficiency of remediation plan to include quick set up and mobilization, fast contaminant removal, and utilization of existing on site construction material and equipment.
- Liability: The effectiveness of the remediation technology to comply with regulations and ability to transfer liability to other entities (for example landfills).

- Complexity: Less complex systems are attractive to real estate investors since they reduce the dependency on technology modifications, intellectual property, and the engagement with field specific contractors that take up time and cost to set up and mobilize.

4. Results and Algorithms

Real Estate driven environmental remediation projects are usually motivated by high cost, high energy, high mechanical expenditures, and fast treatment technologies that mimic construction sites. In contrast, Regulation drivers are lower cost, more environmentally friendly, and less mechanically intensive applications. Risk drivers are more socially acceptable technologies that prioritize halting the contaminant migration and providing further damage protection.

Figure 1 displays the compiled results of this study in a graphical algorithm that serves as a quick and easy reference guide of comparatively cross-examined technologies and project drivers.

4. 1. Influence of Risk Drivers on Remediation Technology Applications

The influence of risk drivers on remediation technology applications are determined from most applicable to minimally applicable, as follows:

Highly Applicable:
- Reactive Walls/Zones: Useful in containing hard to treat underground and groundwater plumes from migrating and posing risks to various other systems. Can be less energy intensive and less costly than pump and treat operations.
- Stabilization & Solidification: Attractive due to fast response stabilization and immobilization of contaminants especially in emergency response situations.
- Soil Vapor Extraction: More attractive than Vacuum Enhanced Recovery for removing VOCs and semi VOCs from a risk perspective since Soil Vapor Extraction is less mechanically intensive, easier to control off gases, and quick onsite set up and mobilization.

Moderately Applicable:
- Pump & Treat: Even though cost prohibitive and energy intensive, Pump and Treat can be one of the very few methods that allow for hydraulic manipulation to control groundwater plume migration.

- Dig & Haul: Cost prohibitive in many instances; however, timely excavation of contaminated soils provides reassurance that air dispersion and water contamination will be prevented.
- Vacuum Enhanced Recovery: Even though applicable, the highly mechanical set up poses risks of off gas explosions and other mechanical failures reducing attractiveness especially from public perspective.
- Bioventing: Bioventing relieves the concerns originating from high pressure air sparging. For long-term interventions, utilizing oxygen as a treatment method is publically acceptable.
- Bioremediation: Both regulators and the public embrace this method that was initially based on natural attenuation principles. Minimal risk can be incurred if bioaugmentation is introduced.
- Phytoremediation: As in bioremediation, this natural method of contaminant uptake is highly favorable. Risk concerns emerge due to contaminated plants (stalks and leafs) consumed by animals/birds and may require proper disposal offsite.

Minimally Applicable:
- Air Sparging: Potential to jeopardize foundations nearby, migration of plume due to air pressure, and potential to increase off gases in underground structures, conduits, and basements

4. 2. Influence of Regulation Drivers on Remediation Technology Applications

The influence of regulation drivers on remediation technology applications are determined from most applicable to minimally applicable, as follows:

Highly Applicable:
- Vacuum Enhanced Recovery: The dual phase extraction capability of this technique reassures regulators that both vapor and dissolved contaminants are addressed and provides a faster way for the client to reach compliance.
- Reactive Walls/Zones: An innovative technique that provides an in situ alternative for pump and treat and reduces hazardous disposal offsite. Can integrate many remediation media to address organic, inorganic, and elemental contaminants.
- Stabilization & Solidification: A time efficient technique that allows for emergency response. In addition, it prevents future contamination and reduces violations costs.

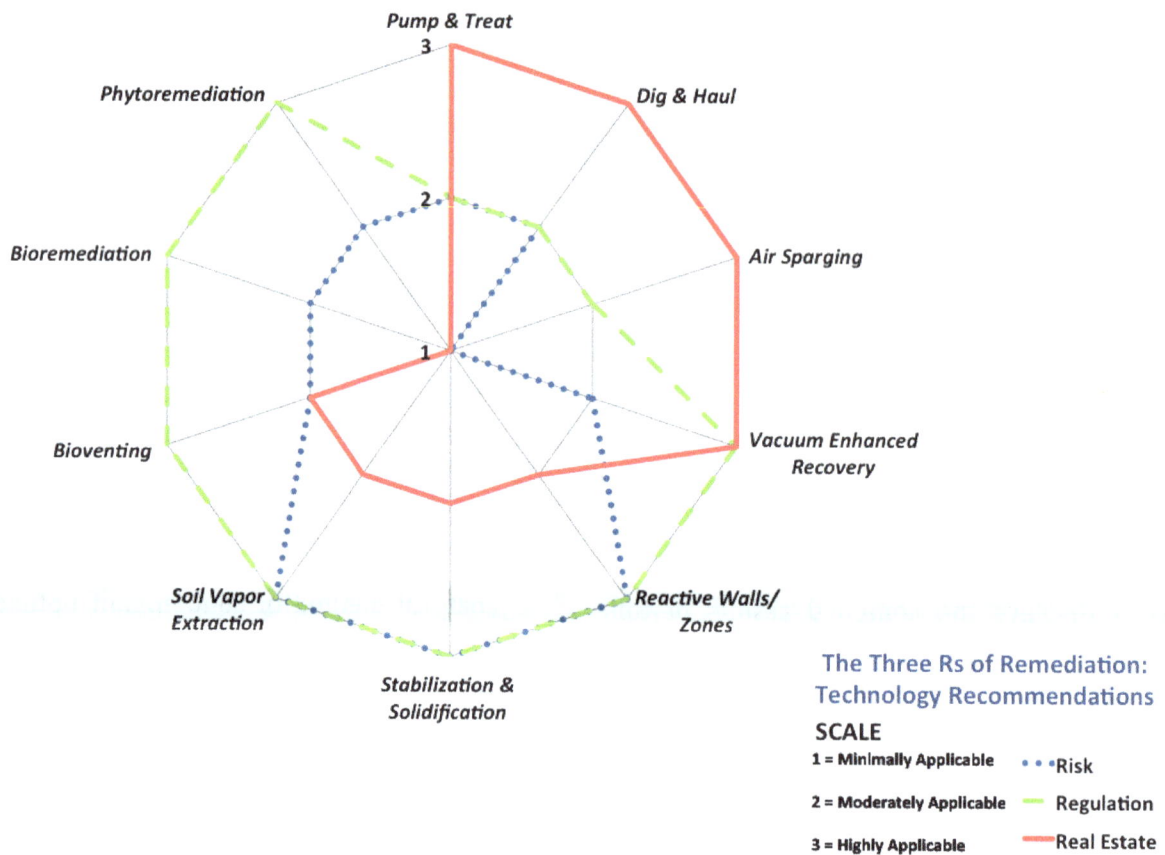

Figure 1. The Three Rs of Remediation Technology Recommendation Graphical Algorithm.

- Soil Vapor Extraction: A time proven method that allows for quick set up and mobilization when treating VOCs and semi VOCs.
- Bioventing: Regulators favor techniques that allow for low-risk natural methods such as bioventing, also known as soil venting.
- Bioremediation: A trendy method that allows for land and resource preservation using the basis of natural attenuation. Also, this technology provides clients/violators with mitigation negotiating possibilities.
- Phytoremediation: Added benefits of land stabilization, hydraulic cycle support and soil enrichment highlights this method of remediation.

Moderately Applicable:

- Pump & Treat: A high energy and high cost old school method has left regulators more impressed by reactive zone on site innovative methods especially when pumped water is

disposed of as hazardous waste offsite or overwhelming local sewer plants (which require coordination with local authorities).

- Dig & Haul: Considered an effective yet old school method of remediation, regulators like to see technology innovation that reduces offsite hazardous waste transfer.
- Air Sparging: Some concerns emerge if foundations or basement/underground structures are nearby.

Minimally Applicable:

- Not applicable since all identified technologies in this study are accepted by regulatory agencies.

4. 3. Influence of Real Estate Drivers on Remediation Technology Applications

The influence of real estate drivers on remediation technology applications are determined from most applicable to minimally applicable, as follows:

Highly Applicable:

- Pump & Treat: Even though costly, real estate investors are happy to save time due to large coverage area and the option to store and dispose contaminated water offsite.

- Dig & Haul: Another costly technique, this method is highly applicable to contaminated soils allowing investors to use the same construction set-up and excavator machinery to remove contaminated soils and dispose of them in a timely manner in approved landfills.

- Air Sparging: The high pressure air sparging to stimulate biodegradation and assist the contaminants in volatizing is an attractive technique in larger project sites where there is no fear of jeopardizing residential basement structures/infiltration and the mobilization of plume to neighboring land.

- Vacuum Enhanced Recovery: Intensive mechanical and energy set up, yet dual phase extraction of both liquid and gas contaminants simultaneously covering large sites makes this attractive when cost is not an issue compared to time.

Moderately Applicable:

- Reactive Walls/Zones: Cost and time of trenching before utilizing trenches for treatment is not as attractive to real estate developers.

- Stabilization & Solidification: Solidification is more attractive in this driver if the material could be solidified and kept onsite without generating any risks.

- Soil Vapor Extraction: Vacuum Enhanced Recovery is usually preferred since it is a dual phase extraction process applicable to larger areas saving time.

- Bioventing: More applicable to low concentration of contaminants that are less volatile. Added advantage of biodegradation is not realized due to time requirements of biostimulation lag period.

Minimally Applicable:

- Bioremediation: Long period of acclimatization and contaminant degradation. In addition cannot be implemented on land to be redeveloped and constructed.

- Phytoremediation: Long period of plant growth and contamination uptake. In addition cannot be implemented on land to be redeveloped and constructed.

4. 4. Sample Case Studies

When approaching a newly discovered contaminated area or spill, professionals can be overwhelmed with the number of technologies and techniques available to remediate any given site. After all, the FRTR matrix provides approximately 60 treatment technologies. The objective of the 3Rs approach presented in this paper is to facilitate a broader method to initial selection of remediation techniques – independent of the nature of the contaminant or the environment affected. Figure 2 provides a questionnaire based decision matrix for guidance and to further demonstrate the application of each driver on choice of technology, Table 1 provides a list of case studies categorized by driver and technology listed by reference and location. Four case studies, included in Table 1, are further evaluated as follows:

Case Study 1: Risk

Removed in 2004 from the National Priority List of superfund sites, the South 8th Street Landfill in West Memphis, Arkansas [10] is a great example of a "Risk" driven remediation approach. The South 8th Street Landfill has completed the third five-year review with a determination that the site remains protective of human health and the environment and that the site controls prohibiting excavation and drilling within the specific landfill areas will prevent future exposure pathways. The 16-acre landfill site and two and a half acre oily sludge pit was a "Risk" driven project since the site is adjacent to the Mississippi River and is on the two-year flood plain. The groundwater table within the alluvial aquifer beneath the site sits just a few feet to 20 feet below the ground surface. It was important to protect the Wilcox aquifer that provides drinking water supply to the City of West Memphis approximately two to four miles from the site. The contaminants of concern included the oily sludge hydrocarbons, municipal and industrial waste, lead, and carcinogenic poly aromatic hydrocarbons to name a few. The site is treated with stabilization and solidification, which as shown in Figure 1 is a highly applicable method for risk driven projects due to the rapid "ability to stop the bleeding," lower cost in preventing contaminant migration, and public acceptance.

Case Study 2: Real Estate

The Vertac Superfund Site in Jacksonville, Arkansas in EPA Region 6 [11] is a great example of a "Real Estate" driven remediation plan. The site used since 1948 and purchased by Vertac Chemical Corporation in 1978 is now remediated with the third five-year review concluding that the site remains protective of human health and the environment as published on May 13, 2014. The 193 acres site was contaminated with chlorinated hydrocarbons with dioxin contamination found in both the soil and drummed waste. The site is not a direct threat to the public, and the underlying aquifer is not used for public water supply or domestic use. Approximately 1,000 residents live within one mile of the site. The site is also adjacent to industry and an air force base. Per Figure 1, the 3R driver focuses on "Real Estate" highly applicable technologies to include both "Dig & Haul" and "Pump & Treat". Excavation of contaminated soil was completed in the late 90s and groundwater extraction wells were installed to eliminate to retract the contaminant plume from the ground water. Other techniques were also used such as demolition of buildings, off-site incineration and decontamination, that align with the more expensive "Real Estate" priority of time effectiveness, compliance with regulation, and less complex systems to enhance repurposing efficiency. The industrial/commercially-zoned site is now repurposed to include the city's recycling center, and police and fire department training facilities.

Case Study 3: Regulation

Since all the techniques presented in this study are acceptable by regulatory agencies, perhaps some of the best case study demonstrating regulatory and penalty driven remediation projects, are dry cleaning establishments. For larger establishments, due to the associated cost and liability of remediation, many of these establishments after bankruptcy or abandoning the site become superfund sites and get placed on the national priority list depending on their risk to human health and the environment. However, this is not the case for many smaller businesses such as dry cleaners. A notable press release by the EPA in 2002 demonstrated the importance of penalties when it comes to compliance [12]. The EPA cited 11 dry cleaners in New Jersey proposing a total of $37,850 in fines. A total of 114 dry cleaners were also cited in New York and New Jersey combined [12]. The EPA has created many mechanisms to better educate and cooperate with dry cleaning establishments to reach compliance. One of the main objectives is to protect public health and the environment from percholoroethylene (a suspected carcinogen and irritant), and other toxic air pollutants. As a result many dry cleaners are now encouraged to use more

environmentally friendly products to meet compliance and protect the workers and the public.

Case Study 4: Multiple Approaches and Multiple Drivers

In many cases, a cookie cutter approach to site management is not possible since many drivers compete for priorities in addition to the always-critical cost and time components. In such scenarios multiple technologies may be applicable. From Table 1, multiple case studies demonstrate a combination of regulatory and risk drivers. Similarly, a number of case studies include a combination of technologies such as "Soil Vapor Extraction" as well as "Air Sparging" which go hand in hand in removing VOCs. It is important to note that multiple approaches can be applicable and many decisions can fall back on the experience of the remediation engineer, contractors, and what they are more familiar in practicing. The proposed 3Rs approach is intended to serve as a reference that better organizes the technologies and options and provides guidance to engineers for determining a priority driver outside the given cost and time factors that are present in most projects.

5. Conclusions

The field of remediation engineering is highly complex and multidisciplinary. From assessing the classification of contaminants, to determining the parameters for lab testing, to investigating soil types and elevations, and calculating water flow and pump curves. This complexity necessitates the introduction of easy to use reference matrixes for sound and expedited decision-making. There is no cookie-cutter approach in remediation engineering; no two projects are alike. This study introduces a matrix that guides engineers to acknowledge specific drivers and technologies that better perform independent of the nature of the contaminant and the environment affected. In an effort to facilitate the initial set of decisions required to select adequate remediation technologies, this paper presents an approach that focuses on three prevalent drivers, the 3Rs: Risk, Regulation, and Real Estate.

Selecting the correct technology to address the fate and transport of contaminants does not only align with time and cost allotments. While time and cost will remain major drivers, the success of contaminant removal is equally dependent on the tools and technologies applied to the geospatial and geopolitical environment. There are multiple treatment technologies and hundreds of treatment technology combinations of which at least 60 are organized and referenced in the FRTR. To better focus the paper, ten remediation technologies that are well established in the field are identified, defined, and used for illustration purposes. Further research can extend the 3Rs method to include other remediation technologies in addition to the coupling of technologies and drivers through the proposed 3Rs approach. A remediation engineer can better navigate available and valuable information when better guided by a broadly applicable approach that defines the priorities for a successful outcome.

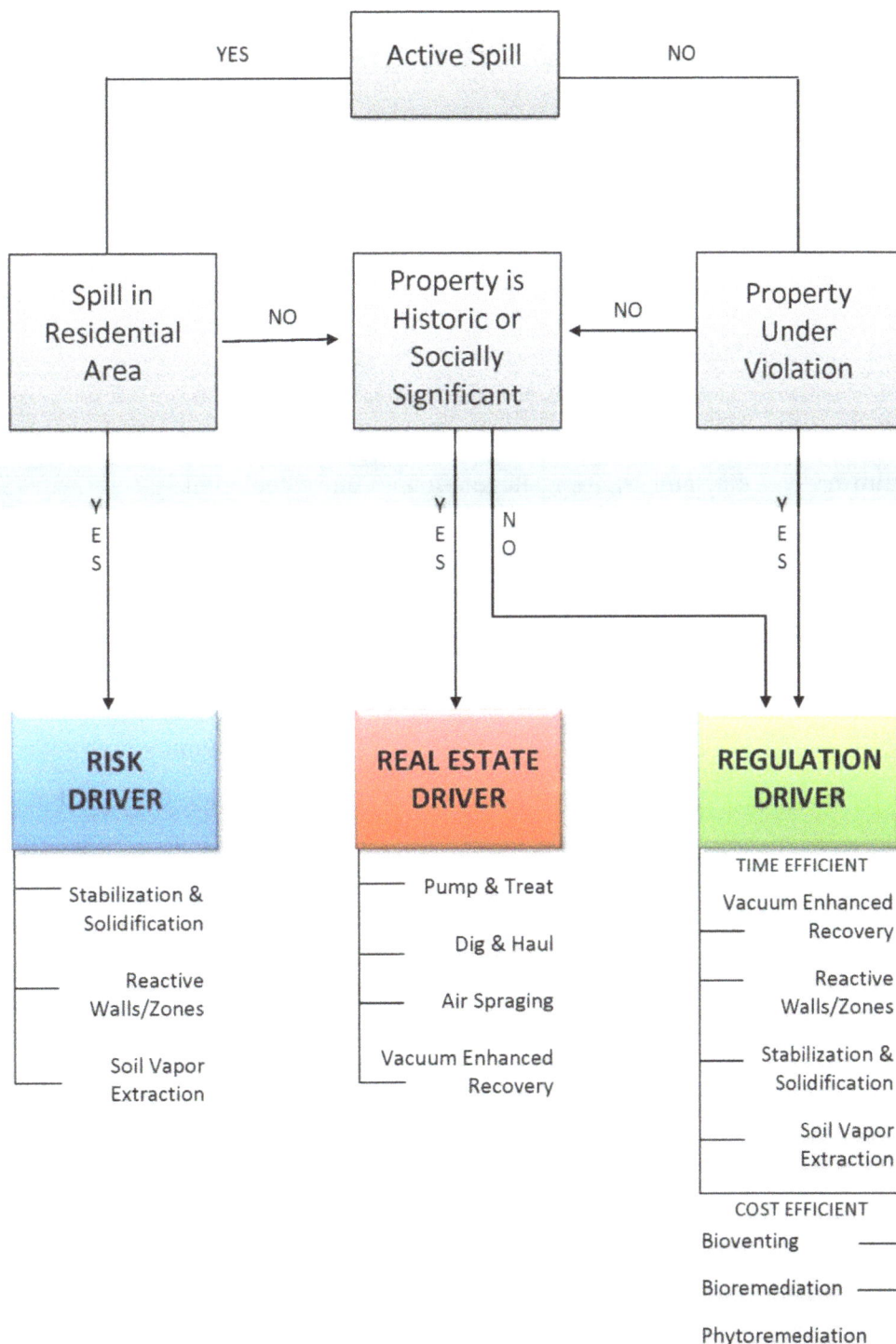

Figure 2. Questionnaire Algorithm for Guidance During Potential Border Driver Scenarios.

Table 1. Case Studies Demonstrating Remediation Driver by Technology and Location.

PROJECT	LOCATION	DRIVER	TECHNOLOGY	REFERENCE
Aberdeen Proving Ground	Edgewood, MD	Regulation	Phytoremediation	[13]
Amoco Petroleum Pipeline	Constantine, MI	Real Estate (Voluntary)	Air Sparging and Vacuum Enhanced Recovery	[14]
Baird and McGuire	Holbrook, MA	Regulation	Pump and Treat	[15], [16]
Big Tex Grain Site	San Antonio, TX	Real Estate	Dig and Haul	[17]
Butterworth Landfill	Grand Rapids, MI	Real Estate	Dig and Haul	[18]
Commencement Bay	Tacoma, WA	Risk	Soil Vapor Extraction	[14]
Federal Creosote	Manville, NJ	Regulation	Dig and Haul	[19]
French Limited	Crosby, TX	Regulation and/or Risk	Bioremediation/Stabilization	[14]
Highlands Acid Pit	Highlands, TX	Regulation and/or Risk	Dig and Haul; Phytoremediation	[20]
Highway 71/72 Refinery	Bossier City, LA	Real Estate	Dig and Haul	[21]
Hill Air Force Base, Site 280	Ogden, UT	Regulation and/or Risk	Bioventing	[14]
Hill Air Force Base, Site 914	Ogden, UT	Regulation and/or Risk	Soil Vapor Extraction and Bioventing	[14]
Iceland Coin Laundry	Vineland, NJ	Regulation	Bioremediation	[22]
Jibboom Junkyard	Sacramento, CA	Real Estate	Dig and Haul	[23]
Love Canal	Niagara Falls, NY	Risk	Pump and Treat; Stabilization and Solidification	[24]
Lowry Air Force Base	Denver, CO	Regulation and/or Risk	Bioventing	[14]
Luke Air Force Base	Glendale, AZ	Risk	Soil Vapor Extraction	[14]
MacGillis and Gibbs	New Brighton, MN	Real Estate	Dig and Haul; Stabilization and Solidification; Pump and Treat	[25]
Midvale Slag	Midvale, UT	Real Estate/Risk	Dig and Haul; Stabilization and Solidification	[26]
Old Esco Manufacturing	Greenville, TX	Risk	Dig and Haul	[27]
Parsons Chemical/ ETM Enterprises (Superfund)	Grand Ledge, MI	Risk	Stabilization and Solidification	[14]
Quarry Market	San Antonio, TX	Real Estate	Dig and Haul	[28], [29], [30]
South 8th Street Landfill	West Memphis, AR	Risk	Stabilization and Solidification	[10]
The Many Diversified Interests, Inc. Site-3617	Houston, TX	Real Estate	Dig and Haul	[31]
The Vertac	Jacksonville, AK	Real Estate	Dig and Haul; Pump and Treat	[32]
Times Beach	Times Beach, MO	Risk	Reactive Walls/Zones; Dig and Haul	[33]
Vienna PCE	Vienna, WV	Regulation	Soil Vapor Extraction	[34]

Secondary to cost implications, the case studies demonstrate that the underlying objective for a successful implementation plan can be attributed to either protecting public health, complying with regulations, profiting from potential real estate value, or a combination of the above. The comparative analysis concludes that "Real Estate" environmental remediation projects are usually driven by high cost, high energy, high mechanical expenditures, and fast acting treatment technologies that mimic construction sites. "Regulation" drivers are lower cost, more environmentally friendly, and less mechanically intensive. "Risk" drivers are more socially acceptable technologies that prioritize halting the contaminant migration and ensuring protection from further harm.

By addressing the Risk, Real Estate, and Regulation impacts of the project, the design process can more efficiently address the fate and transport of contaminants and associated hazards. Understanding the R driver of the remediation project -by focusing on Risk, Regulation, or Real Estate project incentives- will help engineers formulate a more systemic approach resulting in a more socioecologically friendly and publicly acceptable contaminant removal outcome.

References

[1] EPA. (2004). Cleaning Up the Nation's Waste Sites: Markets and Technology Trends [Online]. Available: http://www2.epa.gov/sites/production/files/2015-04/documents/2004market.pdf

[2] S. S. Suthersan, *Remediation Engineering Design Concepts*. Boca Raton, Florida: CRC Press, 1997.

[3] J. V. Deuren, T. Lloyd, S. Chhetry, R. Liou and J. Peck, (2002). Remediation Technologies Screening Matrix and Reference Guide, Version 4.0. *FRTR* [Online]. Available: http://www.frtr.gov/matrix2/top_page.html.

[4] T. Gingrich, "Groundwater remediation: The evolution of technology," *Pollution Engineering*, vol. 39, no. 8, pp. 24, 2007.

[5] EPA. (2012.) A Citizen's Guide to Solidification and Stabilization [Online]. Available: https://clu-in.org/download/Citizens/a_citizens_guide_to_solidification_and_stabilization.pdf

[6] F. Barnett, S. Lynn and D. Reisman. Technology Performance Review: Selecting and Using Solidification/Stabilization Treatment for Site Remediation [Online]. Available: http://nepis.epa.gov/Exe/ZyNET.exe/P1006AZJ.TXT?ZyActionD=ZyDocument&Client=EPA&Index=2006+Thru+2010&Docs=&Query=&Time=&EndTime=&SearchMethod=1&TocRestrict=n&Toc=&TocEntry=&QField=&QFieldYear=&QFieldMonth=&QFieldDay=&IntQFieldOp=0&ExtQFieldOp=0&XmlQuery=&File=D%3A%5Czyfiles%5CIndex%20Data%5C06thru10%5CTxt%5C00000015%5CP1006AZJ.txt&User=ANONYMOUS&Password=anonymous&SortMethod=h%7C-&MaximumDocuments=1&FuzzyDegree=0&ImageQuality=r75g8/r75g8/x150y150g16/i425&Display=p%7Cf&DefSeekPage=x&SearchBack=ZyActionL&Back=ZyActionS&BackDesc=Results%20page&MaximumPages=1&ZyEntry=1&SeekPage=x&ZyPURL

[7] R. A. Selg, "Hazardous waste cost control management," *Cost Engineering*, vol. 35, no. 8, pp. 4, 1993.

[8] P. L. Brooks, L. J. Davidson and J. H. Palamides, "Environmental compliance: You better know your ABCs," *Occupational Hazards*, vol. 55, no. 2, pp. 41, 1993.

[9] R. B. Philp, *Ecosystems and Human Health Toxicology and Environmental Hazards*. Boca Raton, Florida: CRC Press, 2001.

[10] EPA. (2015a). Third Five-Year Review Report for South 8th Street Landfill Superfund Site West Memphis, Crittenden County, Arkansas. [Online]. Available: http://www.epa.gov/region6/6sf/arkansas/south8_street/ar_south8_street_3rd-5yr_review.pdf

[11] EPA. (2015b). Region 6 Congressional District 2 Vertac Superfund Site Jacksonville, Arkansas Update. [Online]. Available: http://www.epa.gov/region6/6sf/pdffiles/vertac-ar.pdf

[12] EPA. (2002b). EPA Cracks Down on Dozens of Area Dry Cleaners. [Online]. Available: http://yosemite.epa.gov/opa/admpress.nsf/8b770facf5edf6f185257359003fb69e/018bc082993ac06685257165006ad605!OpenDocument

[13] EPA. (2014a). Aberdeen Proving Ground (Edgewood Area Site). [Online]. Available: http://www.epa.gov/reg3hwmd/npl/MD2210020036.htm

[14] FRTR. (2007). Abstracts of Remediation Case Studies. vol. 11. [Online]. Available: https://frtr. gov/pdf/volume11%20Final%202007.pdf

[15] EPA. (2013a). Treatment Technologies for Site Cleanup: Annual Status Report (Eleventh Edition) [Online]. Available: https://clu-in.org/download/ remed /asr/11/asr.pdf

[16] EPA. (n.d.) On-Site Incineration at the Baird and McGuire Superfund Site: Holbrook, Massachusetts [Online]. Available: https://clu-in.org/download/ remed /incpdf/baird.pdf

[17] EPA. (2012a). Big Tex Grain Site Ready for Reuse Determination [Online]. Available: http://www. epa.gov /superfund/programs/recycle/pdf/big-tex-rfr.pdf

[18] EPA. (2010b). Return to Use Initiative 2004 Demonstration Project [Online]. Available: http:// www. epa.gov/superfund/programs/recycle/pdf/ butterworth.pdf

[19] EPA. (2012b). National Remedy Review Board Recommendations for the Federal Creosote Superfund Site. [Online]. Available: http://www.epa. gov/superfun d/programs/nrrb/pdfs/fedcreos. pdf

[20] EPA. (2014c). Highlands Acid Pit, Highlands, Texas [Online]. Available: http://archive.orr.noaa.gov/ book_ shelf/388_Highland.pdf

[21] EPA. (2012e). Return to Use Initiative 2011 Demonstration Project [Online]. Available: http:// www.epa.gov/superfund/programs/recycle/pdf/ rtu11-highway7172.pdf

[22] EPA. (2012d). EPA Proposes Innovative Way to Clean Up Iceland Coin Laundry Site [Online]. Available: http://yosemite.epa.gov/opa/adm-press. nsf/8b770facf5edf6f185257359003fb69e/ f208f9133ad7e4be852571e20051e9c9!OpenDoc-ument & Start = 9 & Count = 5&Expand=9

[23] EPA. (2014b). Celebrating Success: Jibboom Junk-yard Sacramento, California [Online]. Available: http://www.epa.gov/superfund/programs/recy-cle/pdf/jibboom-junkyard-success.pdf

[24] A. S. Phillips, Y. T. Hung and P. A. Bosela, "Love Ca-nal Tragedy," *Journal of Performance of Constructed Facilities,* vol. 21, no.4, pp. 313-319, 2007.

[25] EPA. (2010a). Cleanup and Mixed-Use Revitaliza-tion in the Twin Cities [Online]. Available: http:// www.epa.gov/superfund/programs/recycle/pdf/ macgillisgibbscase.pdf

[26] EPA. (2011a). Cleanup and Mixed-Use Revital-ization on the Wasatch Front [Online]. Available: http://www.epa.gov/superfund/programs/recy-cle/pdf/midvale-2011-case-study.pdf

[27] EPA. (2014d). Old ESCO Manufacturing Superfund Site [Online]. Available: http://www.epa.gov/re-gion6 /6sf/pdffiles/old-esco-tx.pdf

[28] J. Hiller, "Quarry Village is Sold: Dallas Firm Buys the Mixed-Use Property," *San Antonio Express News,* 2011.

[29] T. L. Silva, "Alamo Quarry Market is aspiring to new heights," *San Antonio Business Journal,* 2004.

[30] J. C. Garner and W. I. Shank, "Historic American Buildings Survey," in *Alamo Roman and Portland Cement Company,* 1968.

[31] EPA. (2015). Many Diversified Interests, Inc. Su-perfund Site [Online]. Available: http://www.epa. gov/region6/6sf/pdffiles/mdi-tx.pdf

[32] EPA. (2013b). Public-Sector Land Uses and Super-fund Redevelopment [Online]. Available: http:// www.epa.gov/superfund/programs/recycle/pdf/ vertac-case-study.pdf

[33] EPA. (2010c). Times Beach Site [Online]. Avail-able: http://www.epa.gov/region07/cleanup/ npl_files/mod980685226.pdf

[34] EPA. (2002a) EPA Superfund Record of Decision: Vienna Tetrachloroethene [Online]. Available: http:// www.epa.gov/superfund/sites/rods/full-text/r0302060.pdf

14

Atmospheric Pollutants and the Occurrence of Bromeliads in Electric Power Distribution Network

Gleiciane Fernanda de Carvalho Blanc[1], Bernardo Lipski[1], Juliano José da Silva Santos[1], Karime Dawidziak Piazzetta[1], Juliane de Melo Rodrigues[1], Luciana Leal[2]

[1]Institutos Lactec, Department of Environmental Resources
Centro Politécnico da UFPR. Rodovia BR-116, km 98, n° 8813, Curitiba, Paraná, Brasil, 81531-980
gleiciane.carvalho@lactec.org.br; bernardo.lipski@lactec.org.br; juliano.santos@lactec.org.br;
karime.piazzetta@lactec.org.br; juliane@lactec.org.br
[2]Copel Distribuição S.A.
Rua José Izidoro Biazetto, n° 158, Curitiba, Paraná, Brasil, 81200-240
luciana.leal@copel.com

Abstract - *The occurrence of atmospheric bromeliads (Tillandsia spp. – Bromeliaceae) in the structures of electric power distribution networks is a problem for energy concessionaires. It causes the impression of abandonment and energy distributors do not have enough information to understand the impacts caused by this infestation. A research project is underway to investigate possible factors contributing to bromeliad occurrence in these structures. One stage of the project is to evaluate the correlation between atmospheric pollutants and the occurrence of bromeliads in urban electric network cables. This was made by using principal component analysis to evaluate the correlation between atmospheric pollutants concentration obtained by dispersion modelling and field data of bromeliad abundance. The results indicate a correlation between pollutants and infestation of bromeliads in regions with higher vehicular than industrial emissions in the study area (Ponta Grossa city, Paraná State, Brazil).*

Keywords: Air quality, vehicular emissions, *Tillandsia spp.*, principal component analysis.

1. Introduction

The study of atmospheric pollution became important for researchers and society with the development of large urban centres and the industrial revolution. Since then, air quality has been changing and impacting population health and the environment. One tool used to investigate pollutant behaviour and control and manage air quality is mathematical modelling of atmospheric pollutant dispersion. Models have been widely used in scientific and engineering projects and are often applied to find solutions to environmental problems [1].

The occurrence of atmospheric bromeliads (*Tillandsia spp.* – Bromeliaceae) in electric power distribution network structures is a problem for energy concessionaires. The presence of bromeliads provides a negative visual effect and causes the impression of abandonment and lack of asset maintenance. In addition, energy distributors carry the costs for cleaning infested cables and do not have the information to define best maintenance practices nor to understand the impacts caused by this infestation. Situations such as this have already been reported in Brazil – [2] and [3] – and in other countries – [4] and [5]. However, there still has been no in-depth research on the subject.

Studies developed in the city of Curitiba, State of Paraná, Brazil, related the bromeliad occurrence to pollution from vehicular sources – [3] and [6]. The first study [3] used the atmospheric pollutants concentration obtained by dispersion modelling as basis to comparison with data obtained in 26 points identified in the city with bromeliads presence in electric power distribution cables. The results showed that the most part of the selected points were close to locals with high vehicle traffic and the dispersion modelling presented high pollutants concentration by mobile sources in those points. The study concludes that might have a relationship between vehicle emissions and the occurrence of bromeliads in the energy cables.

The second study [6] focus on the assessment of bromeliad *Tillandsia recurvata* as potencial bioindicator of urban atmospheric pollution. Five sampling point were selected and classified according vehicle traffic. In those points, bromeliads were collected and analysed its abundance and metals accumulation (Fe, Cd, Cr, Cu, Pb and Zn). The results showed that the bromeliad abundance and the metals accumulation is bigger in areas with higher vehicle traffic. The study's conclusion is that the bromeliad has potencial as bioindicator in urban areas, especially related with metals accumulation.

A research project in partnership with the local energy concessionaire has been developed with the objective of investigating possible factors that contribute to bromeliad fixation in energy distribution network structures. One stage of this project, presented here, was to evaluate the correlation between atmospheric pollutants and bromeliad occurrence in electric network structures. Using the same approach as study [3], with atmospheric pollutant dispersion modelling, this study aims to evaluate the correlation with bromeliad occurrence in a different area and with more field data.

2. Theoretical Framework

The presence of atmospheric bromeliads (*Tillandsia spp.* - Bromeliaceae) in aerial electric power distribution networks has been reported in urban areas of several Brazilian cities and in other countries in the South American continent. Although bromeliads appear to exist harmoniously with energy networks, it is assumed that in addition to visual aspects, bromeliads can cause problems related to safety and energy supply quality, as well as hamper the visual inspection of

networks, precluding preventive defect verification by maintenance teams [3].

The relationship between species of the genus *Tillandsia* and energy distribution networks is long known and of wide geographic occurrence, although little studied. Infestations dating back to 1951 in the state of Santa Catarina, Brazil, have been reported [2]. In addition, the establishment and growth of *Tillandsia* spp. have been studied in Argentina [4], Panama [5], Mexico [7] and Costa Rica [8].

The genus *Tillandsia* consists of aerial species capable of inhabiting naked rocks, exposed branches or even dead trees and wooden poles. *Tillandsia* species are usually called weeds and can reach significant degrees of infestation on trees and shrubs in altered environments [9] [10].

Several bromeliad species have been studied as bioindicators of atmospheric pollution, mainly by heavy metals. Some of the studies reported in the literature – [11]-[21] – evaluated *Tillandsia* species as potential bioindicators of air pollution in several countries, such as Colombia, Argentina, Mexico and Bolivia. In Brazil, there are only a few studies that have reported the use of *Tillandsia* species as air pollution bioindicators in urban areas – [22]-[25] and [6].

Tillandsia recurvata is the most common species present in electric power distribution networks and can be found from Chile to Argentina in South America, to the Caribbean and Central America, reaching Mexico and Southeast North American [6]. In North America and Europe, *Tillandsia recurvata* is commonly known as "ball moss" although it does not belong to the group of mosses, or bryophytes [3].

Piazzetta [6] presented and summarized the *Tillandsia recurvata* characteristics that show its potential as an atmospheric pollution bioindicator in urban environments with some degree of air quality alteration (Figure 1); once the species colonizes several substrates, it obtains water and nutrients directly from the atmosphere. *Tillandsia recurvata* roots function only for fixation, guaranteeing their independence from soil.

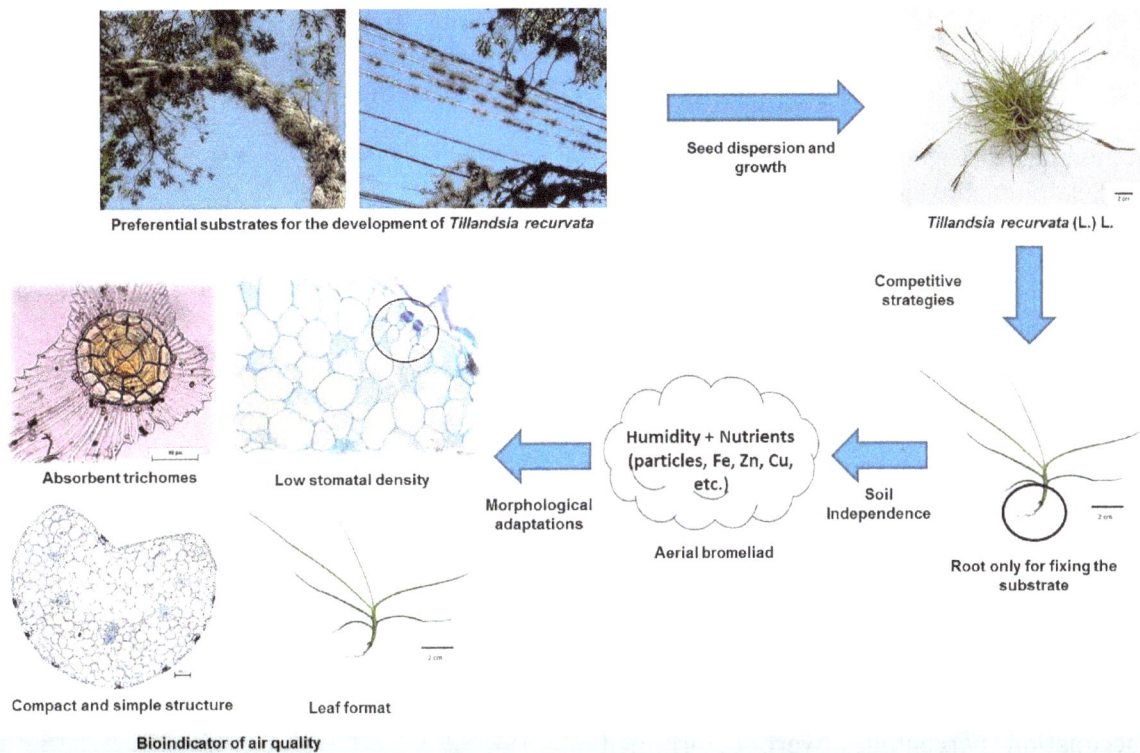

Figure 1. Schematic diagram illustrating *T. recurvata* characteristics that show potential as an atmospheric pollution bioindicator in urban environments with some degree of air quality alteration (adapted from Piazzetta [6]).

3. Methodology

The AERMOD View 9.1 model was used to simulate the atmospheric pollutant dispersion in the study area. AERMOD is a steady-state Gaussian plume model, developed by the American Meteorological Society (AMS) and the U.S. Environmental Protection Agency (EPA) as a formal collaboration group called AMS/EPA Regulatory Model Improvement Committee. The EPA adopted the AERMOD as the preferred regulatory model.

The model is applicable to rural and urban areas, flat and complex terrain, surface and elevated releases, and multiple sources (including, point, area and volume sources). The modelling system consists of one main program (AERMOD) and two pre-processors (AERMET and AERMAP). AERMET uses meteorological measurements, representative of the modelling domain, to compute certain boundary layer parameters used to estimate profiles of wind, turbulence and temperature in the AERMOD. AERMAP is used to create receptor grids and it uses gridded terrain data to calculate a representative terrain-influence height. The AERMOD provides the pollutants' concentration at ground level on each receptor in averaging time and the maximum concentration for the period [26].

3.1. Study Area and Model Data Inputs

The area defined for the study was the urban area of the city of Ponta Grossa (S25°05′42″, W50°09′43″), State of Paraná, Brazil, which presents the highest bromeliad infestation in the state's electric power distribution network.

The AERMOD model requires meteorological, topographic, emission source and receptor data. For meteorological data, five years of hourly data were considered, from January 1, 2010 to December 31, 2014, obtained from an existing automatic meteorological station in the city. The following variables were used: temperature (°C), relative humidity (%), atmospheric pressure (hPa), radiation (W/m^2), precipitation (mm), speed (m/s) and wind direction. Cloud cover data (in tenths) were obtained from the nearest conventional meteorological station with daily data. It was used the daily cloud clover data for all hours of the day.

Topographic data were obtained from contour curves of the Hydrographic Base of the State of Paraná. Contour curves are at 1:50,000 scale with a resolution of 20 metres. In the modelled area, altitude varies from approximately 770 to 1090 metres.

Eighteen industries were considered as fixed sources of pollutant emissions and twenty-one streets with high vehicle circulation as mobile sources. For each source, emission rates were considered for the following pollutants: nitrogen oxides (NOx), carbon monoxide (CO), sulphur dioxide (SO2), total particulate matter (PM) and total hydrocarbons (HCT).

Simulation of atmospheric dispersion in Ponta Grossa was based on a uniform grid of receptors defined by an area of 430 km^2, 20.0 km long and 21.5 km wide, totalling 1804 receptors spaced every 500 metres horizontally and vertically. Sites with bromeliad infestation in the electric network were identified in the field and were used as discrete receptors for pollutant concentration calculations.

3.2. Field Data

Between May and June 2016, bromeliad infestation sites were identified in the study area by dividing the city into squares. These sites were georeferenced and bromeliad abundance was evaluated; occupation percentages were determined according to biomass. The higher the biomass, the greater the percentage. The classes considered were: 0 to 10, 10 to 30, 30 to 50, 50 to 70 and 70 to 100%.

3.3. Statistical Analysis

Principal component analysis was used to evaluate the correlation between field and model results. The data matrix was constructed so that rows represented sites with bromeliad infestation and columns represented the five concentration classes of each pollutant and the classes of bromeliad biomass.

4. Results and Discussion

In the study area, 171 sites were visited and in 99 bromeliads were identified as adhering to electric power distribution network structures. A map with a site infestation index characterizes the percentage of bromeliad biomass occupations (Figure 2).

Pollutant maximum hourly concentrations and average concentrations for the five-year period obtained from modelling the 171 sites are presented in Figure 3, and the stationary and mobile sources are considered separately.

The highest carbon monoxide concentrations are obtained in the city centre and along the highways and avenues of greater flow, the regions strongly influenced by vehicular emissions. For stationary sources, despite modelling considerations including industries with CO

emissions, their contribution is significantly lower (approximately 10%) than emission contributions from vehicular traffic (approximately 90%).

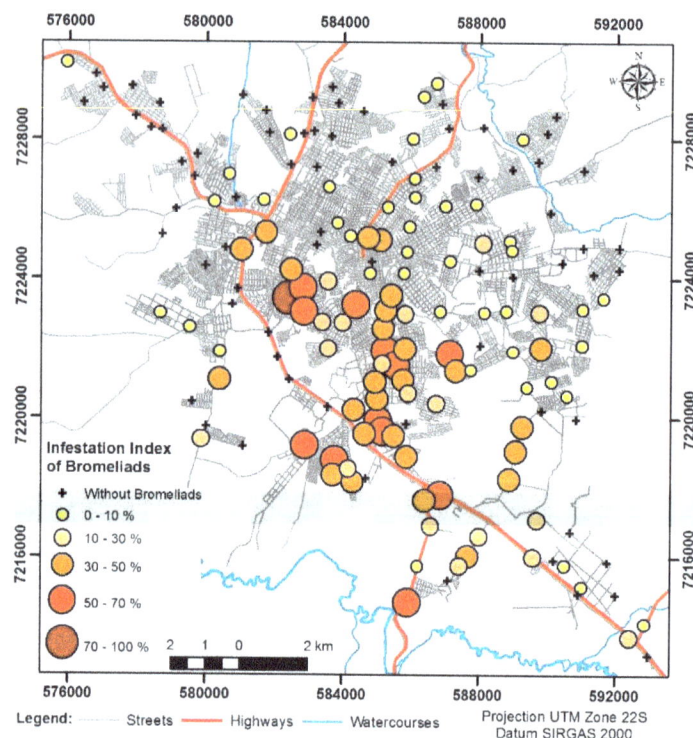

Figure 2. Bromeliad infestation index in the electric power distribution network.

For total hydrocarbons, both stationary and mobile emission sources contribute to the simulated concentrations. As shown in Figure 3, concentrations from fixed sources, mainly metallurgical industries, are concentrated in the area close to their point of emission, while original vehicular emission contributions are evident in central Ponta Grossa where there is a greater flow of vehicles as well as in the areas along highways and avenues with greater flow.

The contribution of stationary sources is also verified in the regions around emission points for the particulate material. Vehicular emissions make no significant contribution to PM concentrations. The concentrations obtained are distributed equally in the central area of the city as well as in the most remote areas and in the regions near the highways. It should be noted that in addition to stationary and mobile sources considered in the modelling, there are diffuse sources, such as resuspension and secondary formation, not addressed in this study, which may explain the low concentrations of this pollutant obtained in the simulations.

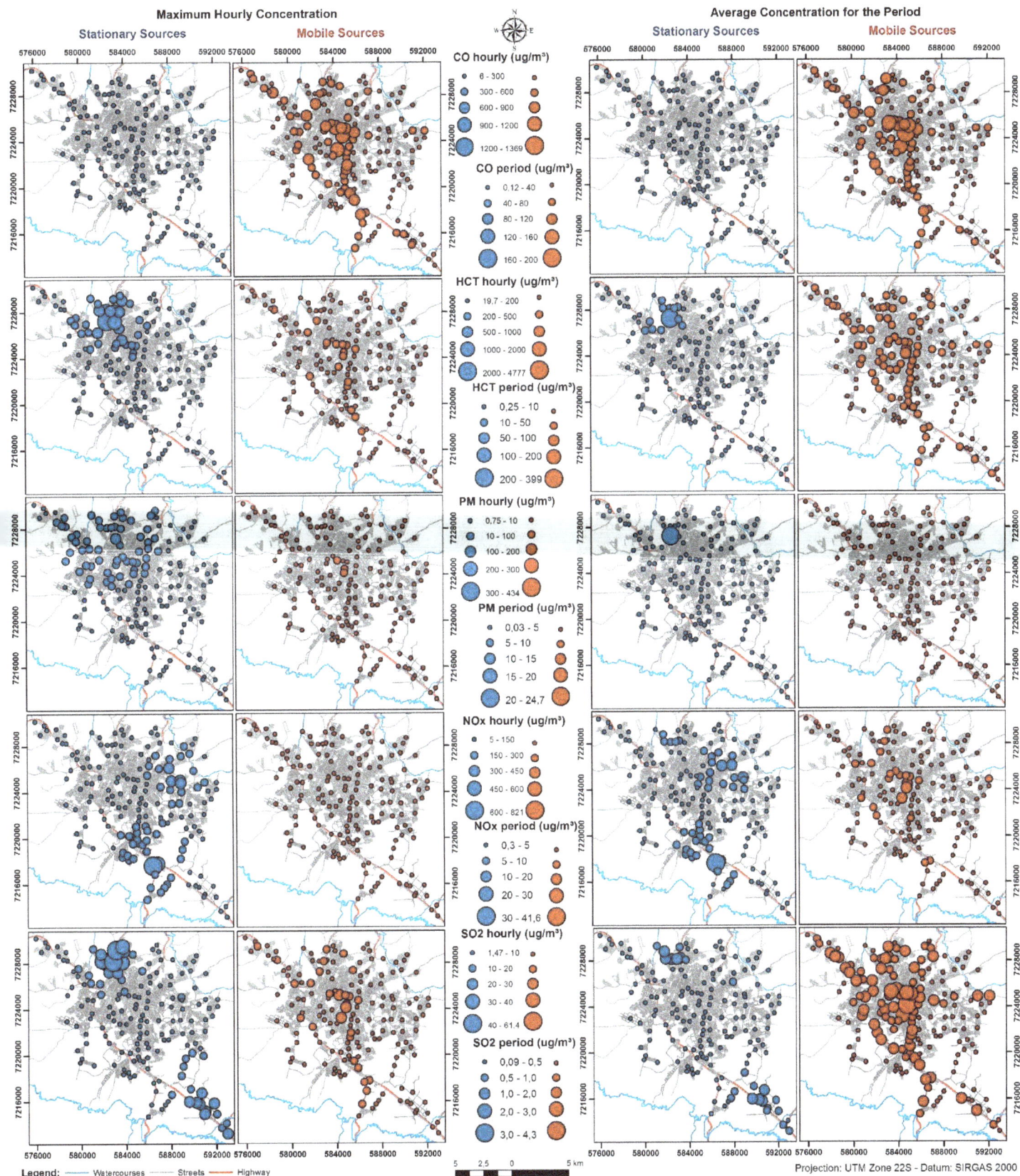

Figure 3. Pollutants maximum hourly concentrations and average concentrations for the five-year period obtained by modelling. This method considers stationary and mobile sources separately at each discrete receptor.

For nitrogen oxides, there is a greater contribution from stationary sources, especially those related to agribusiness and food. The point of highest concentration was located near the emission source. Although there were other fixed emission sources of this pollutant, its concentration at discrete receptors was not significant. Thus, a gradual reduction in NOx concentration is observed with distance from emission sources. For mobile sources, the city's central region registers higher concentrations in the five-year average due to vehicle flow as well as proximity to fixed sources of some regions near highways that cut/surround the city.

Figure 3 shows two different situations for sulphur dioxide. In the hourly maximums, the largest contribution comes from point sources, with the highest concentrations located near points of emission. In the five-year period average, the highest contribution is from mobile sources with the highest concentrations located at city centre points and along the highways and avenues of intense vehicle flow.

The first two principal components obtained for modelled concentrations and bromeliad occurrence explained 60% of the data (Figure 4). In Figure 4, the circle size indicates the correlation obtained and the colours indicate the variable being correlated or grouped. The factor score presents the variables' peaks witch were grouped in principal component. The higher the peaks, the greater the correlation, independent of being positive or negative.

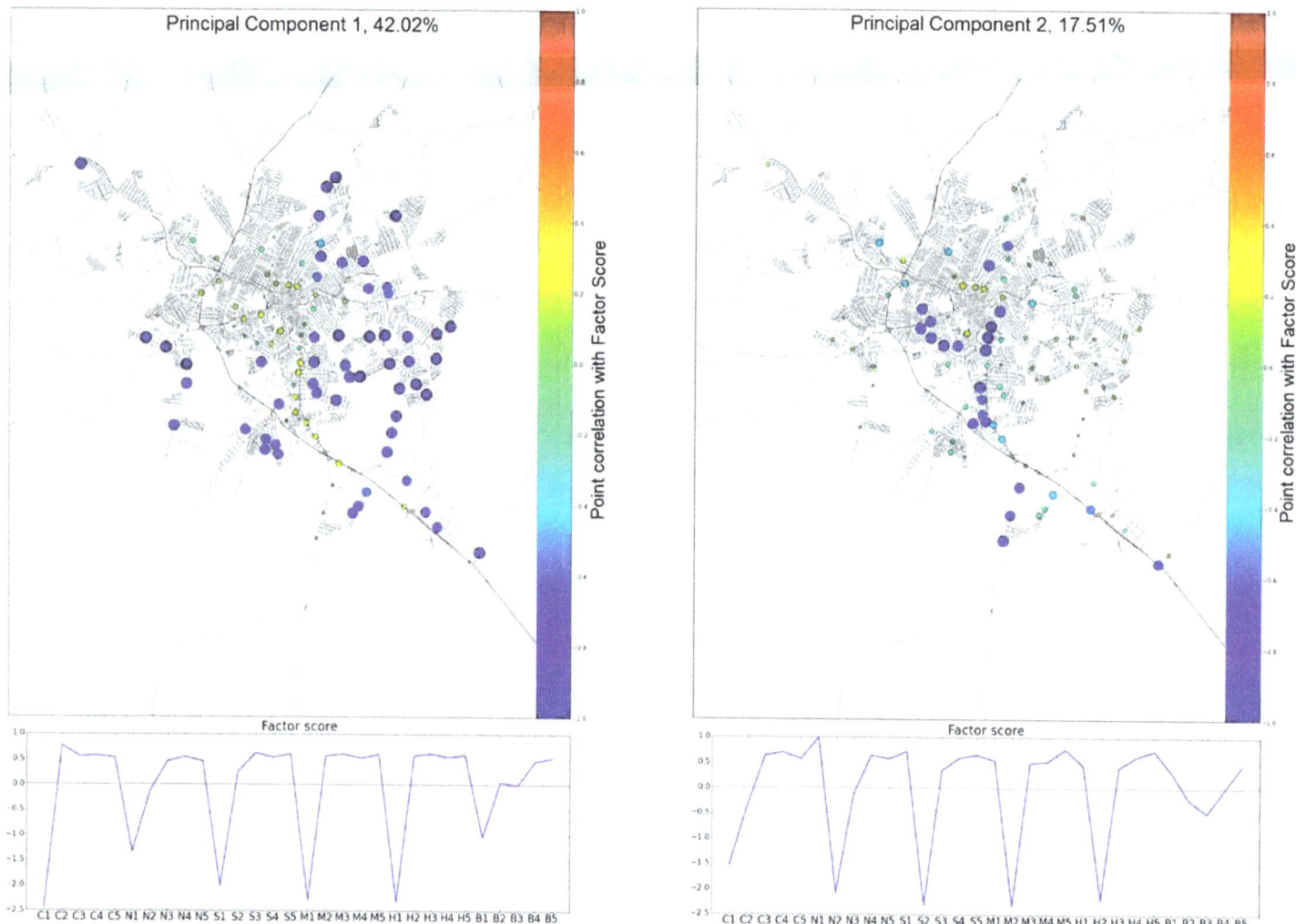

Figure 4. Results of the first two principal components analysis. In this chart, the variable names have been abbreviated for easy viewing. Letters indicate the variable and numbers indicate the class to which it belongs. For example, C1 has C, referring to the pollutant CO, and 1, the first class analysed, being the one of smaller concentrations. Logically, C5 is the class with the highest CO values. Thus, C represents CO, N represents NOx, S represents SO2, M represents PM, H represents HCT and B represents the infestation index of bromeliads.

The first principal component explains more than 40% of the data. The main correlations occurred between the lower classes of pollutants, or those with the lowest concentrations, and the lowest class of bromeliad infestation. These highest correlations are highlighted in the south-central city region in the spatial analysis.

The second principal component accounts for approximately 17% of the data and together with the first, accounts for almost 60% of the variance. This component grouped the first class of CO with the second class of pollutants, NOx, SO2, PM and HCT, and the third class of the bromeliad infestation index.

It should be taking in to account that each principal component represents one variance result, witch means that explain how the data behaves in relation to the total value. The principal component analysis is highly dependent of the data input and the classes adopted. The error on this analysis is associated with the data input and it may diffuse in a multiplicative way until the result. In this study, five classes were used, so any change in these classes might affect the result.

Using spatial analysis major correlations are observed in the main access to the city centre from the highway in the south and in some points in the western region of the city's urban area. This result indicates a correlation between pollutants and bromeliad infestation in a region with greater vehicular emissions than industrial emissions.

If nutrient demand is a factor of extreme importance for the growth of this species, plants present in the distribution network can benefit from the dust raised by vehicle traffic and grow faster. The intense vehicle traffic, especially of trucks, not only releases gases from the combustion but also creates air vortices that circulate taking dust from the ground to the plants. The association of light, water and nutrients, without any of the limiting factors (excess or lack of water, lack of nutrients, shade) being observed, creates the perfect microenvironment for such plants to develop in the electric power distribution network.

5. Conclusion

For the evaluation of the correlation between atmospheric pollutants and bromeliad occurrence in electric network structures was compared field data about the bromeliad infestation and the pollutants concentration by atmospheric dispersion modelling. The results present that there is a correlation between atmospheric pollutants and bromeliad infestation in the electric power distribution network. The correlation is more prominent among the lower classes of pollutants, or those with the lowest concentrations, and the lowest bromeliad infestation class.

In addition, results indicate a correlation between pollutants and bromeliad infestation in regions with greater vehicular emissions than industrial emissions in the city of Ponta Grossa, Paraná, Brazil. These regions are in the city centre and along the highways and avenues of greater flow. However, it is not possible to isolate the vehicle contributions from the industrial ones.

This finding is in agreement with the conclusions obtained by Santos [3], because there is a relationship between vehicle emissions and the occurrence of bromeliads in the electric power distribution network. It is also in agreement with Piazzetta [6]; the presence of *T. recurvata* is indicative of air quality change because it is not significantly impacted by the addition of substances to the atmosphere, but occurs in greater abundance in these areas and accumulates substances proportionally.

Acknowledgements

The authors would like to thank the Brazilian National Electric Energy Agency (ANEEL) and Copel Distribuição S. A. for funding the research and developing the project that includes this work. They also thank the contributions of the other researchers involved in the project.

References

[1] M. El-Harbawi, "Air quality modelling, simulation and computational methods: a review," *Environmental Reviews*, vol. 21, no. 3, pp. 149-179, May 2013.

[2] R. Reitz, *Bromeliads and the endemic malaria-bromeliad*, Itajaí, Brazil: Herbário Barbosa Rodrigues, 1983. (in Portuguese)

[3] J. J. S. Santos, "Study of *Tillandsia spp* (Bromeliaceae) epiphytes in electric power distribution lines of Curitiba," M.S. thesis, Institute of Technology for Development, Curitiba, Brazil, 2014. (in Portuguese)

[4] A. B. Abril and E. H. Bucher, "A comparison of nutrient sources of the epiphyte *Tillandsia capillaris* attached to trees and cables in Cordoba, Argentina," *Journal of Arid Environments*, vol. 73, no. 3, pp. 393-395, Mar. 2009.

[5] S. Wester and G. Zotz, "Growth and survival of *Tillandsia flexuosa* on electrical cables in Panama," *Journal of Tropical Ecology*, vol. 26, no. 1, pp. 123-126, Jan. 2010.

[6] K. D. Piazzetta, "Evaluation of potential of *Tillandsia recurvata* (L.) L., Bromeliaceae, as bioindicator of urban air pollution," (in Portuguese), M.S. thesis, Federal University of Technology, Paraná, Curitiba, Brazil, 2015.

[7] M. E. Puente and Y. Bashan, "The desert epiphyte *Tillandsia recurvata* harbours in nitrogen-fixing bacterium *Pseudomonas stutzeri*," *Canadian Journal of Botany*, vol. 72, pp. 406-408, 1994.

[8] J. Barrat, "Study reveals hazards of the high-wire life for bromeliads," *Bromeliana*, vol. 49, no. 7, pp. 1-4, Oct. 2012.

[9] D. H. Benzing and J. Sheemann, "Nutritional piracy and host decline: a new perspective on the epiphyte-host relationship," *Selbyana*, vol. 2, no. 2/3, pp. 133-148, Sep. 1978.

[10] F. K. Claver, J. R. Alanis and D. O. Caldis, "*Tillandsia spp.*: epiphytic weeds of trees and brushes," *Forest Ecology and Management*, vol. 6 no. 4, pp. 367-372, Oct. 1983.

[11] E. Schrimpff, "Air pollution patterns in two cities of Colombia, S. A. according to trace substances contented of an epiphyte (*Tillandsia recurvata* L.)," *Water, Air, and Soil Pollution*, vol. 21, no. 1, pp. 279-315, Jan. 1984.

[12] G. M. A. Bermudez, J. H. Rodriguez and M. L. Pignata, "Comparison of the air pollution biomonitoring ability of three *Tillandsia* species and the lichen *Ramalina celastri* in Argentina," *Environmental Research*, vol. 109, no. 1, pp. 6-14, Jan. 2009.

[13] A. Z. Garcia, C. M. Coyotzin, A. R. Amaro, D. L. Veneroni, L. C. Martínez and S. G. Iglesias, "Distribution and sources of bioaccumulative air pollutants at Mezquital Valley, Mexico, as reflected by the atmospheric plan *Tillandsia recurvata* L.," *Atmospheric Chemistry and Physics*, vol. 9, no. 17, pp. 6479-6494, Sep. 2009.

[14] M. L. Pignata, G. L. Gudino, E. D. Wannaz, R. R. Plá, C. M. González, H. A. Carreras and L. Orellana, "Atmospheric quality and distribution of heavy metals in Argentina employing *Tillandsia capillaris* as a biomonitor," *Environmental Pollution*, vol. 120, no. 1, pp. 59-68, Nov. 2002.

[15] E. D. Wannaz, H. A. Carreras, C. A. Pérez and M. L. Pignata, "Assessment of heavy metal accumulation in two species of *Tillandsia* in relation to atmospheric emission sources in Argentina," *Science of the Total Environment*, vol, 361, no. 1-3, pp. 267-278, May 2006.

[16] H. A. Carreras, E. D. Wannaz and M. L. Pignata, "Assessment of human health risk related to metals by the use of biomonitors in the province of Córdoba, Argentina," *Environmental Pollution*, vol. 157, no. 1, pp. 117-122, Jan. 2009.

[17] J. H. Rodriguez, S. B. Weller, E. D. Wannaz, A. Klumpp and M. L. Pignata, "Air quality biomonitoring in agricultural areas nearby to urban and industrial emission sources in Córdoba province, Argentina, employing the bioindicator *Tillandsia capillaris*," *Ecological Indicators*, vol. 11, no. 6, pp. 1673-1680, Nov. 2011.

[18] E. D. Wannaz, H. A. Carreras, G. A. Abril and M. L. Pignata, "Maximum values of Ni^{2+}, Cu^{2+}, Pb^{2+} and Zn^{2+} in the biomonitor *Tillandsia capillaris* (Bromeliaceae): Relationship with cell membrane damage," *Environmental and Experimental Botany*, vol. 74, pp. 296-301, Dec. 2011.

[19] E. D. Wannaz, H. A. Carreras, J. H. Rodriguez and M. L. Pignata, "Use of biomonitors for the identification of heavy metals emission sources," *Ecological Indicators*, vol. 20, pp. 163-169, Sep. 2012.

[20] S. Goix, E. Resongles, D. Point, P. Oliva, J. L. Duprey, E. Galvez, L. Ugarte, C. Huayta, J. Prunier, C. Zouiten and J. Gardon, "Transplantation of epiphytic bioaccumulators (*Tillandsia capillaris*) for high special resolution biomonitoring of trace elements and point sources deconvolution in a complex mining/smelting urban context," *Atmospheric Environment*, vol. 80, pp. 330-341, Dec. 2013.

[21] M. A. Martínez-Carrillo, C. Solís, E. Andrade, K. Isaac-Olivé, M. Rocha, G. Murillo, R. I. Beltrán-Hernández and C. A. Lucho-Constanyino, "PIXE analysis of *Tillandsia usneoides* for air pollution studies at an industrial zone in Central Mexico," *Microchemical Journal*, vol. 96, no. 2, pp. 386-390, Nov. 2010.

[22] C. A. Nogueira, "Air pollution evaluation of metals in the metropolitan region of São Paulo using the bromeliad *Tillandsia L. usneoides* as biomonitor," Ph.D. dissertation, Institute of Nuclear and Energetics' Researches, University of São Paulo, São Paulo, Brazil, 2006. (in Portuguese)

[23] A. F. Godoi, R. H. M. Godoi, R. Azevedo and L. T. Maranho, "Pollution and the vegetation density: BTEX in some public areas of Curitiba-PR, Brazil,"

Química Nova, vol. 33, no. 4, pp. 827-833, Mar. 2010. (in Portuguese)

[24] A. M. G. Figueiredo, C. A. Nogueira, M. Saiki, F. M. Milian and M. Domingos, "Assessment of atmospheric metallic pollution in the metropolitan region of São Paulo, Brazil, employing *Tillandsia usneoides* L. as biomonitor," *Environmental Pollution*, vol. 145, no. 1, pp. 279-292, Jan. 2007.

[25] N. A. Vianna, D. Gonçalves, F. Brandão, R. P. Barros, G. M. A. Filho, R. O. Meire, J. P. M. Torres, O. Malm, A. D. O. Júnior and L. R. Andrade, "Assessment of heavy metals in the particulate matter of two Brazilian metropolitan areas by using *Tillandsia usneoides* as atmospheric biomonitor," *Environmental Science and Pollution Research*, vol. 18, no. 3, pp. 416-427, Mar. 2011.

[26] U. S. EPA, (2004, September), AERMOD: Description of Model Formulation. [Online]. Available: https://www3.epa.gov/scram001/7thconf/aermod/aermod_mfd.pdf.

15

Environmental-Friendliness in the Manufacturing Field

Ikuo Tanabe

Nagaoka University of Technology, Department of Mechanical Engineering

1603-1 Kamitomioka, Nagaoka, Niigata, Japan 940-2188

tanabe@mech.nagaokaut.ac.jp

Abstract - In the 21st century, as the importance to manufacture products with environmental-consciousness has been highlighted, manufacturers face a constant challenge to conserve energy, resources and reduce polluting waste. Moreover, in the case of the manufacturing field, large quantities of machine tool lubricating oil, cutting oil, coolant and electricity are used to achieve a smooth drive, precision machining and forced cooling. This represents a significant environmental issue as environmentally-friendly decisions in the technology design, development and production stages are rarely taken. The current research attempted to contribute to the alleviation of the environmental impact of manufacturing. Here, calculation models for a simple environmental impact check and the "Double-ECO model" for technology development were proposed. The proposed methods were then applied and evaluated in a machine tool case that used strong alkaline water mist for forced cooling during machining. It was concluded that; (1) Strong alkaline water mist forced cooling method had a significant cooling performance over the machining-generated heat. (2) the mist was eco-friendly, (3) the calculation models and the Double-ECO model technology can contribute to improve environmental-friendliness.

Keywords: Forced Cooling, Machine Tool, Strong Alkaline Water, Eco-Friendly, Double-ECO Model.

1. Introduction

In the 21st century, the importance to manufacture products in an environmentally-conscious way has been highlighted [1]. In this regard, manufacturers face a constant challenge to conserve energy, resources and reduce polluting waste. Nowadays, there are many researches related to the environmental impact of human activity, as well as countermeasures to reduce it; however, these are still insufficient [2], [3]. Particularly, in the field of manufacturing, most machine tools highly depend on cutting and cooling oils to achieve high accuracy. This represents a large environmental problem, since in most cases the cutting and cooling oils are misused, introduced into the environment and generate undesired pollution [4]. Consequently, the importance of developing new manufacturing ideas that take into account parameters such as high accuracy, high quality and a low environmental impact had been underlined. Hence, manufacturers will be in the need of daring plans, unique ideas and new technologies [5]. Therefore, even though the driving concept of this research was "Ecology", it also included concepts related to mechanical performance such as high precision machining and concepts related to cost management; such as, suitable cost of a machine tool, low running cost and low maintenance fee were included. Along this concepts, the advances on production engineering technology were also considered. For instance, machine tool thermal deformation countermeasures were attempted in order to achieve high accuracy and quality but it was at the expense of using costly equipment and a large quantity of electrical energy that were not enough to satisfy quality parameters [6]. To address this, environmental impact calculations was proposed. Additionally, the "Double-ECO model" was also proposed in order to environmental-friendliness. This model has been previously explored as a method that reconciles "Economy" and "Ecology" during the technology development process [7]. In this paper, both environmental-friendliness improving methods were applied on a milling machine that utilizes strong alkaline water mist to achieve a forced cooling during machining.

2. Environmental-Friendliness Calculation Models and the Double-ECO Model

A model to calculate the environmental impact in the manufacturing field was devised based on the Lifecycle Assessment (LCA). Currently, LCA is one of the major methodologies to calculate the environmental impact of a technology throughout its entire lifespan [8]. However, as the total lifecycle assessment sometimes is rather extensive, manufacturers that have considerable time constraints often neglect LCA considerations. Thus, the current research only considers exhaust carbon dioxide (CO_2) due to electricity and oil consumptions using two calculation models in the form of a simple LCA. The amount of exhaust CO_2 due to electricity consumption was calculated through a relationship between the electricity consumption per hour and CO2 emissions.

The amount of exhaust CO_2 (*EC*) due to electricity consumption EC_E (kg-CO_2) was calculated as follows.

$$EC_E = 0.468 \times W_E \qquad (1)$$

Where, 0.468 is a reference conversion value for kg-CO_2/kWh and W_E is the amount of used electricity (kWh) [9]. The electricity considered was the electricity used by the equipment at all working times.

The amount of exhaust CO_2 (*EC*) due to oil consumption EC_o (kg-CO_2) was calculated as follows.

$$EC_o = (44 \div 12) \times DO \times EHE \times CE \qquad (2)$$

Where, *DO* is the *disposed oil* (kℓ), *EHE* is the emitted heat energy (40.2 GJ / kℓ) and *CE* is the amount of carbon emission (19.22 t-C / TJ) [9].

It was thought that the two calculation models should be used for a simple environmental-friendliness check run during the technology design and development stages. In addition, the Double-ECO model technology was also proposed for promoting environmental-friendliness. The general idea for the Double-ECO model is explained through the following flowchart shown in Fig. 1. Conventional environmentally-friendly technology development stages have environmental protection as first purpose, while their second purpose is cost-profit. Due to this order of priorities, given the different national laws and

policies, such technologies suffered a diminished environemntal-friendliness due to cost.

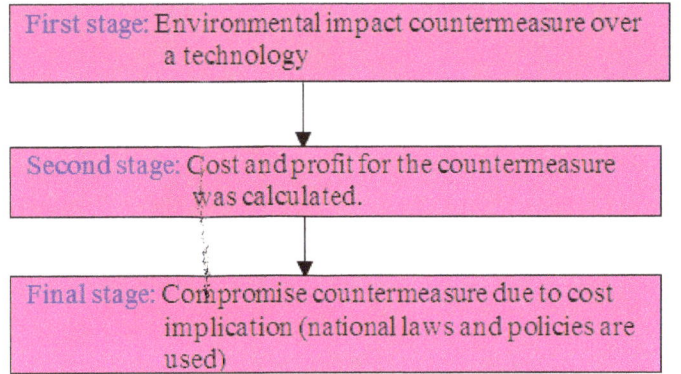

First stage: Environmental impact countermeasure over a technology

Second stage: Cost and profit for the countermeasure was calculated.

Final stage: Compromise countermeasure due to cost implication (national laws and policies are used)

※Final stage differs largely from the first due to compromised countermeasures

(a) Conventional environmentally-friendly technology

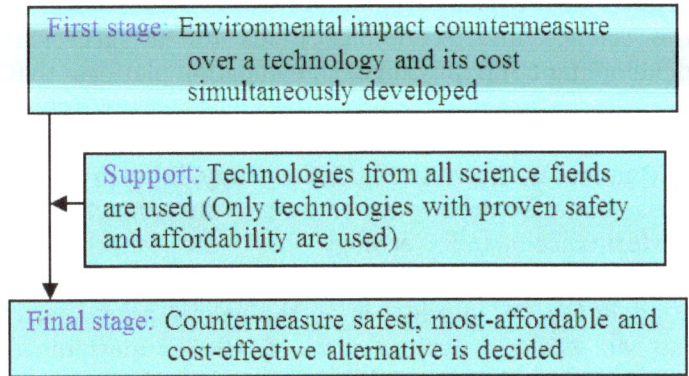

First stage: Environmental impact countermeasure over a technology and its cost simultaneously developed

Support: Technologies from all science fields are used (Only technologies with proven safety and affordability are used)

Final stage: Countermeasure safest, most-affordable and cost-effective alternative is decided

※Final stage do not differ from the first.

(b) Double-ECO model generated technology

Figure 1. Explanation regarding the general idea of the Double-ECO model technologies.

Here, the first purpose would most likely be neglected because of the high cost that they represent. Thus, environmentally-friendly technologies are advancing at a slow pace. In contrast, the "Double-ECO model" technologies prioritize simultaneously environment protection and cost-profit considerations so that no compromises are made at any time. Although it is very difficult to achieve this, recent environmentally-friendly technologies had made remarkable progress. Subsequently, the model proposed solves the technical problems with the incorporation of technology alternatives as countermeasures over a defined case, these countermeasures are selected from all the science fields available so that in the end the safest, most-affordable

and cost-effective alternatives can be selected. At this point, the largest profit that is feasible is guaranteed so the countermeasure is not compromised.

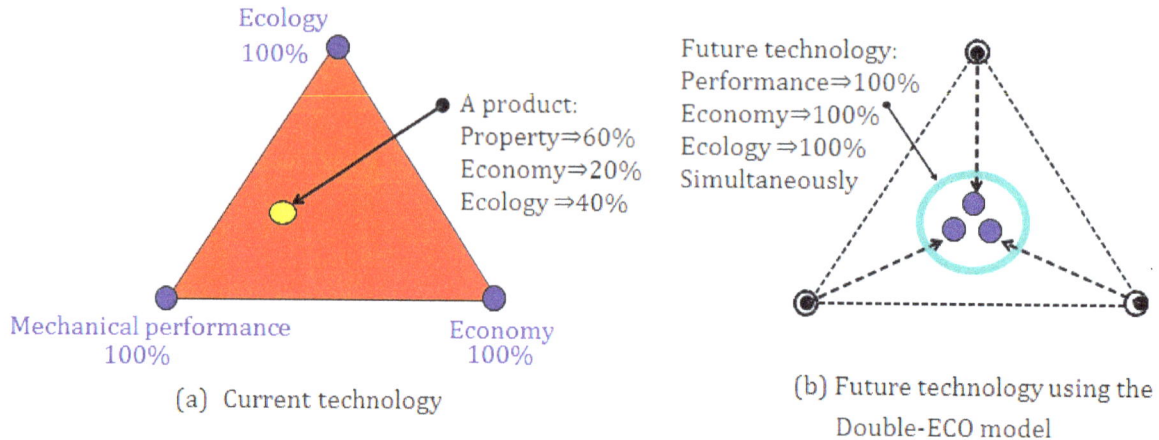

(a) Current technology

(b) Future technology using the Double-ECO model

Figure 2. Depiction of performance evaluation of future technologies based on the Double-ECO model.

In this regard, the Double-ECO model technologies would demand an extensive knowledge of the science and economics fields for a successful development. A depiction of an evaluation platform that includes cost, mechanical performance and ecology as parameters, for future technologies using the Double-ECO model is shown in Fig. 2. Currently, most products are manufactured either on the mechanical performance and ecology or the mechanical performance and economy edges of the evaluation triangle (i.e. "A" product is on the yellow point which has an ecology performance of 40 %, mechanical performance of 60 % and an economy of 40 %). It can be observed that ecology, mechanical performance and economy often clash in the manufacturing field.

However, in the near future, through the Double-ECO model products would optimally be able to reach 100% performances on all parameters. At that stage, all manufactures would willingly become environmentally-friendly. Here, it is necessary to consider that a definition of said parameters is necessary (i.e. what is an environment-friendliness of 0% or 100%) which can be done through polling and statistics-driven weighting of certain parameters such as energy consumption, resulting emissions, toxicity of by-products, risk potential, etc. This particular weighting method has been explored in depth by the BASF Eco-Efficiency Analysis Methodology [10]. Moreover, a definition of the parameters that offers independence from direct weighting that relies on subjective methods (i.e. polling) would certainly be material for further research.

3. Evaluating the Environmental-Friendliness Calculation Models and the Double-ECO Model through an Experimental Case

The two calculation models for checking environmental-friendliness and the Double-ECO model were evaluated in this chapter. Computer numerical control (CNC) milling was performed inside a vessel with strong alkaline water mist in order to investigate the effect of water evaporation in strong alkaline water.

3.1. Corrosion Resistance of Materials in Strong Alkaline Water

The strong alkaline water with pH value above 12.5 has high interfacial permeability, dissolving, emulsification, and separation properties. For these properties, it is well suited for washing, sterilization, corrosion prevention. Moreover, when strong alkaline water is kept for a long time in the air, it loses its alkaline property and becomes normal water (pH 7.0). This remarkable fact has turned alkaline water into a cleaning agent with a diminished environmental impact. In this section, the reactions of the various materials in strong alkaline water are tested. The specification of the device for making alkaline water is shown in Table 1. In the corrosion engineering field, a logarithmic value of metal ion concentration lesser than (-6) means a material does not corrode at equilibrium state; here, according to the corrosion characteristic of strong alkaline water, steel could not corrode in alkaline water with pH above 10 [11]. Similarly, Nickel based alloys shows no chemical reaction in the pH range 8.5~13.0. Titanium alloys also shows no effect under a pH of 13.0.

Table 1. Specification of the system for making strong alkaline water and safety of strong alkaline water.

Generation Method	Closed generation type	Assistant material		Potassium carbonate
Value of pH	pH 12.5	Safety of health	Smell	Nothing
Quantity of generation	10 ℓ/h		Touch	△Wear gloves & glasses
Voltage & Power	100 V & 300W		Inhale	Wear mask
Size	495W×430D×1100H		Drink	×

○ :No problem △ :Avoid × :Prohibition

Table 2. The results of the materials tested in strong alkaline water with pH 12.5 (for two months).

Work piece materials		Condition inside strong alkaline water	Tool materials	Condition inside strong alkaline water
Ti (pure)		○	High speed tool	○
			Carbide(S30T,T725X)	○
Ti6Al4V		○	Cermet (NS530)	○
			Ceramics (LX11)	○
Inconel 718		○	CBN (KBN525)	○
			Diamond (DA2200)	○
Steel (S45C)		○	Coating materials of tool	Condition inside strong alkaline water
Aluminium		×	TiN	○
Copper		Changed to dark brown	TiC	○
			DLC	○
Brass		Changed to dark green	TiAlN	×
			TiAlCr	×

From the aforementioned facts, it was considered that it was possible to operate said materials underwater.

In the experiment, the tested materials were steel, titanium alloy, nickel alloy, copper, aluminium, brass and carbide (tool material) as shown in Table 2. These materials were put in test tubes containing water with three pH values, pH7.0, pH10.0, pH12.5 and kept in a room at a constant temperature of 20±1 ºC and a 60% moisture for two months (Table 2). The alkaline water was changed once a week to keep constant pH values.

The result of the experiment regarding to alkaline resistance of the materials are shown in Table 3. The results showed that there was no corrosion on the tested materials that were kept inside strong alkaline water for two months except for aluminium. From this result, it was confirmed that the underwater cutting process for titanium alloys and nickel alloys can be applicable for a forced tool cooling effect. In the case of aluminium, it was thought that it was necessary to take

appropriate precautions as it corroded in strong alkaline water. Finally, in the case of copper and brass, a change of color occurred. Due to the observed high corrosion resistance of the mentioned materials, it was concluded that strong alkaline water could be used for the forced cooling under the Double-ECO model aided by the two calculation models proposed.

3.2. Cooling Properties of Strong Alkaline Water Mist

The cooling properties of strong alkaline water mist were measured in this section. The experimental set-up is shown in Fig. 3. Here, the nozzle consisted of an air tube (1x7 mm) and a tube (7x7 mm) of strong alkaline water. A sensor for measuring heat transfer coefficient and a manufactured nozzle were set in the centre of the vessel (556×386×310 mm). The sensor for measuring heat transfer coefficient consisted of a ceramic heater (5x5x1.75 mm), two steel plates (5×5×0.06 mm) and 4 thermocouples. The ceramic heater was placed between two steel plates and had a power input E (7.8 W).

At this time, temperature on the steel plate was about 100°C in the air. Two thermo-couples measured temperatures T_{w1} and T_{w2} on the centre of each steel plate. Another two thermocouples measured temperatures T_{M1} and T_{M2} of the strong alkaline water mist at a 5 mm distance from each steel plate. Heat transfer coefficient α was calculated by equation (3).

$$\alpha = \left(\frac{E/2}{A\,(T_{w1} - T_{M1})} + \frac{E/2}{A\,(T_{w2} - T_{M2})} \right) \times 0.5 \qquad (3)$$

Where A (5x5 mm) is the steel plate area. The heat transfer coefficient average on both steel plates was calculated.

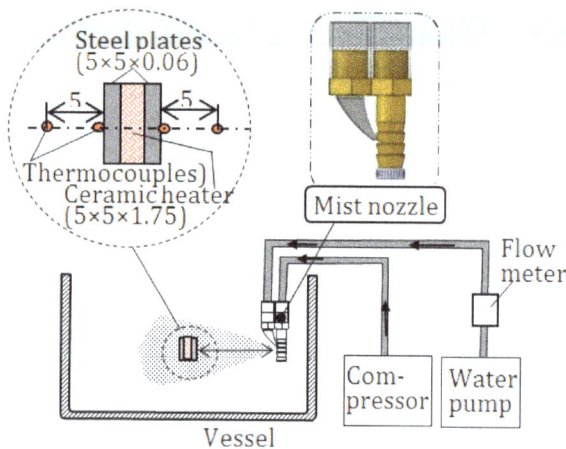

Figure 3. Experimental set-up for measuring the heat transfer coefficient of strong alkaline water mist.

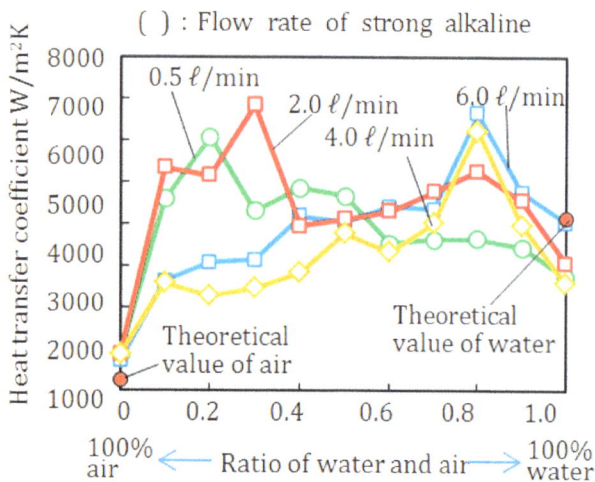

Figure 4. Relationship between the heat transfer coefficient and the ratio of air and strong alkaline water.

The relationship between the heat transfer coefficient and the mixture ratio of air and strong alkaline water is shown in Fig. 4. In this case, the length L from output nozzle to measuring point was 225 mm. The parameter considered here was the total flow rate of strong alkaline water. In this figure it can be observed that one two strong alkaline water stream variations were used; the mist condition (fine strong alkaline water) which has a very large forced cooling effect due to the heat of vaporization, and the other one is the fluid condition (fine air pockets) which has a very large heat transfer coefficient because it presents a high speed. These results clearly show that the forced cooling using strong alkaline water has a significant cooling performance. Here, the heat transfer coefficient measurements should be considered as the apparent heat transfer coefficient as the sensors surface thermal radiation might have played an error factor role. Thus, it can be said that thermal deformation during machining can be effectively cooled down by using a strong alkaline water mist. Additionally, strong alkaline water mist can be used for forced cooling as a technology inside the Double-ECO model and the two proposed calculation models due to its considerable heat transfer coefficient performance.

3.3. Strong Alkaline Water as Cutting Liquid: Cutting Properties during Milling

Cutting fluids or MQL (minimum quantity lubrication) are commonly used for forced cooling during machining processes [12], [13]. However, most cutting fluids pollute the environment and most MQL agents' cooling performance is inferior to other cooling alternatives. In this section, a CNC milling machine was used for experimentation to observe and evaluate the effects of using strong alkaline water mist as cutting liquid. The evaluation consisted in three parameters: tool temperature during cutting, surface roughness and the tool life parameters [14]. In this regard, the entirety of the experimental data and further explanation about it can be consulted in a parallel research [14].

First, the evaluation of tool temperature during cutting was done using the experimental set-up shown in Fig. 5. Here, the cutting area was surrounded by a strong alkaline water and microbubbles mist and two thermocouples placed near the cutting tool tip. The cutting conditions described a middle milling pattern as shown in Table 3. On the other hand, the mist had a 113.3 ℓ /min amount of air and the total strong alkaline water flow rate was 0.82 ℓ /min and the L mm from output nozzle to measuring point was 50 mm.

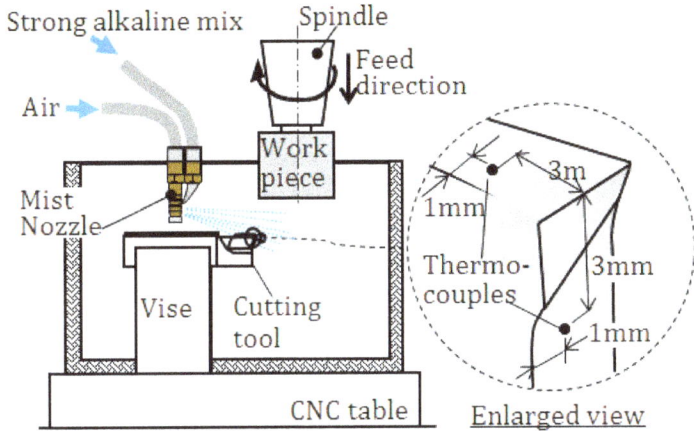

Figure 5. Experimental setup for measurement of tool temperature.

Table 3. Cutting conditions for tool temperature measurement.

Cutting conditions		
Cutting speed 80 m/min	Feed speed 0.25 mm/rev	Depth of cut 0.4 mm
Work piece		
Material : S50C		
Tool (Bite)		
Rake angle: 5°		Coated carbide

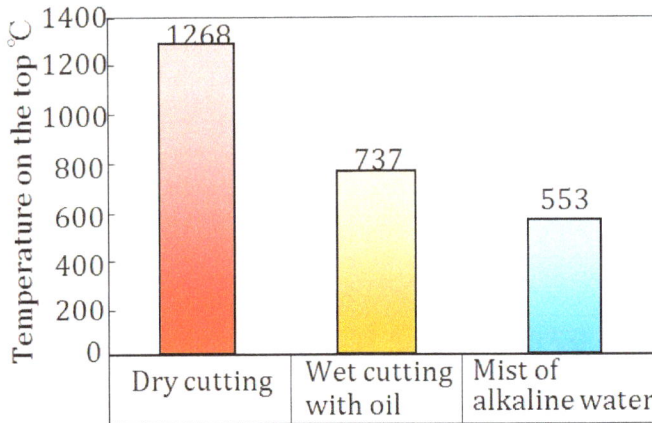

Figure 6. Experimental results for temperature on the top.

In Fig. 5, as conventional milling process would make the use of thermocouples difficult, the workpiece was fixed to the spindle and the tool was fixed to the CNC machine table vise. The tool tip temperature was estimated using the finite element method (FEM). The experimental data obtained through the thermocouple measurements the tool tip allowed a calculation of the tool heat transfer coefficient and the amount of generated heat. As a result, it was possible to obtain an approximation of the tool tip temperature that was considered the representative temperature [11].

Temperatures on the top of the tool are summarized in Fig. 6. Dry cutting and wet cutting with oil are also shown for reference. Material of the used tool was coated carbide and its optimum temperature for cutting is about 800°C. Temperature on the top of the tool using mist of strong alkaline water was 44% that of dry cutting and 75% that of oil wet cutting. Thus, strong alkaline water mist was deemed as an effective method for tool cooling.

Second, the evaluation of tool life during cutting was done using a conventional milling set-up shown in Fig. 7 (Dry cutting and oil wet cutting were also performed for reference). Here, the objective cutting area was surrounded with a strong alkaline water mist. The cutting conditions described a middle milling pattern as shown in Table 4. On the other hand, the mist had a 113.3 ℓ /min amount of air and the total strong alkaline water flow rate was 0.82 ℓ /min and the L mm from output nozzle to measuring point was 50 mm. An End-mill with 2 throw away inserts was used as milling tool due to the simplicity to measure limit of tool life that it presents. The results of the tool life tests are shown in Fig. 8. Here, it was observed that that the tool life of the tool using mist of strong alkaline water was 2.5 times of that of dry cutting and 1.4 times that of oil wet cutting.

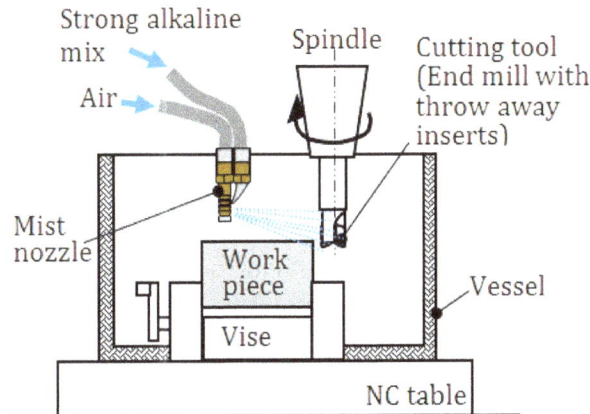

Figure 7. Experimental set-up for measurement of tool life and surface roughness.

Table 4. Cutting conditions for tool life.

Cutting conditions		
Cutting speed 100 m/min	Feed/ tooth 0.15 mm/tooth	Width of cut 3 mm	Depth of cut 2 mm
Work piece		
Material : S50C		
Tool (End mill with 2 throw away tips)		
Rake angle: 5°	Coated carbide	

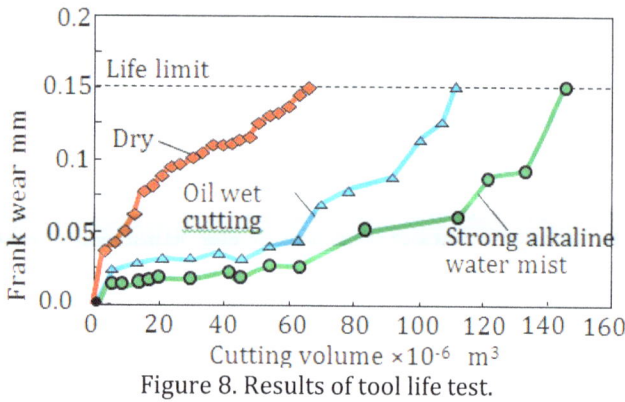

Figure 8. Results of tool life test.

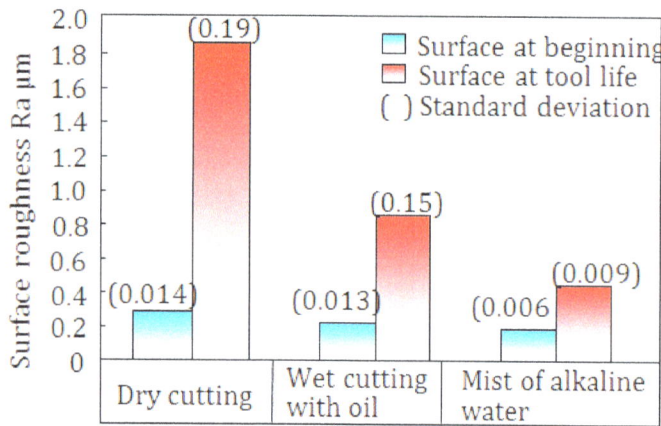

Figure 9. Results of surface roughness.

Hence, it is thought that a mist of strong alkaline water was an effective method for extending the tool life during machining. Similarly, in the case of the cutting using mist of strong alkaline water, defects were not observable in the microscope. Here, surface roughness test results shown in Fig. 9 demonstrate that strong alkaline water mist achieved a better surface finish when compare to existing cutting methods.

3.4. Evaluation and Considerations Regarding Machining using Strong Alkaline Water Mist using the Proposed Models

First, the environmental impact of the proposed cooling technology was assessed based on a comparison of the amount of exhaust CO_2 of the proposed and conventional cooling as shown in Fig. 10. In the case of the conventional method, it was considered that the power used by the coolant pump on the milling machine during conventional wet cutting was 1.2 kW and a working year being composed of 250 days and one 8-hours shift per day. Furthermore, the amount of CO_2 emissions due to electricity consumption, EC_E, was calculated by using equation (1). W_E was the amount of used electricity (kWh) used in the coolant pump and 0.468 the conversion value for kg-CO_2/kWh.

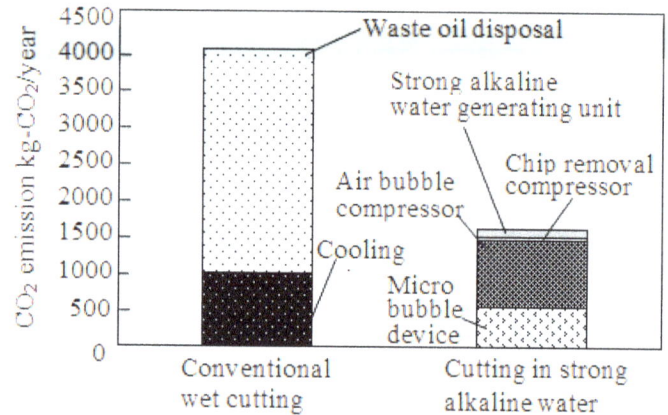

Figure 10. Comparison of CO_2 emissions.

Table 5. Cooling expenses general overview.

Category	Conventional	Proposed
Initial cost	Alkaline water devices ○	Cooling device ○
Running cost	Electricity ×	Electricity × Oil supply ×
Maintenance	Few times ◎	Constant ×
Note: Symbols = ◎ : Cheap, ○: Moderate, ×: Expensive		

Here, the calculated amount of CO_2 emissions from the coolant pump was 1123.2 kg-CO_2. Subsequently, the amount of CO_2 emissions due to oil consumption, EC_o, was calculated by using equation (2). In this case, the amount of oil was assumed to be 340 ℓ and disposal times to be 2 times a year. However, milling machines require a monthly oil fill-up which was assumed to be 30 ℓ a month (30 $\ell \times 12$ months=360 ℓ). Hence, the total amount of disposed oil was 1040 ℓ. Where, the emitted heat energy EHE was 40.2 GJ / kℓ

and the amount of carbon emission CE was 19.22 t-C / TJ. Here, the calculated amount of CO_2 emissions from oil consumption was 2946.3 kg-CO_2. Therefore, the total amount of CO_2 emitted was 4069.5 kg-CO_2.

In the case of the strong alkaline water method, it was considered that the power used by the air compressor, microbubble device and chip removal pump on the proposed cutting were 0.95 kW, 0.56 kW, 0.0132 kW respectively. Here, the usage time was considered to be a working year being composed of 250 days and one 8-hours shift per day. Additionally, the strong alkaline water generating unit had a power of 0.75 kW and a 25 h usage time per year. Furthermore, the amount of CO_2 emissions due to electricity consumption, EC_E, was calculated by using equation (1). Here, the calculated amount of CO_2 emissions from electricity consumption was 1434.6 kg-CO_2 (64.7 % reduction in a year). Here, as no oil consumption was generated the calculations from equation (2) were omitted. Thus, it can be considered that this method is not only effective in cooling the machine tool but also **capable of reducing the impact to the environment.** Finally, a comparison between the expenses involved in the proposed cooling and the conventional cooling during machining is shown in Table 5. It can be noted that the proposal is more affordable, given that the initial costs would be considerably less because of the low market price of alkaline water. Thus, the proposed system goal of simultaneously reaching a "highly cost-effective" and "environmentally-friendly" technology was achieved.

4. Conclusions
1. The Double-ECO Model was used as this model reconciles "Economy" and "Ecology" by exploring technological trends and setting the same degree of priority to these categories and mechanical performance during the technology development process. Here, calculation models for checking environmental-friendliness were also effective in creating an evaluation reference.
2. The application of the Double-ECO Model into real production engineering problems and technology was proposed and presented through the "Forced cooling using mist of strong alkaline water" researches that demonstrate its potential.
3. It was concluded from the experimental results of the proposed mist technology, that improvements in the environmental pollution, mechanical

properties and cost parameters were achieved through the proposed research.

References
[1] I. Tanabe and M. T. Hong "Cutting with an environment-friendly cooling method using water evaporation," *Transactions of Japan Society of Mechanical Engineers*, vol. 67, no. 664, Series C, pp. 4011–4016, 2001. (In Japanese)

[2] e-Gov Japan, *Ordinance related to calculation for carbon dioxide equivalent greenhouse gas emissions with their business activities of specified emitters*, Article 2, 2013. [Online]. Available: http://law.e-gov.go.jp/htmldata/H18/H18F15002002003.html

[3] I. Tanabe, H. M. Truong, K. Yoshii "Turning with environment-friendly cooling method using water evaporation," *Transactions of the Japan Society of Mechanical Engineer*, series C, vol. 66, no. 643, pp. 1026-1030, 2000. (in Japanese)

[4] I. Tanabe and H. M. Truong "Cutting with an environment-friendly cooling method using water evaporation," *Trans. of Japan Society of Mechanical Engineers*, vol. 67, no. 664, series C, pp. 4011–4016, 2001. (In Japanese)

[5] I. Tanabe, K. Yamanaka, J. Mizutani and Y. Yamada "A new design of lathe structure for reducing thermal deformation (Design of zero-center on three directions, self-compulsory cooling and design of thermal synchronism)," *Transactions of Japan Society of Mechanical Engineers*, series C, vol. 65, no. 639, pp. 4508-4513, 1999. (In Japanese)

[6] I. Tanabe, H. S. Ye, T. Iyama and Y. Abe "Control of tool temperature using neural network for machining work-piece with low thermal conductivity," *Transactions of Japan Society of Mechanical Engineers*, series C, vol. 77, no. 776, pp. 1556-1564, 2011. (In Japanese)

[7] I. Tanabe, "Double-ECO model technologies for and environmentally-friendly manufacturing," *Procedia CIRP: 23rd CIRP Conference on Life Cycle Engineering*, vol. 48, 495-501, 2016.

[8] M. Inkbeiner, A. Inaba, et al., "The New International Standards for Life Cycle Assessment: ISO 14040 and ISO 14044," *The International Journal of Life Cycle Assessment*, vol. 11, pp. 80-85, 2006.

[9] Enviroment agency, *Law (enforce No. 3) for Global Warming Countermeasures in Japan-(Exhaust coefficient list)*, 2006.

[10] P. Saling, A. Kicherer, B. Dittrich-Krämer, et al., "Eco-efficiency Analysis by BASF: The Method," *The International Journal of Life Cycle Assessment,* vol. 7, no. 4, pp. 203-218, 2002.

[11] S. Shiddaira, *Material science for corrosion and corrosion resistance.* AGNE technological center, pp. 30-32, 255-257, 287-288, 1995. (In Japanese)

[12] S. Shintani, *Hand book for cutting fluid.* Kogyo Chosakai Publishing Co., Ltd., 2004, pp. 33-48. (In Japanese)

[13] M. Okada, A. Hosokawa, N. Asakawa, Y. Fujita and T. Ueda "Influence of minimum quantity lubrication on tool temperature in end milling of difficult-to-cut materials having low thermal conductivity," *Transactions of Japan Society of Mechanical Engineer,* series C, vol. 78, no. 792, pp. 3093-3103, 2012. (In Japanese)

[14] I. Tanabe "Development of forced cooling using mist of strong alkaline water for restraining thermal deformation on a machine tool," *MM-science Journal,* pp. 521-526, 2014

16

Development and Validation of Noise Maps for the Container Terminals at the Port of Long Beach

I-Hung Khoo, Tang-Hung Nguyen

National Center of Green Technology & Education, College of Engineering,
California State University Long Beach,
1250 Bellflower Blvd, Long Beach, CA 90840, USA
i-hung.khoo@csulb.edu

Abstract- Noise emissions from various transportation modes including seaports have become a major concern to environmental and governmental agencies in recent years due to the impact they have on the community. The Los Angeles-Long Beach port complex is the nation's largest ocean freight hub and its busiest container port complex. As the container sector has the highest growth potential, the levels of noise generated by container traffic and handling activities may present a problem. The purpose of this study is to model the noise of container terminals at the port of Long Beach with the following specific objectives: (1) to determine, using noise mapping, the level of noise generated by the cargo handling and transport activities at the container terminals. A noise model of the port and its surroundings will be created, and validated with field measurements; (2) to assess the noise impact and identify the key noise source in the area; (3) to determine, through field measurements, the noise and activity variations during the period of study. The noise model will be a very valuable tool for the city and port authorities in making planning decisions and to predict future noise impact on the port and its surroundings.

Keywords: Noise mapping, transportation noise, container terminals, seaport.

1. Introduction

Noise is a prevalent pollutant that affects all aspects of life around the globe. Noise can affect health, interrupt activities, and disrupt normal cognitive process [17, 24]. During the past few decades, the mobility of people and goods has increased, and with it the amount of traffic and the environmental noise [1]. So noise emissions from various transportation modes including seaports have become a major concern to environmental and governmental agencies. The European Union leads the rest of the world in recognizing the negative impact of environmental noise and issuing legislature to assess and reduce the noise [5]. As a result of EU directive 2002/49/EC, noise studies have been done at major European cities such as Paris, Brussels, Ireland, and Bologna [9, 16] as well as major seaports such as Hamburg, Copenhagen, and Livorno [15]. Noise studies have also been done at Asian cities such as Chojun, Korea and Tainan, Taiwan [13, 25]. These studies utilize the modern technique of noise mapping to analyze the noise distribution. The United States lags behind the European countries in terms of noise mapping. Currently community noise mapping is not mandated by the U.S. federal or state governments [10] and noise studies in the U.S. are limited to highway and airport noise. The noise prediction models used in these studies are limited in scope and do not include noise from sources other than the infrastructure under study. For example, the Traffic Noise Model (TNM) provided by the Federal Highway Administration (FHWA) can only model traffic noise and not rail or industrial noise [8]. In addition, there is no known noise mapping study of major US seaports. The Los Angeles-Long Beach port complex is the nation's largest ocean freight hub and its busiest container port complex. The combined Ports have had a constant rate of growth every year which exceeded that of the national average and they are designated as an Inter-modal Corridor of Economic Significance [14]. So there is a need for an appropriate noise study to ensure that the noise levels in the port and the surrounding areas do not exceed a reasonable level. This paper is intended to describe a study of the noise from the container related activities at the port of Long Beach by creating a noise model and noise maps of the port. The container terminals were selected as the target of this study due to the fact that 82% of all cargo handling equipment operated at the Port are used at container terminals. The noise map approach used in this study has several advantages over the past noise studies conducted at the Port of Long Beach which were limited to the monitoring of noise levels at a few selected locations or as part of the environmental impact statements [21]. One key advantage is that the noise map approach provides a geographical view of the noise distribution in and around the port areas; this can help the port authorities assess the noise situation in the port and the surroundings. The noise maps can also be used to evaluate the noise impact and identify the key noise source. The results of this study also give an insight into the relative contribution of different types of sources (such as truck traffic, rail traffic and industrial noise) to the overall noise.

As part of the study, field noise and activity data were also collected at different locations around the port near the container terminals. The data were compiled into charts which provide detail insights into the noise and activity variations by hour, day of the week, and month at the different locations. The noise charts are used to supplement the noise maps which show only the annual average values and not the variations. Using the charts, the noise and activity peaks can be identified.

2. Research Methodology

To understand the noise distribution and its impact, the modern approach of noise modeling and mapping is commonly used [2, 13, 23]. Noise mapping is the geographic presentation, via a map, of data related to outdoor sound levels and sound exposure. It takes into account the contribution of all noise sources as well as the effects of obstacles and terrain. Its focus is on the long-term averaged noise and not sporadic intermittent noise. The production of noise maps can be broadly divided into the following steps:

1) Create digital models of the buildings, screens, and topography.
- The ground topography (ground contours and buildings) must be accurately provided since sound propagation is strongly affected by the ground contours and obstacles between sources and receivers.
2) Collect the source power level and characteristics of all noise sources. These can be measured under normal working conditions, or obtained from the manufacturer or noise database. When quantifying the noise level, the unit of dBA is used. This represents the sound level in decibel with an A-weighted filter applied, which correlates well with subjective reactions to noise.

The operation of container terminals involves the following principal noise sources which are the focus of this study:
- **Grantry cranes:** The gantry cranes are used to load and unload containers from the ships. They are equipped with a large electric motor, located at high level, to lift the container up and down.
- Ship generators: The ship generators are large diesel generators that are used to produce the power required for onboard activities when ships are at berth.
- Trucks and trains: These are the two main methods of transporting containers to and from the container terminals.
- RTG cranes: The RTG crane is a mobile crane, equipped with a diesel motor, used for stacking containers within the container yard.
- Forklifts and yard tractors: These are used for moving the containers around the container yard.

Note that there are other sources of noise at the port such as warning sirens on cranes, ship's horns sounded on departure, train crossing warning bells, as well as the impact of containers on other surfaces. The noise levels resulting from these sources, although of concern to residents, do not have much effect on measured average noise levels due to their short duration and intermittent operation and are not included in this study.

3) Collect the operational information of all the noise sources at the port. The information required include the number of each type of noise source that is in operation, their locations, and the period of time that they are active.
4) Calculate the noise levels using noise propagation models to create the noise contours. In this study,

the RLS-90, Schall-03, and ISO 9613-2 standards are used for calculating the road, rail, and industrial noise propagation respectively. Because noise levels can vary significantly over a short period, they are usually described in terms of an average level that has the same acoustical energy as the average of all the time-varying events. This energy-equivalent sound/noise descriptor is called Leq. A common averaging period is hourly, but Leq can also describe any arbitrary duration. The hourly Leq is denoted as dBA Leq(h).

5) Verifying the noise model using field measurements.

The following is the detailed discussion on the steps involved in developing the noise model of the Port.

2.1. The Spatial and Geographical Information for the Port

There are altogether 7 container terminals at the Port of Long Beach spread around different piers (Figure 1):

1. Pier A: SSA
2. Pier C: SSA
3. Pier E: California United Terminals
4. Pier F: Long Beach Container Terminal
5. Pier G: International Transportation Service
6. Pier J: Pacific Container Terminals
7. Pier T: Total Terminals International

Figure 1. Port of Long Beach Terminals (Port of Long Beach 2009).

In this study, the ground topography was obtained from the USGS digital elevation model. The 1/3-arc second National Elevation Dataset (NED) was used,

which has a resolution of approximately 10 meters. The current NED, however, still does not include the elevation of the recent extension to Pier J. So this data needs to be digitized manually.

Next, the features of the port such as roads, rails, buildings were digitized manually using the high resolution 0.6m orthoimagery of the entire port from the USGS. During the digitization process, Google, Bing, and AAA maps were used to identify the names of the roads so that they can be input into the computer model as well. Finally, the heights of the buildings and major structures at the port were entered into the computer model. These were obtained from Google Earth and through field observation.

The complete digitized spatial model of the Port with elevation contour is shown in Figure 2. The upper boundary of model is slightly north of Anaheim St. The left boundary is the edge between the Port of Long Beach and the Port of Los Angeles. The right boundary is the Los Angeles River, although some of the roads and buildings around the Long Beach Marina were included. Note that the roads and buildings around downtown Long Beach were not digitized since the container activities do not extend to that area. The area however is included in the noise propagation simulation.

Figure 2. Digital spatial model of the Port with elevation contour.

2.2. Field Data Collection

The data collected in the field include the average noise level (Leq) and activity information for the trucks, trains, and cargo handling equipment. The annual average value of the measured noise will be used in validating the noise map results. The activity data is needed for compiling the operational information of the noise sources, which will be explained in detail next. The data were also compiled to provide the hourly, daily, and monthly averages. As the noise maps only show the annual average noise, the hourly, daily and monthly noise averages will be useful in understanding the noise variations.

For collecting the field data, representative locations were selected around the Port. Due to restrictions from the terminal operators of the Port, it was not permitted to take the measurements inside the container terminals. So, eight measurement locations were selected outside the terminals but close enough to the various key noise sources such as the trucks, cargo handling equipment, and trains. These eight locations are shown on the map in Figure 3. They include truck entrances, areas next to the container yards where container handling activities occur, and also the railroads. The locations are spread around the Port in order to give a reasonable sampling of the noise level.

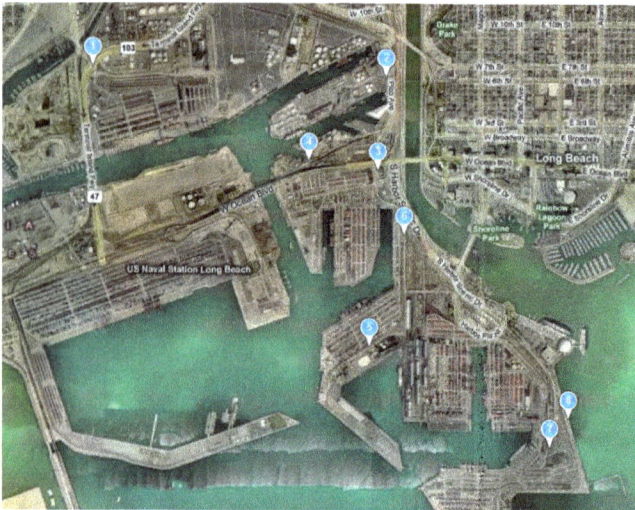

Figure 3. Data collection locations.

The noise and activity data were collected from November 2009 to June 2010. This timeframe covers the peak container activity period around November, as well as the slower periods thereafter; hence, the variations can be studied. The field data was collected for 4 to 6 hours each day during weekdays. Some data were also collected during the weekends. Due to limited number of research personnel, in order to get a good sampling, the research assistant moved around the different locations and took readings of 20-30 minutes each. The measurements were done during daylight

hours between 8am to 4pm with a focus on the peak hours around 8am and 1pm where the activity is the highest.

2.3. Compiling the Sound and Operational Information of the Noise Sources

The volume of noise generated by the container activities is determined by the quantity of each noise source and their noise power and characteristics. The contribution of the noise source to the overall noise distribution then depends on the location of the source and their movement pattern plus the surroundings. So, the noise characteristics and the operational information of the sources need to be obtained and input into the noise model. The noise sources in the port areas can be broadly grouped into two major categories: industrial activities and traffic related activities [18]. The data needed for modeling industrial noise sources include:

- Sound power level and characteristics of every relevant industrial source such as ship generators, cranes, forklifts, yard tractors, etc.
- The number of each source, their location and movements.
- Operating hours of every source for each time period (day, evening, night)

The data required for modeling traffic related noise sources are the following:

- Road traffic data: the type of vehicles (light, medium, heavy, trucks, passenger cars), their number per hour for each time period (day, evening, night), and their average speed.
- Train traffic data: the type of train (cargo, passenger, etc), their number per hour for each time period (day, evening, night), their average speed.
- Location of roads.
- Location of rail tracks.

The sound power level and characteristics for the industrial sources can be obtained by direct sound measurements or by using values from available noise source databases. In this study, due to field access restriction, the noise database SourceDB is used. For the traffic related noise sources, the noise characteristics are already built into the calculation standards.

The remaining operational information such as the number of noise sources, their location and working hours are derived from the field data and data from the port authority [19, 20]. One useful information available from the Port is the air emission study that is conducted annually. The Air Emission Inventory Report contains details of the equipment and vehicles used at the port and their operational information. The Port uses the operational information of the equipment and vehicles to calculate the pollutants emitted; this information can be used for our noise study as well. This information together with the field activity data collected were used in the noise model. Below are the discussions on modeling the noise emissions from trucks, trains, ships, and cargo handling equipment.

Trucks

For the trucks and trains, the calculation standards have built-in assumption on the noise emission characteristics of the vehicles, so only the operational information is required.

The RLS-90 standard is used to calculate the noise from the truck traffic. The standard allows the simulation of road noise by modeling standard vehicle types such as cars and trucks. In this study, we are focusing on the container activities where only trucks are involved in the movements of containers on the road. So cars are excluded in the calculation. The parameters required for the model include the number of trucks per hour and their speed for each road segment used by the trucks in transporting containers in and out of the port. The Emission Report contains the average truck numbers for each of the major roads for different period of the day, including evenings and nights. These are adjusted using the actual truck data collected at the gates. The resulting number of trucks for each pier for different time period is shown in Table 1.

The truck routes are obtained from the field and the data is then compiled for each road segment and entered into the model. The speed of the trucks is obtained from both field estimation and the air emission report.

Table 1. Average number of trucks by time period for each pier (derived from field data and Air Emission Report).

Pier	AM (6-9am)	MD (9am-3pm)	PM (3-7pm)	NT (7pm-6am)
A	237	1528	663	717
C	161	873	407	557
E	600	2183	858	884
F	406	1542	672	514
G	374	2321	975	938
J	163	1023	435	451
T	326	1897	744	514

Trains

The Schall-03 standard is used to calculate the noise from the train activities. The information needed for the calculations are the number of trains per day, the length of the train, and the speed. The standard also takes into account the different noise level for different types of trains such as express, freight, commuter trains etc. Freight train is selected for this calculation. The average number of trains per day carrying containers and the average length of the trains are obtained from the Air Emission Report. The number however is for the twin ports of LA and Long Beach. So, the average container volume for each port is used to divide the train numbers between the two ports. Next, the trains are distributed among the different container terminals depending on their cargo volume. The truck count recorded in the field for each pier, which is a good estimate of the cargo volume, is used for this purpose. Table 2 shows the data for the train activities. (Note that Pier C and E do not have rail activities). The information is entered for each rail segments serving the different terminals. The average speed of the trains is assumed to be 20mph based on field estimation.

It is to be noted that the standard requires the number of trains to be an integer value. So the number is rounded to the nearest integer and the length of the train is adjusted proportionally so that the noise effect remains the same.

Table 2. Average number of trains per day & average number of container ships active per hour for each pier (derived from Emission Report and truck data).

Pier	Average # of trains per day	Average length of train (meters)	# of container ships active per hour
A	3	1744	1.24
C	-	-	0.67
E	-	-	1.74
F	3	1760	1.16
G	3	2648	1.44
J	2	1751	0.83
T	3	2165	1.48

Ships and Cargo Handling Equipment

The ISO 9613-2 standard is used for calculating the noise from the ships and the cargo handling equipment. Since it is a general standard, the noise power and spectrum of the source and its operational information need to be provided. In this study, due to restrictions, it was not possible to measure the noise characteristics of these sources in the field. So, the noise database, SourceDB, is used instead. SourceDB is an industrial noise database which contains the noise characteristics of approximately 1,100 sources in over 70 different industries including those needed for our study. The database was developed for the EU IMAGINE (Improved Methods for the Assessment of the Generic Impact of Noise in the Environment) project and has been used in the EU NoMEPorts (Noise Management in European Ports) noise mapping activity which is similar to our project. The sound characteristics are available for all the sources needed in this study: ships, dockside cranes, RTG cranes, forklifts/sidepicks/top-handlers, yard tractors. The sound spectrum of the equipment is specified in 1/3 octave band from 25Hz to 10kHz. To reduce the calculation time, the forklifts/sidepicks/top-handlers were grouped together due to their similarity. **Because of access restrictions, the activities for the** ships and cargo handling equipment could only be recorded for a few of the piers and the data covered only a part of the overall activities. So, instead the operational information for these sources was derived from the air emission report and then adjusted using the field data.

For the container ships, the emission report lists the total number at berth for the entire port. This number is scaled to the current year's level using the cargo volume statistics and then distributed among the terminals based on the cargo volume of each pier using the observed truck activities as indicator. The emission report also indicates the average time at berth. Multiplying this with the number of ships will give the total hours of operation for the ships. This is then converted to the number of ships active per hour at each pier. The data is shown in Table 2. It can be seen that the number of container ships at berth per hour at each pier is less than 2, which is within the capacity of the piers. On average, 1.4 dockside cranes are needed to load/unload each ship. So the ship data is multiplied by 1.4 to get the number of dockside cranes active per hour at each pier.

The air emission report also lists the make and model of each piece of cargo handling equipment (forklift, RTG crane, side-pick, top handler, yard tractor) in use at each terminal, and their annual hours of operation. Once again, the values are scaled to the current year's level using the cargo volume statistics and then adjusted for each pier using the observed truck activities as indicator for the cargo volume. The

final values are shown in Table 3 as the number of equipment active per hour for each pier for different time period. The same time distribution from the truck data is used here to divide the activities into the different time periods.

To complete the calculation, the locations of the sources need to be specified. Line sources are used to represent the ships and the cargo handling equipment. The location of the ships, dockside cranes, and some yard tractors will be next to the berth. The forklifts, RTG cranes, and yard tractors will be located in the container yard; several line sources are needed depending on the number of rows of containers in the yard.

Table 3. Number of cargo handling equipment active per hour by pier and time period (derived from Emission Report and truck data).

RTG Cranes			
Pier	AM (6-9am)	MD (9am-3pm)	PM (3-7pm)	NT (7pm-6am)
A	2.1	6.8	4.5	1.9
C	0	0	0	0
E	2.7	8.8	5.9	2.4
F	1.3	4.2	2.8	1.1
G	2.0	6.5	4.4	1.8
J	0.7	2.1	1.4	0.6
T	1.9	6.1	4.1	1.7
Forklifts/side-picks/top-handlers			
Pier	AM (6-9am)	MD (9am-3pm)	PM (3-7pm)	NT (7pm-6am)
A	3.6	11.7	7.8	3.2
C	1.5	4.7	3.1	1.3
E	1.9	6.1	4.1	1.7
F	1.3	4.3	2.9	1.2
G	3.2	10.3	6.9	2.8
J	3.0	9.8	6.5	2.7
T	1.9	6.0	4.0	1.6
Yard tractors			
Pier	AM (6-9am)	MD (9am-3pm)	PM (3-7pm)	NT (7pm-6am)
A	15.9	51.2	34.0	14.0
C	5.0	16.1	10.7	4.4
E	10.0	32.1	21.4	8.8
F	7.6	24.2	16.1	6.6
G	14.8	47.6	31.6	13
J	9.5	30.4	20.2	8.3
T	16.0	51.3	34.1	14.0

3. Noise Maps for the Port of Long Beach

With the spatial information of the port and the operational information of the noise sources, the noise model is created using the noise modeling software SoundPLAN from Braunstein & Berndt. The software then calculates the noise over the area of interest using the appropriate noise propagation standards to produce the noise maps. A noise receiver grid spacing of 10 meters was used in the calculation. The overall noise maps of the Port are shown in Figure 4 and Figure 5 respectively for the day period (6am-10pm) and night period (10pm-6am). These time periods are defined as per the RLS90 and Schall03 calculation standards. The noise values displayed on the noise maps correspond to the annual average values. From the noise maps, it can be observed that the highest noise is concentrated along the 710 freeway and the major roads. The container yards also have significant noise.

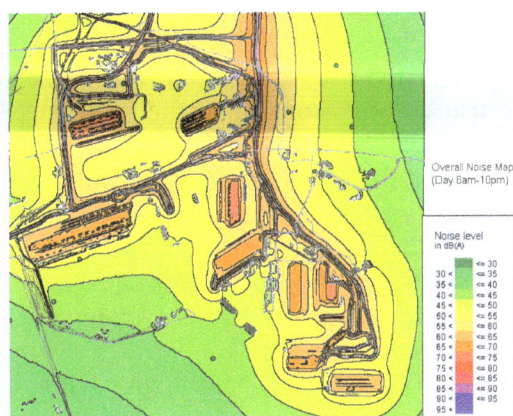

Figure 4. Overall noise map for day period (6am-10pm).

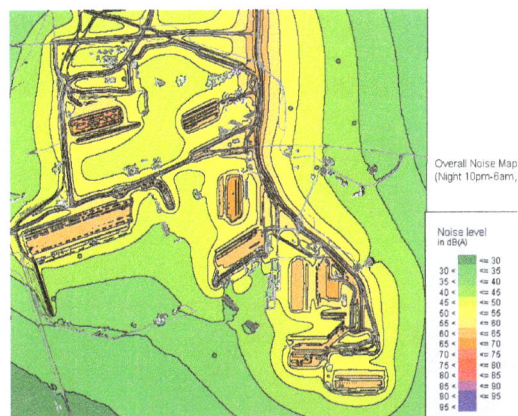

Figure 5. Overall noise map for night period (10pm-6am).

4. Validation of the Noise Model

It is very important when creating the noise model to ensure the reliability and accuracy of the input data such as the noise characteristics of the sources and their operational information. Inaccuracies in the data would result in errors in the noise maps which could have far reaching consequences such as incorrect action plans. In this study, the validation is done by comparing the daytime noise map (Figure 4) with actual field measurements at selected location throughout the Port (as described in section 2.2). It can be assumed that if the daytime noise map is accurate, the night time result is accurate as well since it is generated from the same noise model. The comparison is shown in Table 4. The average difference between the predicted and measured noise for all 8 locations is -2.85 dB with a standard deviation of 3.59 dB.

Table 4. Comparison of noise map values with actual field measurements.

Location	Noise level from noise map (dB) (1)	Average noise level measured in the field (dB) (2)	Difference (dB) = (1) – (2)
1	67.2	71.6	-4.4
2	67.3	70.9	-3.6
3	62.5	72.8	-10.3
4	72.6	71.8	+0.8
5	67.2	68.8	-1.6
6	62.5	65.9	-3.4
7	64.9	66.1	-1.2
8	66.7	65.8	+0.9

Following the procedure outlined in the Good Practice Guide published by the European Commission WG-AEN [7], it is estimated that the uncertainty in the vehicle speed would result in an error of up to 3dB in the calculated noise map results. Additionally, because short term measurements were used to determine the long-term noise levels, an error up to 2dB is expected in the extrapolated measurement values [22]. So, a total error of 5dB is possible in the difference between the noise map value and the measured noise value. Thus, the calculated mean difference of -2.85 dB is well within this expected range.

Looking at the results in Table 4, it can be seen that all the errors are reasonably low except for Location 3. Location 3 is 100 meters south of a truck entrance. The trucks come in from the north and enter the terminal without passing by location 3. So in the noise simulation, the location is not next to any activity. However, the field measurement is picking up the noise from the non-container traffic traveling on the road next to the observation location. This explains why the field noise measurement is so much higher than that predicted by the noise map. This location should not be used for validating the noise map result. Without location 3, the average difference between the predicted and measured noise is -1.78dB with a standard deviation of 2.12dB.

In conclusion, the noise model is reasonably accurate. For validating the results, it is very important to use measurements at locations close to the container related activities and away from any non-container activities that are not included in the noise map simulation.

5. Evaluation of Noise Impact using Noise Maps

The noise maps can be used to evaluate the noise impact of the port container activities on high priority or sensitive areas. Since the Port is located in a predominantly industrial area, there are few sensitive areas in its immediate vicinity. The closest ones are the non-industrial areas across the river, to the east of the Port, and the Queen Mary Hotel next to the cruise terminal. It is to be noted that there is no specific guidelines or regulations on port noise in the United States. The Long Beach Municipal code (Section 8.80) specifies the noise standards for various districts but allows the limit to be raised if the ambient noise already exceeds the existing limits. So it is not suitable for our study. Instead, the Community Noise Exposure guidelines of the Los Angeles City municipal code, the assessment method of the adjacent Port of Los Angeles, will be used here to assess the noise impact [4].

- The area to the east of the LA River is the closest non industrial area to the port. Due to its distance from the Port, the noise level is low as indicated on the noise map in Figure 4. The noise level for the area next to Cesar Chavez park on the eastern edge of the LA River do not exceed 60dB during the day period. This is within the Community Noise Exposure guidelines of the LA municipal code, which consider 50-70 dB to be normally acceptable for playgrounds and parks. During the night period, it is below 55dB. The noise drops steadily further east towards the city. Here the port activity noise will be insignificant compared to the urban city noise.

- The Queen Mary Hotel is situated on the Port next to the cruise terminal. The noise level at the location is well within acceptable limits. Its noise level is 55dB during the day period and 50 dB during the night period. The Community Noise Exposure guidelines of

the LA municipal code consider 60-65 dB to be normally acceptable for multifamily homes.

6. Determination of the Most Significant Source

To analyze which source is dominant, the noise maps are generated for each type of source acting alone, i.e. truck activities only (Figures 6), ships and cargo handling activities only (Figures 7), and train activities only (Figures 8). From these noise maps, it is obvious that the truck movement activity is the highest source of noise, while train activity contributes the least to the overall noise. So if any noise mitigation is required, the focus should be on the truck activities.

The noise impact can also be assessed for each type of activity. The FHWA has standards and regulations related to traffic noise. These are identified as Noise Abatement Criteria (NAC) for a Type 1 federally funded highway improvement project. Although these standards are not directly applicable to this study since it is not a Type 1 funded highway improvement project, they are still useful for evaluating the noise impacts from the truck traffic. Looking at the noise map for the truck activities only, it can be seen that the noise is concentrated on the roads and radiates outwards. It can be observed from the noise map that the noise level is within the 71dB limit for developed land 50 feet away from the major roads (not counting the Freeway), following the Caltrans/FHWA Noise Abatement Criteria for Category C activities [3].

The noise from the cargo handling activities can also be evaluated using the LA municipal code for industrial equipment noise. The code stipulates a limit of 75dB 50 feet away. Using the noise map for the cargo handling activities, it can be observed that the noise is well within the limit, 50 feet away from container yards.

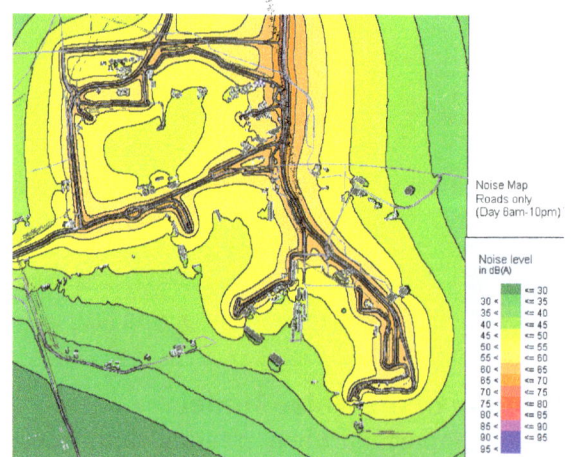

Figure 6. Noise map with only truck traffic.

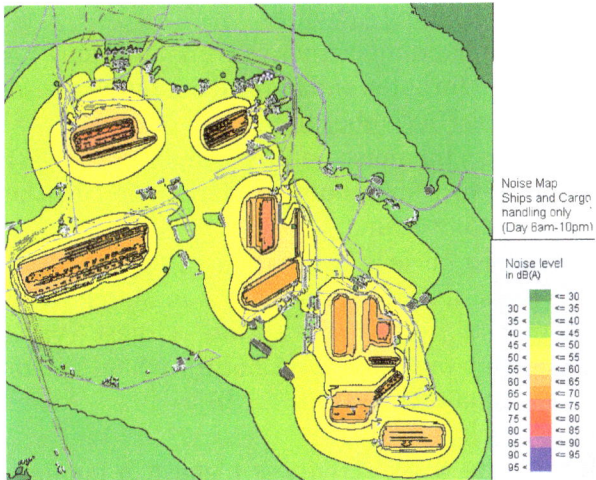

Figure 7. Noise map with only ships and cargo handling activities.

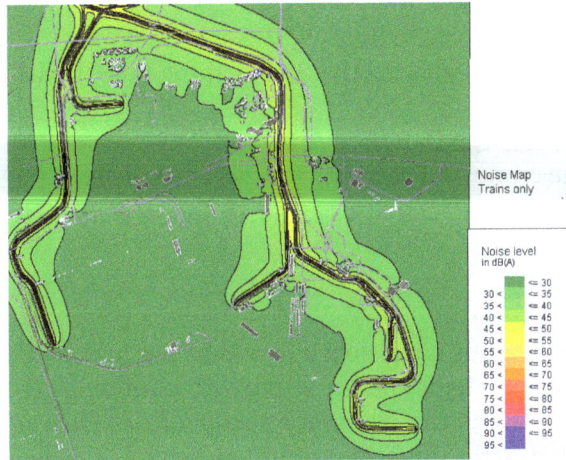

Figure 8. Noise map with only train activities

7. Analysis of the Noise and Activity Variations

The noise maps only display the annual average noise values. To help understand the noise variations, the field data can be used. As mentioned earlier, the noise and activity data were collected at 8 different locations around the Port (see Figure 3). The activity data include the truck, rail, crane, forklift, and yard tractor activities. The truck activity is quantified as the number of truck movements per hour. The rail activity is quantified as the fraction of the hour that the rail is active. The crane, forklift, yard tractor activities are quantified as the number of equipment that is active per hour. Note that the forklifts, side-picks, and top-handlers are grouped together as forklifts. The field data were compiled to provide the hourly, daily, and monthly averages for the noise levels and amount of activities for each location. The detailed analysis of the

data is summarized below. A few of the key charts are also presented.

Hourly noise

- The average hourly noise measured at each location is shown in Figure 9. On average, the noise peaks around 8am (70.3dB) and tapers off after that to a minimum of 67.8dB around noon. It then peaks again around 1pm (70.3dB) and 2pm (70.4dB), and tapers off again after that. This is consistent with the operating characteristics of container terminals at the port.

- The highest noise is at Location 1 around 2pm (75.8dB) and lowest noise is at location 6 around 3pm (57.6dB). Location 6 experienced the largest variation throughout the day (11.6dB), while Locations 2 and 3 have the smaller variations, 2.9dB and 2.8dB respectively.

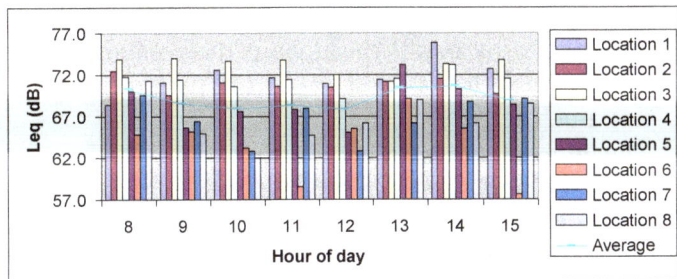

Figure 9. Hourly noise variation for each location.

Hourly Truck Activities

- The average number of truck movements observed at each location per hour is shown in Figure 10. On average, the PM truck activity is higher than the AM truck activity. The peak truck activity is around 1pm. The lowest truck activity is around noon.

- Location 7 has the highest truck activity around 3pm, followed by Location 5 around 1pm. These are truck entrances.

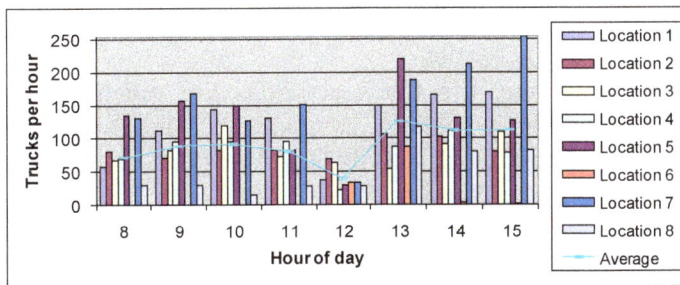

Figure 10. Hourly truck activity for each location.

Hourly Train Activities

- The rail is most active around 1pm at Location 1. Overall, Location 1 also has the most rail activities. This is consistent with the fact that Location 1 is at the beginning of the rail lines that serve most of the port.

Hourly Cargo Handling Activities

- Overall, the cargo handling activities are the highest around 9am and lowest around noon. The AM period has slightly more activities compared to the PM period.
- The highest crane and yard tractor activities are at Location 6, around 9am. The highest forklift activity is at Location 5, also around 9am.

Daily Noise

- The average daily noise at each location is shown in Figure 11. On average, the noise is very much higher during the weekdays compared to the weekends. The noise peaks slightly on Wednesday (71.8dB), but varies only by 0.8dB throughout the weekdays. The lowest noise is on Sunday (64.1dB) where there is not much activity.
- The highest noise is at Location 1 on Wednesday (75.7dB) and the lowest noise on a weekday is at location 8 on Monday (66.2dB). Location 5 experienced the largest variation throughout the weekdays (2.6dB), while Location 2 has the smallest variations, 1dB.

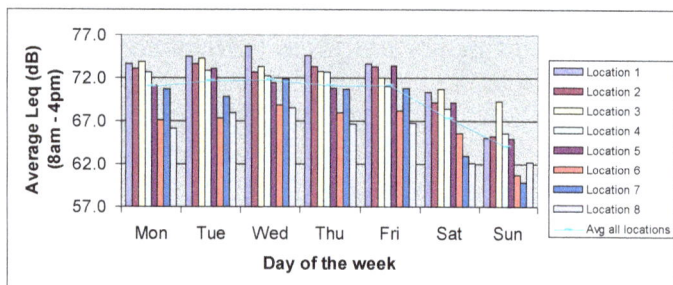

Figure 11. Noise variation by day of the week for each location.

Daily Truck Activities

- On average, the truck activities are much higher during the weekdays compared to the weekends. The highest truck activity is on Friday and the lowest activity is on Sunday.
- Location 7 on Friday has the highest truck activity.

Daily Train Activities

- The rail is most active on Thursday at Location 1. This is followed by Location 2 on Friday.

Daily Cargo Handling Activities

- The cargo handling activities peak on Friday for the cranes and forklifts, and bottom out on Wednesday.

Monthly Noise (December to June)

- On average, the noise peaks in January (72.4dB) and drops off to a minimum in March (66.1dB) before rising steadily again.
- The highest noise is at Location 1 in January (76.9dB) and lowest noise is at location 7 in March (60.6dB). Location 5 experienced the largest variation throughout the months (12.6dB), while Location 3 has the smallest variation, 4.1dB.

Monthly Truck Activities

- Overall, the truck activity is highest in January and lowest in March.
- Location 7 has the highest truck activity in January and February. Next is Location 5 in January.

Monthly Train Activities

- The rail is most active in April for both Locations 1 and 2.

Monthly Cargo Handling Activities

- The crane and yard tractor activities peak in January.
- The forklift activities peak in January as well as in June.

8. Conclusion

The Port of Long Beach is one of the major nodes in the logistic chain and an important economic center in the region. As the container sector of the Port of Long Beach has the highest growth potential, the levels of noise generated by cargo transportation and handling activities are especially of interest. In this pilot study, the noise distribution at the container terminals at the Port was modeled by means of noise mapping. The noise maps generated present the noise distribution in and around the port areas and give an insight into the relative contribution of different groups of sources (e.g. road traffic, rail traffic and industrial noise). It was determined that the truck movements were the main contributor of container activities noise in the Port, followed by cargo handling, and then rail. The noise maps were also used to evaluate the noise impact on sensitive non-industrial areas near the port, as well as the relative impact of specific type of sources. It was found that the noise did not exceed the relevant guidelines. To supplement the noise maps, the field data collected in this study were compiled to provide the hourly, daily, monthly noise and activity variations. It was observed that the average noise was highest around 8am and 1pm and lowest around noon. The noise levels during the weekdays were also very much higher than the weekends. During the period of study (December – June), it was observed that the noise peaked in January and dropped off to a minimum in March before rising steadily again.

Noise mapping is a very valuable tool allowing port authorities not only to assess the current noise situation in the port, but also to examine the potential impact of future development plans. If noise abatement measures are recommended, the noise model can then simulate the new scenario to see if the desired reduction is achieved. All these will be part of the next phase of our research on port noise and the environment.

Acknowledgement

This work was funded by the California Department of Transportation through METRANS. The authors wish to thank the port authorities of the Port of Long Beach for their invaluable assistance in this research project. Special thanks to Mr. Rick Cameron, Director of Environmental Planning, Mr. Don Snyder, Director of Trade Relations, and Mr. Mitch Poryazov, Assistant Terminal Services Manager. Finally, the authors wish to acknowledge the efforts of student assistants Kiran Rajanna, Dane Christensen, Eduardo Delgado, Benjamin Solinsky.

References

[1] Baaj, M., El-Fadel, M., Shazbak, S., & Saliby, E. (2001). "Modeling noise at elevated highways in urban areas: a practical application" *ASCE Journal of Urban Planning and Development,* pp169-180.

[2] Bourbon, C., Noel, P., Mummenthey, R. (2001) "Brussels Life Project: Noise Mapping As a Tool for Management and Planning Road Traffic Noise in Urban Area" *Proceedings of the International Congress and Exhibition on Noise Control Engineering (INTERNOISE 2001).*

[3] Caltrans. (2006). "Traffic Noise Analysis Protocol for New Highway Construction and Reconstruction Projects".

[4] City of Los Angeles. (2006). "City of Los Angeles CEQA Thresholds Guide".

[5] European Commission. (2002). "Directive 2002/49/EC of the European Parliament and of the council of 25 June 2002 relating to the assessment and management of environmental noise" *Official Journal of the European Communities.*

[6] European Commission. (2003). "2003/613/EC Commission recommendations of 6 August 2003 concerning the guidelines on the revised interim computation methods for industrial noise, aircraft noise, road traffic noise and railway noise, and related emission data" *Official Journal of the European Communities.*

[7] European Commission Working Group - Assessment of Exposure to Noise (WG-AEN). (2006). *Good practice guide for strategic noise mapping and the production of associated data on noise exposure.*

[8] Federal Highway Administration. (2004). *Federal Highway Administration traffic noise model user's guide version 2.5.* Washington D.C.

[9] Garai, M., & Fattori, D. (2009). "Strategic Noise Mapping of the Agglomeration of Bologna, Italy". *Urban Transport XV. Urban Transport and the Environment,* pp519-528.

[10] Kaliski, K., Duncan, E., & James, C. (2007). "Community and regional noise mapping in the United States", *Sound and Vibration,* pp 14-17.

[11] Khoo, I., Nguyen, T. (2011). "A Preliminary Study of Noise at the Port of Long Beach", *METRANS National Urban Freight Conference (NUF).*

[12] Khoo, I., Nguyen, T. (2011). "Developing a Noise Model for Container Terminals at the Port of Long Beach", *2011 Transportation Research Forum.*

[13] Ko, J.H., Chang, S. I., Lee, B.C. (2011). "Noise impact assessment by utilizing noise map and GIS: A case study in the city of Chungju, Republic of Korea", *Applied Acoustics,* pp 544-550.

[14] Lea, J. and Harvey, J.T. (2004). "Data Mining of the Caltrans Pavement Management System (PMS) Database", Technical Report to California Department of Transportation, University of California at Berkeley, CA.

[15] Morretta, M., Iacoponi, A., & Dolinich, F. (2008). "The port of Livorno noise mapping experience". *Journal of the Acoustical Society of America,* 123(5), 3137.

[16] Murphy, E., King, E. A., & Rice, H. J. (2009). "Estimating human exposure to transport noise in cen-

tral Dublin, Ireland". *Environment International,* 35(2), 298-302.

[17] Niemann H., Maschke C. (2004). "WHO report on noise effects and morbidity" *World Health Organization.*

[18] Noise Management in European Ports (2008). "Good Practice Guide on Port Area Noise Mapping and Management".

[19] Port of Long Beach (2005). "Port of Long Beach Air Emissions Inventory – 2005"

[20] Port of Long Beach (2008). "Port of Long Beach Air Emissions Inventory – 2008"

[21] Port of Long Beach (2009). "Middle Harbor Redevelopment Project Final EIS/EIR".

[22] Romeu, J., Genesca, M., Pamies, T., Jimenez, S. (2011). "Street categorization for the estimation of day levels using short-term measurements" Applied Acoustics, pp 569-477.

[23] Stapelfeldt, H., Jellyman, A. (2001). "Noise Mapping in Large Urban Areas" *Proceedings of the International Congress and Exhibition on Noise Control Engineering (INTERNOISE 2001).*

[24] Sust, C., & Lazarus, H. (2003). "Signal perception during performance of an activity under the influence of noise" *Noise and Health Journal, 6,* pp 51-62.

[25] Tsai, K. T., Lin, M. D., & Chen, Y. H. (2009). "Noise mapping in urban environments: A Taiwan study". *Applied Acoustics,* 70(7), 964-972.

17

Analysis of Air Purification in a Woodland by Field Observation and Wind Tunnel Experiment

Yoichi Ichikawa[1], Syunsuke Mukai[1], Masahiro Nishimoto[2], Hideaki Mouri[3], Akihiro Hori[4]
[1]Ryukoku University, Faculty of Science and Technology
1-5 Yokotani, Setae-cho, Otu, Shiga 520-2194, Japan
ichikawa@rins.ryukoku.ac.jp
[2]Polytech Add, Inc.
RBM Tsukigi Square 3F, 1-18-8 Shintomi, Chuo-ku, Tokyo, 104-0041, Japan
nishimoto@polyadd.co.jp
[3]Meteorological Research Institute
1-1 Nagamine, Tsukuba, Ibaraki 305-0052, Japan
hmouri@mri-jma.go.jp
[4]Meteorological and Environmental Sensing Technology Inc.
1-1 Nagamine, Tsukuba, Ibaraki 305-0052, Japan
fudo@mri-jma.go.jp

Abstract- *As part of a comprehensive study on the harmonious coexistence of humans and nature focusing on woodland, we evaluated the role of the Ryukoku forest in air purification. The Ryukoku forest, located at Seta Hill in Shiga Prefecture, Japan, is composed of a mixture of conifer and deciduous trees. There are two mechanisms by which forest air is purified: the deposition of air pollutants onto leaf surfaces and the effect of a tree wind break (air pollutant break). We measured vertical distributions of the nitrogen dioxide (NO₂) concentration inside and outside the Ryukoku forest using passive samplers attached to a 25-m-tall observation tower and to neighbouring university campus buildings, respectively. Wind tunnel experiments that treat atmospheric dispersion among model trees and cannot consider the effect of deposition were conducted to evaluate the effect of a tree wind break. The observed NO₂ concentration at the forest floor showed a 30 percent decrease compared with that above the canopy top. The results of the wind tunnel experiment for four parallel rows of model trees showed a decrease in the concentration of air pollutants inside the forest comparable to the observation in the Ryukoku forest. However, the decrease in the concentration of air pollutants from the top of the canopy to the forest floor was not observed in the wind tunnel experiment. It is deduced that the low NO₂ concentration inside the forest compared with that outside the forest is mainly due to the tree wind break effect and that the variation in the NO₂ concentration with the height from the canopy top is mainly due to the deposition of air pollutants onto leaf surfaces.*

Keywords: Air purification, Forest, Wind tunnel experiment, Nitrogen dioxide.

1. Introduction

It is well known that forests contribute to improving the atmospheric environment because they play a crucial role in air purification by the uptake and deposition of air pollutants. Nowak et al. [1] revealed that trees and forests remove substantial amounts of pollution and can produce substantial health benefits, particularly within urban areas. Field observation studies of forest air pollutants can be found in the literature. For example, Fontan et al. [2] measured vertical profiles of the ozone concentration in a pine forest in the Landes region of southwest France. A positive gradient of the ozone concentration at the vegetation level and at the ground level was observed in relation to a temperature inversion. Lamaud et al. [3] measured ozone fluxes at the same site as Fontan et al. and found that the ozone uptake by the understorey of short vegetation accounts for a significant portion of ozone deposition. Cox et al. [4] measured vertical ozone profiles in a 15.7-m-high forest canopy at the University of New Brunswick woodlot in Fredericton, Canada, and found a 50 percent reduction in the ozone concentration from the canopy top to the breast height. Krzyzanowski [5] conducted a similar observation in a 13-m-high forest canopy in the Lower Fraser Valley, British Columbia, Canada, and found a 31 to 47 percent reduction in the ozone concentration from the top of the canopy to the forest floor. Utiyama et al. [6] found in a 15-m-high canopy of a red pine forest situated in central Japan that the ozone concentration decreased by 35 percent from the top of canopy to the forest floor in autumn, during which no ozone formation was found to occur in the trunk space. Launiainen et al. [7] measured ozone fluxes above and within a boreal Scots pine forest in Southern Finland, where the mean canopy height was 15 – 16 m, and discussed ozone removal by the canopy and forest floor. Savi et al. [8] found, in a 14–m–high evergreen Holm oak forest located within the Castelporziano Estate, Italy, that the forest canopy was a net sink of ozone.

It is important to accumulate observation data at various forest sites while considering vegetation and meteorological conditions. In this study, as part of a comprehensive study on the harmonious coexistence of humans and nature focusing on woodland, we measured the distributions of the nitrogen dioxide (NO_2) concentration inside and outside the Ryukoku forest located at Seta Hill in Shiga Prefecture, Japan. Ozone is sometimes photochemically formed in the trunk space owing to the emission of hydrocarbons from trees [6]. Since NO_2 is much less reactive than ozone, the observation of NO_2 is useful for studying removal mechanism of air pollutants in forests.

There are two mechanisms by which forest air is purified: the deposition of air pollutants onto leaf surfaces and the effect of a tree wind break (air pollutant break). A wind tunnel experiment to determine the effect of the canopy layer on flow turbulence was conducted by Pietri et al. [9]. They used conical model trees made of metal and fine green foam, and quantitatively obtained the wind break effect in the case where the ratios of the tree spacing to tree height were 1, 1.5 and 2. Gromke et al. [10, 11] investigated the effect of trees on the dispersion of pollutants by means of a wind tunnel experiment and a computational fluid dynamics simulation. They placed one or two rows of a fibrous wadding material in street canyon models to simulate avenues of trees. A tracer gas was emitted in the street canyon or between the model trees. Al-Dabbous et al. [12] conducted field measurements to investigate the influence of roadside vegetation barriers on the movement of air pollutants. Janhäll [13] reviewed the

influence of urban vegetation on the deposition and dispersion of air pollutants and introduced some studies on vegetation barriers in street canyons. However, no studies on the effect of trees on the dispersion of air pollutants transported from outside a forest have been reported to the best of our knowledge. We therefore conducted wind tunnel experiments to elucidate the effect of trees on air pollutant break.

2. Field Observation

Seta Hill, Shiga Prefecture, Japan, is near the Kansai metropolitan area, which includes Osaka and Kyoto. Shiga prefecture is almost located in the centre of Japan and Seta Hill is located at the southwest end of Shiga prefecture as shown in Figure 1. Seta Hill includes the Ryukoku forest, a 38 ha woodland area primarily composed of konara oak (*Quercus serrata*, deciduous), longstalk holly (*Ilex pedunculosa*, broad-leaved evergreen), Japanese red pine (*Pinus densiflora*, coniferous) and hinoki cypress (*Chamaecyparis obtusa*, coniferous). The study area is the Ryukoku forest and the neighbouring Ryukoku University campus, located 150 - 160 m above sea level at N35.0 deg., E135.9 deg., approximately 15 km east of central Kyoto. The Ryukoku forest and the surrounding area are also shown in Figure 1. There is a highway at a distance of about 1 km from the university campus but no other major stationary emission sources of air pollutants. The area surrounding the university campus can be considered to be a typical suburb. A 25-m-tall observation tower was set up in the forest. The tower has a fetch longer than 300 m from the nearest forest edge. The height of the canopy top varies with the season. In this paper, the height of the canopy top is assumed to be 15.6 m, which is the highest observation point of NO_2 in the canopy layer. The leaf area index (LAI) of the canopy at the observation tower was measured by an optical sensor (LI-COR LAI-2000 plant canopy analyzer). The LAIs estimated from measurements made above the canopy and at ground level were 4.2 and 3.5 in autumn and winter 2013, respectively.

NO_2 was collected using an Ogawa passive sampler [14], and nitrite ions were extracted from the filter of the passive sampler. The NO_2 concentration was determined from the nitrite ion concentration analysed using a spectrometer (Shimadzu UV2550). Two filters were placed in the body of the passive sampler. We averaged the NO_2 concentrations analysed from each filter. The difference in the NO_2 concentrations between the two filters was approximately 3 percent of the average and

185 out of 193 sample pairs showed differences within 10 percent of the average. Bytnerowicz et al. [15] observed the distributions of ozone and nitrogen dioxide in 32 forests sites in central Europe using passive samplers. Passive samplers have been proven to be useful tools for determining the temporal and spatial variation when studying the exposure of the forests to gaseous air pollution [16].

Map data : http://www.craftmap.box-i.net/

Map data : http://mapps.gsi.go.jp/maplibSearch.do?specificationId=757688

Figure 1. Location of Seta Hill (upper) and the study area (lower) showing the NO₂ observation points in the campus marked by triangles.

Twelve passive samplers were attached to the observation tower at heights of between 0.1 and 25.1 m. Each measurement was performed for 3 days in almost every month from January to December 2013. Furthermore, the vertical distribution of the NO_2 concentration outside the Ryukoku forest was measured at heights of between 1.5 and 30 m using passive samplers attached to university campus buildings at the locations shown by triangles in Figure 1. Wind measurements with a vane anemometer (Vector Instruments W200P) were performed at a height of 5 m using the observation tower in the Ryukoku forest and at a height of 6 m using a mast in the university campus.

3. Wind Tunnel Experiments

Wind tunnel experiments were conducted using facilities at the Meteorological Research Institute of Japan Meteorological Agency. The test section of the wind tunnel was 18 m long, 3 m wide and 2 m high. A grid and rods were placed at the entrance to the test section and 10 mm cubes were placed at the floor of the test section to produce a suburban atmospheric boundary layer. Experiments were conducted with a flow velocity of 5 m/s at the boundary layer height. The height of the boundary layer and the wind-profile power-law exponent were 280 mm and 1/7, respectively. The streamwise turbulence intensity was more than 10 percent near the surface.

Figure 2 shows wooden model trees, 10 mm cubes and a sampling probe for tracer gas. Two or four parallel rows of model trees at 100 mm intervals were set up. In the case of four parallel rows, the trunks of the trees were arranged in a staggered manner as shown in the figure. One model tree was 10 mm wide and 12 mm thick, and comprised a 75-mm-high canopy layer and a 25-mm-high trunk space. Thirty-six holes of 6 mm diameter were bored in the canopy layer with an opening ratio of 14 percent. A wire gauze of 12 meshes per inch was used to simulate the canopy.

Pure propane was used as the tracer. The tracer gas was released parallel to the wind tunnel floor using a gamma-type pipe of 2 mm internal diameter, collected using a sampling probe and analysed using a flame ionisation detector (Beckman Industrial Model 400). The sampling probe was positioned at the centre of the model trees, as shown in Figure 2. The release point of the tracer was situated at a distance of 2000 mm from the sampling point on the windward side. The release heights were 0, 30, 60, 90, 120, 150 and 190 mm.

Measurements of the tracer gas concentration were carried out for both flat terrain and model trees.

The measurement was repeated two times for the flat terrain to examine the accuracy of concentration measurements. The differences in the standard deviations of lateral and vertical dispersions were 3 percent and 1 percent of the average of the two measurements, respectively.

The standard deviation of lateral dispersion for flat terrain was 75 mm at a distance of 2000 mm from the release point. The total width of model trees had to be much larger than 75 mm in order to obtain similar dispersion phenomena among the model trees. In our wind tunnel experiment, five or six model trees per row were set up and the total width of model trees was 500 or 600 mm. Vertical profiles of the concentration of tracer gas are analysed with respect to the height normalized by canopy height. It can therefore be considered that the number and height of the model tree do not affect the interpretation of the dispersion phenomena in this study.

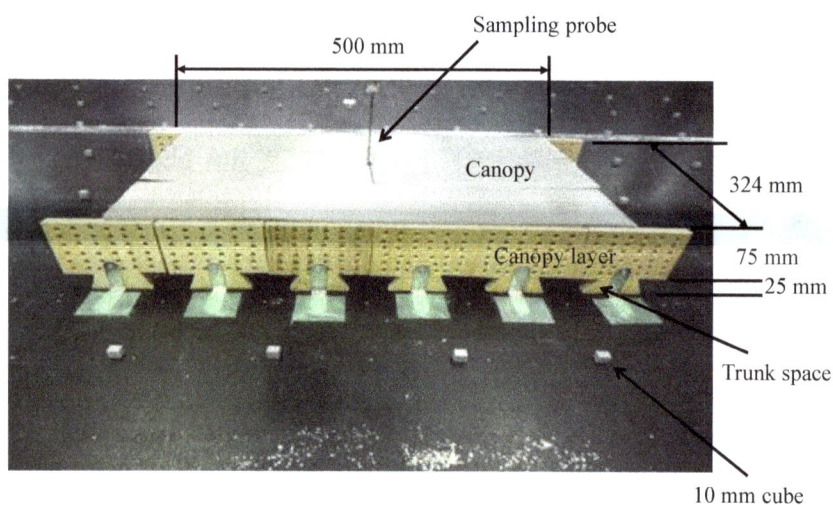

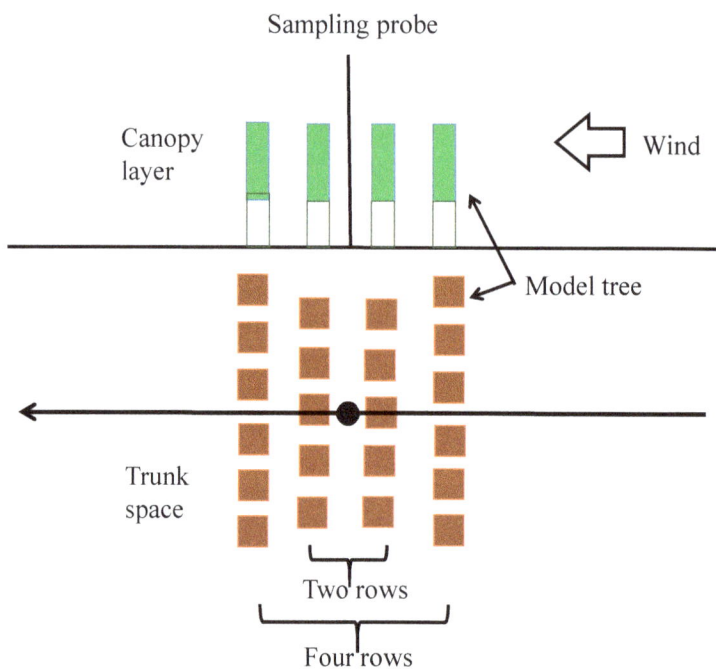

Figure 2. Model trees in a wind tunnel (upper) and the arrangement of model trees (lower).

We discuss here the influence of the opening ratio of a model tree on the dispersion characteristics among model trees. The opening ratio is determined from various factors such as the size of the model tree, number of holes and the diameter of holes. The opening ratio itself is more important than each single factor. In our study, the size of each model tree and the hole diameter were fixed. Figure 3 shows the effect of the number of holes on the vertical profiles of the tracer gas. The number of parallel rows of model trees was two and the release height of the tracer gas was 60 mm. The number of holes and the corresponding opening ratio are shown in the legend. The concentration of the tracer gas below the canopy top for 36 holes was approximately 20 – 30 percent higher than those for zero and 6 holes. The zero hole corresponds to an impermeable canopy layer and is unrealistics. There is a difference of 10 percent between the results for 36 holes and 12 (or 18) holes. The model tree with 36 holes was used in our wind tunnel experiment. Comparison between the results of the wind tunnel experiment using model trees with 36 holes and those of the field observation is discussed later with reference to Figure 9.

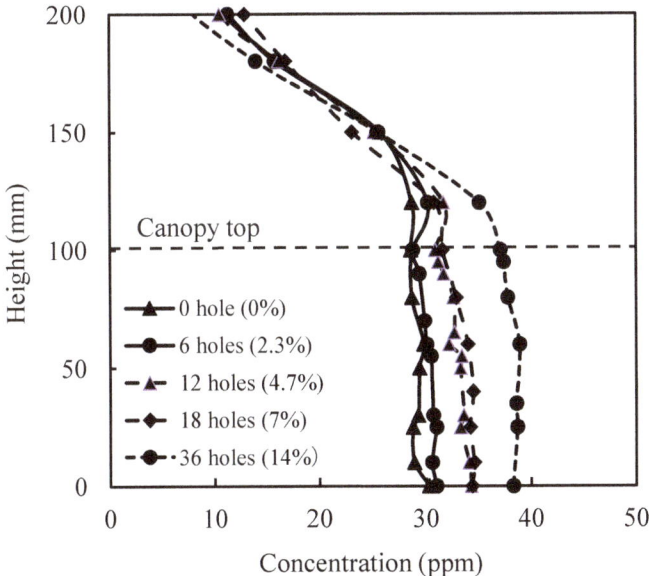

Figure 3. Vertical profiles of tracer gas concentration for two parallel rows of model trees. The release height was 60 mm. The legend denotes the number of holes and the values in parentheses represent the opening ratio.

The flow fields were measured by laser Doppler velocimetry (LDV) combining an argon laser (Spectra Physics model 2017) and a Burst Spectrum Analyser (Dantec Dynamics F60), and visualised by the smoke wire technique. In this technique, smoke is generated by heating a glycerine-covered nickel-chromium wire of 0.2 mm diameter. Since LDV measurements and smoke visualisations under the canopy cover were very difficult, flow fields were acquired without the canopy cover.

4. Results

Figure 4 shows the observed NO_2 concentration plotted against the height from the forest floor in the Ryukoku forest. The observed NO_2 concentration decreased from the top of the canopy to the forest floor throughout the year. The average ratio of the NO_2 concentration at the forest floor to that above the canopy top was approximately 70 percent. The decrease in the NO_2 concentration in the forest is due to not only the deposition onto leaf surfaces but also the effect of a tree wind break. Since wind tunnel experiments cannot consider the effect of deposition, their results show the contribution of the effect of trees on air pollutant break. A forest consists of the canopy cover, canopy layer and trunk space. Figure 5 shows a wind tunnel experimental result showing the effect of the canopy cover on the vertical profile of the tracer gas. The number of parallel row of model trees was four and the release heights of the tracer gas were 60 and 120 mm. The opening ratios in the cases of canopy cover and no canopy cover were 27.9 and 100 percent, respectively. Similar results were obtained regardless of the existence of a canopy. This is because the tracer gas mainly flows in the spaces between trees through holes in the canopy layer and trunk space.

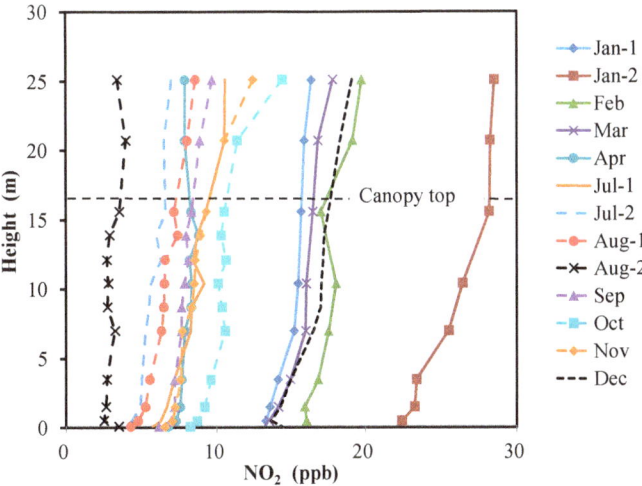

Figure 4. Vertical NO_2 profiles in the Ryukoku forest. The legend denotes the observation months in 2103.

The result of flow visualisation using the smoke-wire technique is shown in Figure 6. The movement of smoke is sketched by the thick arrows. The thin arrows show wind vectors around the height of the canopy top at the centreline of the model trees. The wind velocity at the canopy height was very low and the winds at and below the canopy height blew in opposite directions. Figure 6 shows that less smoke flowed in the spaces between the trees through the canopy cover, although there were instantaneous in- and outflows of smoke. Gromke et al [17] studied the dispersion phenomena in a wind tunnel model of a street canyon and were unable to find a substantial difference in the flow fields between impermeable and permeable tree crowns. Hereafter we will analyse the results of wind tunnel experiments using a canopy cover.

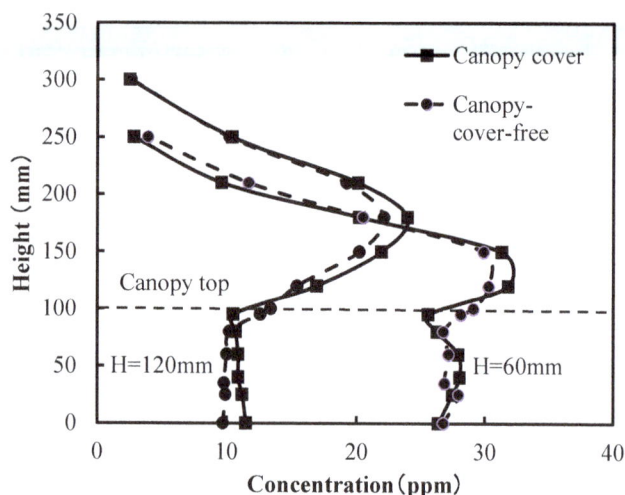

Figure 5. Comparison of the vertical profile of the tracer gas concentration between canopy-cover and canopy-cover-free conditions. H denotes the release height.

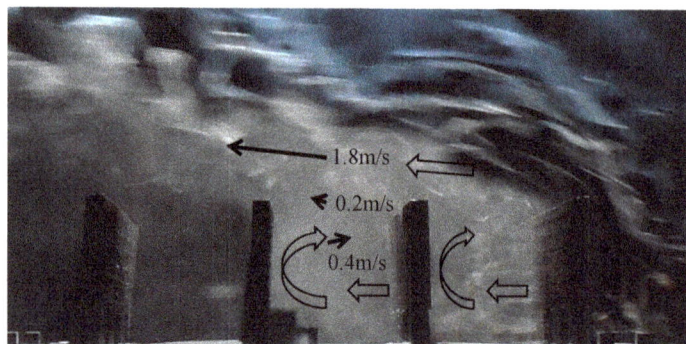

Figure 6. Visualisation of the smoke moving between model trees.

The observed vertical NO_2 profile outside the Ryukoku forest (i.e., at the University campus) was almost uniform as shown in Figure 7. It is considered that this is because exhaust gases released from various heights are transported to the forest. The following integrated concentration was therefore estimated from the results of the wind tunnel experiment for each release height:

$$IC(z) = \int_0^{H_0} C(H,z)w(H)dH / \int_0^{H_0} w(H)dH \qquad (1)$$

Where $IC(z)$ is the integrated concentration of the tracer gas at height z, $C(H, z)$ is the concentration of the tracer gas at height z for the release height H, $w(H)$ is a weighting function and H_0 is the maximum release height .

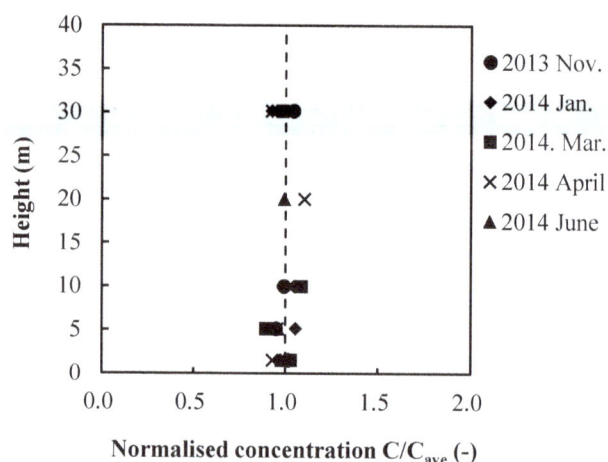

Figure 7. Vertical NO_2 profiles outside the Ryukoku forest. The legend denotes the observation months. C_{ave} is the average NO_2 concentration in the vertical direction.

In the wind tunnel experiments, vertical distributions of the tracer gas concentration for flat terrain were obtained for various release heights. A uniform distribution corresponding to the observed NO_2 concentration outside the forest was obtained by integrating the vertical concentration distributions for the flat terrain. The values of $w(H)$ in Eq. (1) were chosen so that the uniform distribution outside the forest was obtained. The value of $w(H)$ ranged from 1 to 1.6. Similarly, Eq. (1) was used to estimate the vertical concentration distribution of the tracer gas among the model trees. $w(H)$ for the model trees was the same as that for the flat terrain.

The wind tunnel experimental results for four parallel rows of model trees are shown in Figure 8. This figure shows the vertical distribution of the tracer gas for various release heights. An increase in the tracer gas concentration above the canopy top and a tendency for the tracer gas concentration to be uniform below the height of the canopy top are observable. These characteristics are due to the upward flow that occurred on the windward side of the model trees and strong turbulence among the model trees. Another characteristic is the appearance of inflection points at a height of 100 mm, i.e., the canopy height, except for the release height of 190 mm, indicating that there is little exchange of the tracer gas throughout the canopy cover.

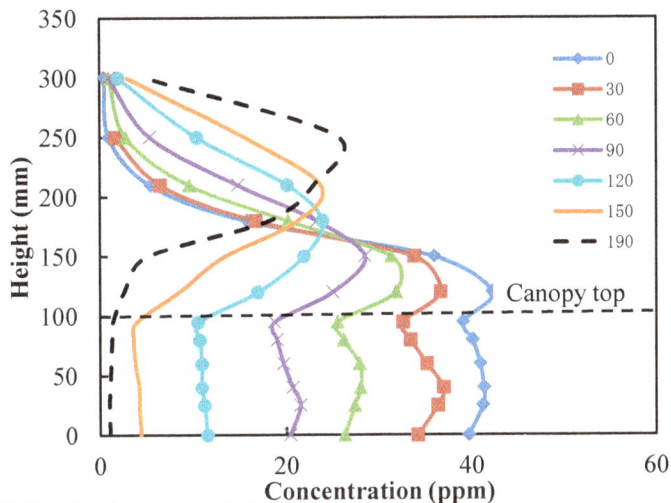

Figure 8. Vertical profiles of tracer gas concentration for four parallel rows of model trees. The legend denotes the release heights in units of mm.

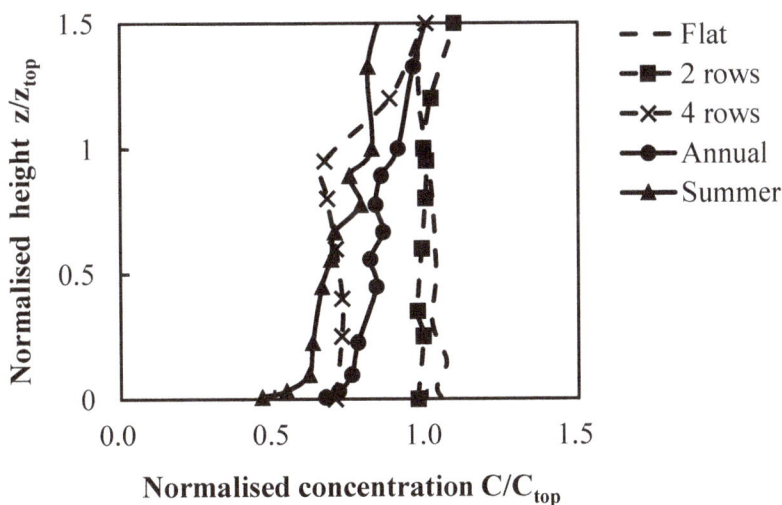

Figure 9. Vertical profiles of integrated concentrations of tracer gas estimated from wind tunnel experiments (broken lines) and NO_2 profiles observed in Ryukoku forest (solid lines). C_{top} is the concentration for flat terrain (or outside the forest) at the height of the canopy top (z_{top}).

The integrated concentration of the tracer gas obtained using Eq. (1) is shown in Figure 9. The concentration was normalised by the tracer gas concentration for flat terrain (or outside the forest) at the height of the canopy top (z_{top}). The vertical axis was normalised by z_{top}. The solid lines were obtained from the observation in the Ryukoku forest, while the broken lines were obtained from the wind tunnel experiments. In the figure, the observed results in the Ryukoku forest show the annual average and the results for the summer period. The results of the wind tunnel experiments for two and four parallel rows of model trees are also included. The concentrations of the tracer gas below the canopy top for the cases of two and four parallel rows decreased by 5 and 30 percent, respectively, compared with that for flat terrain. The annual average of the observed vertical NO_2 profile lay between the wind tunnel experimental results for two and four parallel rows. The wind tunnel experimental results indicated a uniform vertical distribution in the canopy layer and trunk space, while the observed result showed a decrease in the NO_2 concentration from the canopy top to the forest floor. It is deduced that the decrease in the tracer gas concentration among the model trees that appeared in the wind tunnel experiment corresponds to the tree wind break effect, while the concentration gradient appearing in the Ryukoku forest is due to the effects of the plant uptake and deposition of air pollutants. The observed vertical NO_2 profile for the summer period is similar to the wind tunnel experimental result for four parallel rows.

A comparison of the wind velocity observed in the Ryukoku forest with that measured in the wind tunnel is shown in Figure 10. The ratio of the wind velocity inside the Ryukoku forest to that at the campus varied with the season. The result for four parallel rows in the wind tunnel was almost coincident with the results observed from summer to the beginning of autumn when the amount of biomass was large and the LAI was high in the forest.

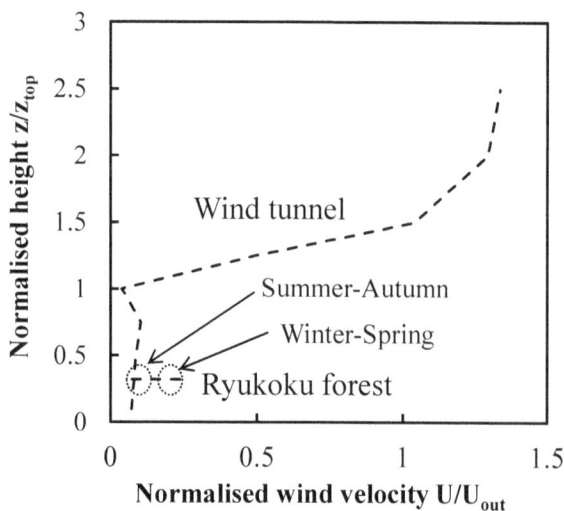

Figure 10. Comparison of wind velocities between the wind tunnel experiments and the observation in the Ryukoku forest. U_{out} is the wind velocity outside the forest and the measurement height corresponds to the mast height in the university campus.

5. Conclusion

A decrease in the NO_2 concentration from the top of the canopy to the forest floor was observed in the Ryukoku forest at Seta Hill in Shiga Prefecture, Japan. The annual average of the ratio of the NO_2 concentration at the forest floor to that at the canopy top was approximately 70 percent. Similar decreases in the ozone concentration in the forest canopy layer have previously been observed and discussed from the viewpoint of deposition on leaf surfaces, uptake by vegetation and chemical destruction. The decrease in the air pollutant concentration in forests is due to not only the deposition of air pollutants onto leaf surfaces but also the effect of a tree wind break. Wind tunnel experiments cannot consider the effect of deposition, and can treat atmospheric transport and dispersion in model trees. This study dealt with the decrease in the air pollutant concentration due to the fluid dynamic effect of a wind break through wind tunnel experiments.

The results of the wind tunnel experiment for four parallel rows of model trees showed a decrease in the concentration of air pollutants inside the forest comparable to the observation in the Ryukoku forest. This is probably because a similar wind break effect was found in the wind tunnel experiment and the observation. However, the decrease in the concentration of air pollutants from the top of the canopy to the forest floor was not observed in the wind tunnel experiment. From the above results, it is deduced that the low NO_2 concentration inside the forest compared with that outside the forest is mainly due to the tree wind break effect and that the variation in the NO_2 concentration with the height from the forest floor is mainly due to the deposition of air pollutants onto leaf surfaces.

Since the wind tunnel experiments were conducted using simple model trees, the effect of the canopy cover on the inflow of air pollutants into the forest was examined. There were no substantial differences in the concentration distribution of the tracer gas between the cases of model trees with and without the canopy cover. Other configurations of model trees remain to be explored in order to simulate the transport and dispersion of air pollutants in the Ryukoku forest in a more rigorous manner. This will be the purpose of our future work.

Acknowledgements

This work was supported by JSPS KAKENHI Grant Number 25340018 and Ryukoku University Short-Term Research Fellowship. The authors are indebted to

Toshimasa Yagi of Meteorological and Environmental Sensing Technology Inc. for his contribution to the flow visualisation in the wind tunnel and to Shiori Hashimoto for her assistance in the wind tunnel experiment.

References

[1] Nowak D. J., Hirabayashi S., Bodine A., Greenfield E., "Tree And Forest Effects On Air Quality And Human Health In The United States," *Environmental Pollution*, vol. 193, pp. 119-129, 2014.

[2] Fontan J., Minga A., Lopez A., Druilhet A., "Vertical Ozone Profiles in a Pine Forest," *Atmospheric Environment*, vol. 26A, pp. 863-869, 1992.

[3] Lamaud E., Carrara A., Brunet Y., Lopez A., Druilhet A., "Ozone Fluxes Above And Within A Pine Forest Canopy In Dry And Wet Conditions," *Atmospheric Environment*, vol. 36, pp. 77-88, 2002.

[4] Cox R. M., Malcolm J. W., "Passive Ozone Monitoring For Forest Health Assessment," *Water, Air, and Soil Pollution*, vol. 116, pp. 339-344, 1999.

[5] **Krzyzanowski J., "Ozone Variation With Height In A Forest Canopy – Results From A Passive Sampling Field Campaign,"** *Atmospheric Environment*, vol. 38, pp. 5957-5962, 2004.

[6] Utiyama M., Fukuyama T., Maruo Y. Y., Ichino T., Izumi K., Hara H., Takano K., Suzuki H., Aoki M., "Formation And Deposition Of Ozone In A Red Pine Forest," *Water, Air, and Soil Pollution*, vol. 151, pp. 53-70, 2004.

[7] Launiainen S., Katul G. G., Grönholm T., Vesala T., "Partitioning Ozone Fluxes Between Canopy And Forest Floor By Measurements And A Multi-Layer Model," *Agricultural and Forest Meteorology*, vol. 173, pp. 85-99, 2013.

[8] Savi F., Fares S., "Ozone Dynamics In A Mediterranean Holm Oak Forest: Comparison Among Transition Periods Characterized By Different Amounts Of Precipitation," *Annals. of Silvicultural Research*, vol. 38, pp. 1-6, 2014

[9] Pietri L., Petroff A., Amielh M., Anselmet F., "Turbulence Characteristics Within Sparse And Dense Canopies," *Environmental Fluid Mechanics*, vol. 9, pp. 297-320, 2009.

[10] Gromke C., Buccolieri R., Sabatino S. D., Ruck B., "Dispersion Study In A Street Canyon With Tree Planting By Means Of Wind Tunnel And Numerical Investigations – Evaluation Of CFD Data With Experimental Data," *Atmospheric Environment*, vol. 42, pp. 8640-8650, 2008.

[11] Gromke C., Ruck B., "Pollutant Concentrations In Street Canyons Of Different Aspect Ratio With Avenues Of Trees For Various Wind Directions," *Boundary Layer Meteorology*, vol. 144, pp. 41-64, 2012.

[12] Al-Dabbous A. N., Kumar P., "The Influence Of Roadside Vegetation Barriers On Airborne Nanoparticles And Pedestrians Exposure Under Varying Wind Conditions," *Atmospheric Environment*, vol. 90, pp. 113-124, 2014.

[13] Janhäll S., "Review On Urban Vegetation And Particle Air Pollution – Deposition And Dispersion," *Atmospheric Environment*, vol. 105, pp. 130-137, 2015.

[14] Ogawa and Company, *Passive Sampler*. Available: http://ogawausa.com/passive- sampler/.

[15] Bytnerowicz A., Godzik B., Frączek F., Grodzińska K., Krywult M., Badea O., Barančok P., Blum O., Černy M., Godzik S., Mankovska B., Manning W., Moravčik P., Musselman R., Oszlanyi J., Postelnicu D., Szdźuj J., Varšavova M., "Distribution Of Ozone And Other Air Pollutants In Forests Of The Carpathian Mountains In Central Europe," *Environment Pollution*, vol. 116, pp. 3-25, 2002.

[16] Cox R. M., "The Use Of Passive Sampling To Monitor Forest Exposure To O_3, NO_2 And SO_2: A Review And Some Case Studies," *Environmental Pollution*, vol. 126, pp. 301-311, 2003.

[17] Gromke C., Ruck B., "Influence Of Trees On The Dispersion Of Pollutants In An Urban Street Canyon – Experimental Investigation Of The Flow And Concentration Field," *Atmospheric Environment*, vol. 41, pp. 3287-3302, 2007.

18

Relationship for the Concentration of Dissolved Organic Matter from Corn Straw with Absorbance by using UV–Visible Spectrophotometer

Shamshad Khan[1], Wu Yaoguo[1], Zhang Xiaoyan[1], Xu Youning[2], Zhang Jianghua[2], Hu Sihai[1]
[1]Department of Applied Chemistry, School of Science,
Northwestern Polytechnical University, Xi'an, 710072, China
shamshadkhan768@yahoo.com;wuygal@nwpu.edu.cn
[2]Xi'an Institute of Geology and Mineral Resources，Xi'an, 710054, China

Abstract- A rapid technique to estimate the concentration of dissolved organic matter from corn straw using the UV–visible spectrophotometer was developed in this study. We tested samples and suitable relationship for the concentration of dissolved organic matter (DOM) with absorbance. The results demonstrated that the relationship between absorbance (254nm) and concentration of DOM is good surrogate to estimate the concentration of the DOM from corn straw. Absorption of DOM verified that it doesn't depend on pH in normal working range. Relatively, this method is very less time consuming, and low cost than the Chemical oxygen demand method and TOC–analyzer method. The absorbance (254nm) with concentration of DOM has linear relationship with high correlation ($R^2 = 0.998$). It is suggested that the absorbance (254nm) should be used as a surrogate for concentration of DOM from corn straw.

Keywords: absorbance, chemical oxygen demand, corn straw, dissolved organic matter, concentration.

1. Introduction

Dissolved organic matter is a complex heterogeneous mixture of organic compounds containing the carboxylic, carbonyl, methoxyl, hydroxyl, and phenolic functional groups and the major source of these organic matters are living organisms deposited on or within soil components that play significant role in plant nutrition and soil environments [1]. If soil contains the sufficient amount of organic matter, there the plants grow are better, produce higher yields; and the nutritional quality of harvested foods and feeds are greater. During the last decay dissolved organic matter is become key parameter in agriculture and environmental fields because of its involvement in mobilization and transportation of acidty, colloids, nutrients, metals and pollutants and serves as a substrate for microbial growth [2–6]. Although various literatures have been considered to the consequence of DOM on the plants and soil environment, but not many study are available on characterization of DOM in water. Primary compositions of elements in DOM are carbon (52-56%), hydrogen (4-5.5%), oxygen (33-39%), and small fractions of sulfur, nitrogen, and phosphorus [7]. Christman et al. [8] were suggested the hypothetical structures for DOM on degradation of the products. However, it is difficult to describe the specific chemical structure of DOM due to its heterogeneous organic matter characters.

Yet now, there is no direct method is accessible to estimate the concentration of DOM in water because of its highly complicated structures [8,9]. Two of the most widely used methods to surrogate the concentration of DOM are the total organic carbon analyzer (TOC) and Chemical oxygen demand methods. Previous method estimates the concentration of organic carbon content in water which proxy the concentration of DOM. Afterward estimates the quantity of oxygen required to oxidize the organic materials which indicate the organic matter in water. But such methods required the much time, expensive chemicals, sophisticated instrument and require a large sample volume for the analysis of concentration of DOM of each sample. These shortcomings have lead to the improvement of spectroscopic methods towards the quantification of DOM concentration.

Ultraviolet (UV) absorption spectrophotometer is generally used to study various properties of the DOC, such as its aromaticity, hydrophobic content, and biodegradability [2,10–12] because DOM have constitute such type of components which have capacity to absorb the ultraviolet light for example unsaturated aliphatic bonds and benzenoid type components [13,14]. Specific absorbance at different specific wavelengths has been used to measure the aromaticity, hydrophobic content, apparent molecular weight and size, and biodegradability [15–18].UV-absorbance is good surrogate the concentration of DOM specifically concentration of aromaticity in DOM at 254 nm absorbance [13,18]. Anton et al. [19] reported that UV-absorbance showed much closed linear relationship (0.99) between concentration of DOM and UV–254 absorbance in throughfall and soil solution samples. Aromaticity fraction in DOM is surrogate to estimate dissolved organic matter by using absorbance 254nm because of the absorbance of organic solutes is directly proportional to their concentration of aromatic compounds [20].This correlation was considered here to develop the relationship for concentrations of DOM in water extracted from corn straw and absorbance.

Based on the above discussion, absorbance of a water sample at specific wavelength 254nm is good proxy to the concentration of DOM. Standard calibration curve should be able to indicate the total concentration of DOM from corn straw with absorbance 254nm. Thus the aim of this study to establish relationship between absorbance 254nm and concentration of DOM from corn straw by using UV–visible spectrophotometer should be able to surrogate the concentration of DOM from corn straw. We will also demonstrate the effect of pH on absorbance at 254nm.

2. Material and Methods

Corn Straw is used as source of DOM in water. Extractions of DOM from corn straw just carried out by follow the method mentioned by Zhongqi et al.[6,21]. In concise, initially corn straw air-dried, and ground to pass through a 1-mm sieve. Just before the absorbance experiments, mix with 40:1 (v/w) water to sample proportion using cold water, periodic shaking it 18 h, suspension were centrifuged (900×g) for 30 min and

filtered through 0.45-μm pore size polycarbonate filters. Aliquots of water samples of corn straw were analyzed with total carbon analyzer (Shimadzu TOC 5000A) after filtration through 0.45-μm pore size polycarbonate filters. Stock solution of 563.8mg L^{-1} DOM was used to prepare the standard solutions containing different amount of dissolved organic matter. UV–visible absorbance measurements were performed on a high precision, double–beam spectrophotometer (model 2550) between 220 and 450nm with the reference of distilled water. A quartz cell with 1.0 cm path length was used. Buffer solutions to maintain pH were not need because a set of 5mg L^{-1} of DOM solutions prepared at different pH have demonstrated that UV–visible spectra were not significantly affected by the 4 to 10 range of pH as shown in Figure 1. The negligible effect of pH on the DOM samples presented in this study is less than observed for soil humic substances [22]. This result also is in agreement of the observed result of Wang and Hsieh who reported that humic acids solutions prepared at different pH have showed that the UV–visible spectra were not affected by the pH at the normal working range. The minute dependency of UV absorbance on pH in the range of 4–10 means that, within this pH range, it is unnecessary to adjust the pH to a constant value to compare results between samples.

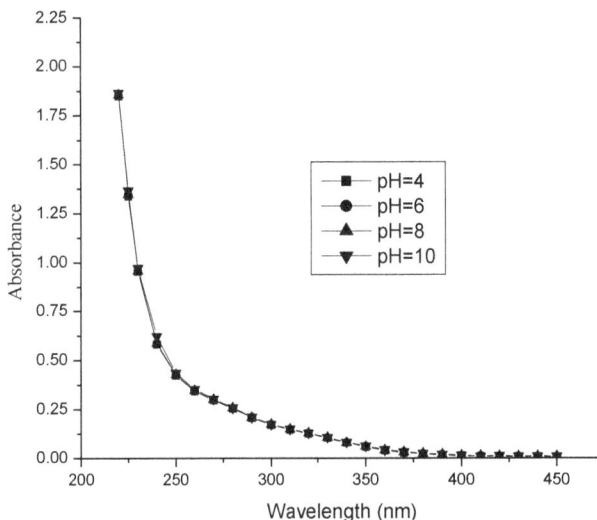

Figure 1. Effect of pH on UV–visible spectra of 5 mg L^{-1} of dissolved organic matter in water from corn straw.

3. Results and Discussion

As can be seen from figure 2, that absorbance increases with increasing the concentrations of DOM.

The absorbance at wavelength length less than 280nm is much greater than that of longer wavelength as a result a perfect slope is obtained at the shorter wavelength (<285nm), when concentration of DOM is higher than this trend is much more visible as seen in the spectra. The absorbance measured at greater wavelength (>400nm) is comparatively less than that absorbance at 280-400nm.This result almost agrees with the experimental results of Wang and Hsieh [23], who reported that the absorbance of humic acids at wavelength length less than 250nm is causes a sharp slope at the longer wavelength (<300), and absorbance at longer wavelength (>400nm) is relatively low as compared with those observed at UV and sub-UV ranges (200 –400 nm).

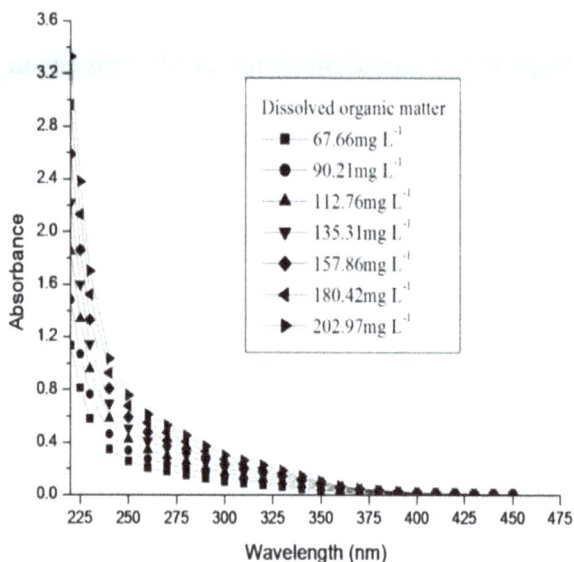

Figure 3. UV–visible spectra of different concentrations of dissolved organic matter in water from corn straw for verification of Beer–Lambert law.

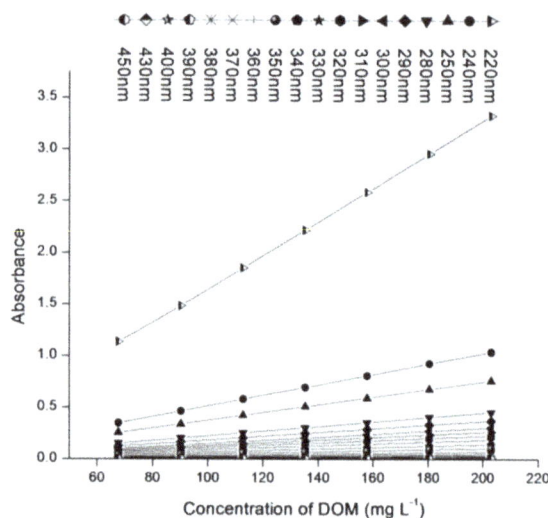

Figure 2. UV–visible spectra of different concentrations of dissolved organic matter in water from corn straw.

For the validation of these experimental results (Figure 2), it should follow Beer–Lambert law because Beer–Lambert law is independent of wavelength. The Beer–Lambert law has been examined at wavelengths 220,240,250,280,290,300,310,320,330,340,350,360,370,380,390,400,430,450nm which showed the linearity with correlation coefficients (R^2) greater than 0.99. It implies that Beer–Lambert law is applicable between range of 220-450nm as shown in figure 3.

In order to justify the UV method, we developed the relation between the absorbance at 254nm and various concentrations of DOM determined by carbon analyzer. As shown in figure 4, the relationship between the absorbance (254nm) and concentration of DOM, the UV absorbance at 254nm is highly correlated (R^2 =0.998; $P<0.0001$) with concentration of DOM determined by carbon analyzer. So it is suggested that the absorbance at 254nm should be used as a proxy for concentration of DOM from corn straw. Such correlation between absorbance and concentration of DOM determined by Huoliang et al [24] for the printing and dyeing waste water as shown in figure 4.

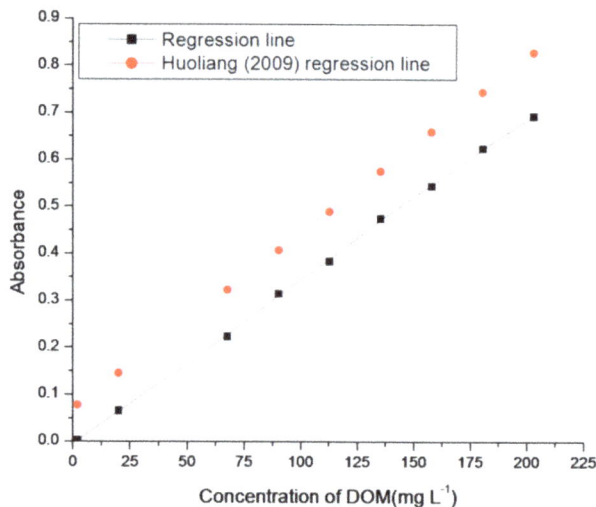

Figure 4. Correlation between absorbance at 254nm and concentration of DOM determined by carbon analyzer method.

The equation of the least squares regression line of figure 4 for corn samples is:

$$DOM \ (mg \ L^{-1}) = 288.078 \times \text{absorbance} \ (254nm) + 1.3 \tag{1}$$

In the above equation 1, 288.078 and 1.3 are representing the slope and y-intercept respectively. The intercept of the regression line shows the presence of 1.3 mg L^{-1} of non-UV absorbing DOM, possibly due to the organic matter containing no chromophores absorbing at 254nm. These could include saturated aliphatic compounds, low molecular weight oily acids, or carbohydrates. Lower limit detection is a significant affecting factor on the precision and accuracy of the measure concentrations in analytical chemistry. We have calculated lower limit of detection 6.248×10^{-5} mg L^{-1} from calibration graph of figure 4 by the following phenomena. The actual absorbance (response points) on Y- axis composing the line usually do not fall exactly on the line for the "fitting" of the calibration graph (Figure 4). Hence, random errors were implied. The parameters for calculating errors due to calibration graph in figure 4 using equation 1 are in shown table 1.

Table 1. Parameters for calculating errors due to calibration graph of figure 4.

Xi	Yi	Yi*	(Yi- Yi*)	(Yi-Yi*)²
2	0.0024	0.0024	0	0
20	0.0650	0.0649	0.0001	1.0×10^{-8}
67.66	0.2231	0.2303	-0.0072	5.184×10^{-5}
90.21	0.3145	0.3086	0.0059	3.481×10^{-5}
112.76	0.3833	0.3869	-0.0036	1.29×10^{-5}
135.31	0.4750	0.4652	0.0098	9.60×10^{-5}
157.86	0.5435	0.5435	0	0
180.42	0.6241	0.6218	0.0023	5.29×10^{-6}
202.97	0.6929	0.7000	-0.0071	5.04×10^{-5}
$\sum$(Yi-Yi*)²= 2.512×10^{-4}				

The standard error of the y-estimate was calculated by Equation 2.

$$Sy = \sqrt{\frac{\Sigma (Yi - Yi^*)^2}{n-2}} = 0.006 \tag{2}$$

Where Yi* is "fitted" y-value for each xi, (calculated from Eq. 1). Thus, Yi- Yi* is the vertical deviation of the found y-values from the line and "n" is number of calibration points.

This uncertainty about the y-values (the fitted y-values) is transferred to the corresponding concentrations of the unknowns on the x-axis and can be expressed by the standard deviation of the obtained x-value. The exact calculation is rather complex but a workable approximation can be calculated with:

$$Sx = \frac{Sy}{b} = 2.083 \times 10^{-5} \tag{3}$$

So Lower Limit of Detection derived from a calibration graph was calculated finally by using following equation:

$$LLD = 3 \times Sx = 6.248 \times 10^{-5}$$

This is very convenient and economical for determination of concentration of DOM from corn straw. Data presented in Figure 5 compare the values of concentrations of DOM estimated by TOC carbon analyzer and UV–visible spectrophotometer at 254nm. A strong linear correlation ($R^2 > 0.99$) exists between these two methods.

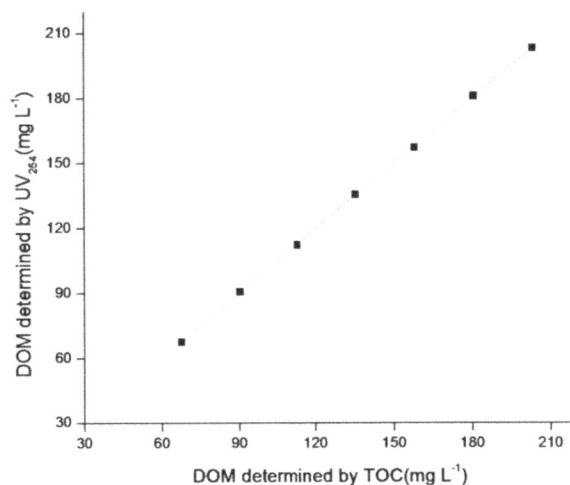

Figure 5. Correlation of the concentrations of DOM by UV$_{254}$ technique versus concentrations of DOM determined by the TOC carbon analyzer.

Other methods like Chemical oxygen method and TOC analyzer method are more expensive, time consuming and need of more sophisticated instrument but in this method no chemical is required and very commonly using device needed like UV–Visible

Spectrophotometer. Another advantage is that in normal working range of pH no need to use buffer solutions because changes of pH within normal working ranges has no effect on this method.

4. Conclusion

A very highly linear correlationship (R^2=0.998) is illustrated between absorbance (254nm) and concentration of DOM in water from corn straw. UV–visible spectrophotometer method is very straightforward, fast and low-priced alternative for assessment of DOM. This relation has advantage over common methods for estimation of DOM with less time consuming. It is also working well in normal working pH ranges (4-10).

Acknowledgements

This work was supported by the Land and Resources Scientific Research of China from special fund (20111020) in the public interest, the project titled "survey and assessment of groundwater pollution in main cities of Northwestern China (1212011220982)" and "NPU Foundation for Fundamental Research (NPU-FFR-JCR20130145)".

References

[1] H.O. Neung, A.P. Brian, A.M. Philip, J.H. Peter, M.B. Sandra, O. Noriaki, M.L. Kavvas, A.B. Brian, R.H. William "The role of irrigation runoff and winter rainfall on dissolved organic carbon loads in an agricultural watershed" Agriculture, Ecosystems and Environment, 179(1), 2013, pp.1–10.

[2] D. Jorg, K. Klaus "Estimation of the hydrophobic fraction of dissolved organic matter in water samples using UV photometry" Water Research, 36(20), 2002, pp. 5037–5044.

[3] R.M.W. Amon, R. Benner "Bacterial utilization of different size classes of dissolved organic matter" Limnology and Oceanography, 41(l), 1996, pp. 41-51.

[4] A.L. Bonnie, M.C. Rose, S.W. Howard "Changes in dissolved organic matter fluorescence and disinfectionbyproduct formation from UV and subsequent chlorination/chloramination" Journal of Hazardous Materials, 264(1), 2014, pp. 411–419.

[5] G. Rachel, J.H. Peter, W. Naomi, F. Christopher "Dissolved organic carbon and trihalomethane precursor removal at a UK upland water treatment works" Science of the Total Environment. 468–469(1), 2014, pp.228–239.

[6] K. Shamshad, W. Yaoguo, Z. Xiaoyan, H. Sihai, L. Tao, F. Yilin, L. Qiuge "Influence of dissolved organic matter from corn straw on Zn and Cu sorption to Chinese loess" Toxicological & Environmental Chemistry, 95(8), 2013, pp.1318-1327.

[7] D.A. Reckhow, P.C. Singer, R.L. Malcolm "Chlorination of humic materials: by product formation and chemical interpretations" Environmental science &technology, 24(11), 1990, pp. 1655– 1664.

[8] R.F. Christman, D.L. Norwood, Y. Seo, F.H. Frimmel "Humic substances: II. In search of structure" In: Hayes, M.H.B., Mac-Carty, P., Malcolm, R.L., Swift, R.S. (eds).Publishing Wiley, Chichester, England, 1985, pp. 451-463.

[9] G.G. Choudhry "Humic substances. structural, photophysical, photochemical and free radical aspects and interactions with environmental chemicals" Gordon & Breach Publishers, New York, 1984, pp. 98-106.

[10] C.S. Uyguner, C. Hellriegel, W. Otto, C. Larive "Characterization of humic substances: Implications for trihalomethane formation" Analytical and Bioanalytical Chemistry, 378(6), 2004, pp.1579–1586.

[11] J.P. Croué, M.F. Benedetti, D. Violleau, J.A. Leenheer "Characterization and copper binding of humic and nonhumic organic matter isolated from the South Platte River: Evidence for the presence of nitrogenous binding site" Environmental Science & Technology, 37(2), 2003, pp.328– 336.

[12] K. Hautala, J. Peuravuori, K. Pihlaja "Measurement of aquatic humus content by spectroscopic analyses" Water Research, 34(1), 2000, pp.246–258.

[13] A.U. Eaes, P.R. Bloom "Fulvic acid ultraviolet-visible spectra: Influence of solvent and pH" Soil Science Society of America Journal, 54(5), 1990, pp. 1248-1254.

[14] S.J. Traina, J. Novak, N.E. Smeck "An Ultraviolet Absorbance Method of Estimating the Percent Aromatic Carbon Content of Humic Acids" Journal of Environmental Quality, 19(1), 1990, pp.151-153.

[15] M. Simonsson, K. Kaiser, F. Andreux, J. Ranger "Estimating nitrate, dissolved organic carbon and DOC fractions in forest floor leachates using ultraviolet absorbance spectra and multivariate analysis" Geoderma.124, 2005, pp.157–168.

[16] K. Kalbitz, J. Schmerwitz, D. Schwesig, E. Matzner "Biodegradation of soil-derived dissolved organic matter as related to its properties" Geoderma, 113, 2003, pp.273–291.

[17] J. Peuravuori, K. Pihlaja "Molecular size distribution and spectroscopic properties of aquatic humic substances" Analytica Chimica Acta, 337 (20), 1997, pp.133–149.

[18] Y. Chin, A. George, E. O'Loughlin "Molecular Weight, Polydispersity, and Spectroscopic Properties of Aquatic Humic Substances" Environmental science & technology, 28 (11), 1994, pp. 1853-1858.

[19] B. Anton, S.S. Ronald, M. Axel, W.W. Walter "Estimating dissolved organic carbon in natural waters by UV absorbance (254 nm)" Journal of Plant Nutrition and Soil Science, 159(6), 1996, pp.605-607.

[20] B.A. Bergamaschi, M.S. Fram, C. Kendall, S.R. Silva, G.R. Aiken, R. Fujii "Carbon isotope constraints on the contribution of plant material to the natural precursors of trihalomethanes" Organic Geochemistry, 30(8), 1999, pp. 835–842.

[21] H. Zhongqi, M. Jingdong, W.C. Honeycutt, O. Tsutomu, J.F. Hunt, J.C.M. Barbara "Characterization of plant-derived water extractable organic matter by multiple spectroscopic techniques" Biology and Fertility of Soils, 45(6), 2009, pp.609-616.

[22] P.R. Bloom, J. Leenheer "Vibrational, Electronic, and Highenergy Spectroscopic Methods for Characterizing Humic Substances" In Humic Substances II: In Search of Structure; Hayes, M. H. B., Malcolm, R., Swift, R. S., Eds.; John Wiley & Sons: New York, 1989, Chapter 14, pp. 409-446.

[23] G. Wang, S. Hsieh, "Monitoring natural organic matter in water with scanning spectrophotometer" Environment International, 26 (4), 2001, pp. 205-212.

[24] K. Huoliang, W. Huifang "A Rapid Determination Method of Chemical Oxygen Demand in Printing and Dyeing Wastewater Using Ultraviolet Spectroscopy" Water Environment Research, 81(11), 2009, pp. 2381-2386.

19

A Comparison Study of Water Quality among Different Regions of Sharjah and Ajman in the United Arab Emirates

Sabrina Chelli, Adnan Falah, Rami El Khatib
School of Environment and Health Sciences, Canadian University of Dubai,
P.O. Box 117781, 1st Interchange, Sheikh Zayed Road,
Dubai, United Arab Emirates
rami@cud.ac.ae

Abstract- *A comparison study of water quality among different regions of two major cities in UAE, Sharjah and Ajman, was undertaken. Numerous water samples were collected from five regions of Ajman and four regions of Sharjah near human activity areas, entertainment areas and industrial areas. The samples were taken approximately 10-20 meters from the shoreline, where the depth was 0.5-1.0 meter. Various tests were conducted to identify the levels of inorganic substances and physico-chemical parameters to determine if they were within an acceptable range for living organisms. Based on the analyzed data, the pH was relatively acidic, where it ranged between 5 and 6. Salinity average was 30,886 µs/cm. Total dissolved solids' (TDS) average was around 58,275 mg/L, and the specific conductance average was about 75,356 µs/cm. Many other inorganic substances involving PO_4^{3-}, P, P_2O_5, Fe^{3+}, Cu^{2+}, SO_4^{2-}, Br^-, NO_3^-, and NO_3^- as nitrogen, were also analysed to determine the impact of human activity on water quality. Some of the data collected for such parameters have showed low concentrations which is an indicator for desirable level of pollution.*

Keywords: Water quality, pH, TDS, physico-chemical parameters, Sharjah, Ajman, UAE.

1. Introduction

Water is a crucial compound for humanity [1, 2] However, Dubai's hot and arid climate prevents us from having readily available natural fresh water with rain only few days in a month per year. It is then imperative for the UAE to find a way around this knowing that they have one of the highest water footprints ("the average resident's 550 litres a day for drinking and washing are more than triple the world average") [3] yet have little fresh water available. With today's advanced technology, there are a few methods available to increase the supply of fresh water, either from sea water or from underground aquifers. Desalination and reverse osmosis are two ways to address this issue [4]. However there are some limitations to those methods. Reverse osmosis requires high pressure which is very expensive. Desalination is also costly because the pumps require a lot of power and energy. Jebel Ali is known to have one of the largest desalination industries which also harm the environment by draining sea water from oceans and by pumping it back in, often too concentrated in salts and other inorganic species.

UAE has significantly changed in the last ten years due to the construction of buildings, artificial islands, chemical industries and port activities [4]. Furthermore, these activities have negatively impacted the natural environment, particularly the water. In this report, the effect of the surrounding environment on the water quality in several industrial, human activity and entertainment areas in Sharjah and Ajman will be investigated and discussed.

Nine water samples were collected and analyzed from several locations in Sharjah and Ajman, UAE. They were subjected to water analysis to determine the concentration of inorganic substances such as Fe^{2+}, Cu^{2+}, PO_4^{3-}, P and P_2O_5 and physico-chemical parameters like pH, TDS (total dissolved solids), EC (electrical conductivity) and salinity [5].

2. Materials and Methods
2.1. Chemicals and Instruments

The analysis of water samples was done using the HI 83200 Multiparameter Bench Photometer and the reagents from "HANA Spectroquant kit" [4].

Every water sample was filtered at first using filter paper. The photometer can measure certain concentration range (0-2000 mg/L), in case of sea water, high concentration substances. Thus in order to determine the exact concentration the water sample should be diluted first; and then analyzed in which 99mL of distilled water was added to 1mL of the water sample and was then multiplied by 100.

The tests conducted involved determining the concentration of inorganic substances such as Fe^{3+}, Cu^{2+}, PO_4^{3-}, P and P_2O_5 and also physico-chemical tests which involved electrical conductivity, pH, TDS (total dissolved solids) and salinity [4]. All water samples were analyzed at the same time after collection and were stored under dry conditions at 25°C. The samples were collected about 10-20 meters from the shore at depth of 0.5-1 meters.

3. Results and Discussion

The water samples studied were taken from Sharjah as shown in table 1 and from Ajman as shown in table 2.

Table 1. Beaches analyzed from Sharjah and their surrounding environment.

Number of the Beach	Name of Beach	City	Surrounding Environment
1	Kanat Al Qasha	Sharjah	Entertainment area, jet ski activity, next to coral reefs.
2	Tawan next to Mamzar	Sharjah	Entertainment area, jet ski activity, next to coral reefs.
3	Golden Gate Hotel beach area	Sharjah	Industrial area, next to Zulal water bottling & electricity industry
4	Nayya	Sharjah	Human activity area, port

Table 2. Beaches analyzed from Ajman and their surrounding environment.

Number of the beach	Name of Beach	City	Surrounding Environment
5	Port Ajman	Ajman	Human activity area, port
6	Corniche	Ajman	Entertainment area, beach club
7	Ajman beach	Ajman	Entertainment area, Dana Beach Hotel
8	Al Muntazah	Ajman	Human activity area, fishing port
9	Beginning of Corniche	Ajman	Entertainment area, beach

3.1. Characterization of Sea Water

The phosphate concentrations were found to be within normal range (as they are all low) in all beaches with the exception of Al Muntazah and Tawan next to Mamzar (Table 3) as their phosphate concentrations were slightly higher than the other beaches; and the industry in Sharjah with the beginning of the Corniche of Ajman which had a very low presence of phosphate. The beginning of the Corniche of Ajman indicated negative tests for all three phosphate groups representing a low pollution level. Considering its surroundings of hardly any human activity, it isn't surprising that there are no indications of pollution. However, the fact that phosphate isn't present in water signifies that it is the limiting factor for algal growth. Low concentrations in the water next to the industry in Sharjah were unexpected as contaminated water often has high phosphate levels. Since it was in an acceptable range, the industry seems to manage well since the phosphate levels were low.

Al Muntazah and Tawan however, indicated higher levels of phosphate compared to the other beaches showing a slightly polluted area. There could have been many factors causing that such as "decaying plant matter, fertilizers, mineral treatment chemicals, contaminated well water, acid rain, contamination with soil, ground water runoff, bird droppings, bather wastes, urine and sweat" [6]. Nevertheless in the case of Al Muntazah, the most apparent factors are due to human activity practiced in presence of a fishing port. Moreover, when the sample was collected, the water had an orangey colour to it which already gave a first impression of pollution.

The copper range in water (Table 3) should not exceed more than 1 mg/L [7] as it may cause vomiting and liver damage. In general, it is not seen as a health hazard if in a low concentration. The results below for all 9 beaches indicate low concentrations of copper under 1 mg/L.

Iron concentrations should not exceed 0.1-0.3mg/L [8] in drinkable water. If exceeded, it may cause gastrointestinal problems, heart diseases and sometimes cancer. Seawater should normally contain less than 1.5 to 2 mg/L [8] of iron. The results for iron (Table 3) are all under the abnormal range with the exception of Port Nayya (sample 4), water next to Zulal water & electricity industry (sample 3) and the beginning of the Corniche of Ajman (sample 9) which are a little bit higher than the others. The reason why iron was slightly higher in Port Nayya could be due to the commercial shipping of goods and the activity may lead to iron corrosion in water. Since sample 3 was collected next to an industry in Sharjah, it was anticipated that the iron level would be slightly higher than the others. Lastly, sample 9 wasn't expected to have a level of iron as high as sample 3 and 4. But since it was an entertainment area linked to Jet Ski activity, the saltwater could have caused the jet skis to corrode which could be the justification for this cause.

The pH range of the seawater should be 7.5 to 8.5 [9, 10]. The pH of all the water samples (Table 3) is between 5 and 6 which is slightly acidic. However, beaches 1 and 2 have a slightly more acidic pH than the other beaches. In general, low pH levels are caused by dissolved carbon dioxide and acid generated salts [4]. This can be potentially harmful to the marine life, particularly those that are adapted to live in water of neutral pH.

The normal range of salinity is about 33 to 37 ppt [11, 12]. The data collected (Table 3) respects the range of salinity except for the fishing boat area at Al Muntazah and the beginning of Corniche of Ajman with lower results of 23 and 29 ppt. This means that the sum of solid materials in the water is lower than the normal range. Low salinity may slowly kill corals and other marine creatures present in the water which are sensitive to salinity concentration changes [5]. This may be a point to consider in a long term aspect.

The total dissolved solids' normal seawater range should be 30,000-40,000 ppm [13]. All the beaches have an extremely high range of TDS as they all exceed 40,000 ppm (Table 3). These high concentrations of dissolved solids show that the human activity is dominant in the UAE (construction sites, chemical industries...). It also implies a hazard for marine life. We have noticed that the pH, salinity and the TDS ranges in most samples have shown compatible values with the exception of sample 8 which had revealed a higher pH value. This may be due to higher values of alkaline earth metals (Ca^{2+} and Mg^{2+}) that were not studied in this report. The values of specific conductance [10] amongst different water samples have shown variations that can be due to the fact that some ions may exist in two different forms as either free or complex.

Seawater contains about 2700 mg/L of Sulfate [14]. Most of the beaches (Table 4) are around that range with the exception of Kanat Al Qasha in Sharjah, Port Nayya in Sharjah and the fishing boat area at Al Muntazah in Ajman. The concentration of Sulfate in seawater depends upon the discharge from the soil and the rocks present as well as the leakage of sewers, precipitation and other [8]. The concentration in Kanat Al Qasha and Port Nayya are slightly higher with a concentration of 3000 mg/L whereas the boat area in Al Muntazah is low with a concentration of 2000 mg/L. Higher concentrations of Sulfate may indicate higher levels of pollution from leakage of toxic substances or simply high rates of discharge from the soil and the rocks [4].

A normal range of Bromide in seawater is within a range of 65 mg/L to 80 mg/L [15] and a suitable range for bromine is about 70 ppm [16] which is around 69.92 mg/L. The concentration of bromine in all the beaches (Table 4) was found to be very low if not, non-existent in some cases with concentrations of 0.00-0.25 mg/L. Nitrate concentrations should not exceed 5 ppm in seawater as it could cause severe problems and toxicity in plants [8]. All beaches (Table 4) did not exceed this range with the exception of Al Muntazah in Ajman. The concentration found in this area was 12.5 mg/L which greatly exceeded the appropriate range. This high level of toxicity presents a hazard for aquatic plants and indicates a high level of pollution [4].

Table 3. Physico-Chemical analysis of the concentration of inorganic substances and properties in several areas in Ajman and Sharjah, samples 1-9 (in mg/L).

Sample number	Beach	Location	pH	Specific Conductance (µs/cm)	Salinity (µs/cm)	Total Dissolved Solid (TDS)	Phosphate (PO4²⁻)	Phosphorus (P)	Phosphorus Pentaoxide (P₂O₅)	Iron (Fe³⁺)	Cu²⁺ (µg/L)
Standard Value			7.5 - 8.5	54,000	33 - 37	30,000-40,000 (ppm)	<8	<8	<8	0.1-0.3	< 1 (mg/L)
1	Kanat Al Qasha	Sharjah	5.69	58,500	31,563	59,552	0.4	0.1	0.3	0.00	369
2	Tawan next to Mamzar	Sharjah	5.68	80,000	31,642	59,702	1.4	0.5	1.1	0.04	173
3	Next to Zalal water & electricity industry	Sharjah	5.75	79,100	31,721	59,851	0.1	0.0	0.1	0.00	430
4	Port Nayya	Sharjah	5.58	81,000	32,670	61,641	0.5	0.1	0.3	0.1	422
5	Ajman Port beside the creek of Ajman	Ajman	5.89	86,500	34,094	64,328	1.0	0.3	0.8	0.01	262
6	Comiche of Ajman	Ajman	5.69	82,000	32,591	61,493	0.8	0.3	0.6	0.04	219
7	Ajman Beach	Ajman	5.64	78,100	31,009	58,508	1.0	0.3	0.7	0.03	218
8	Fishing boat area at Al Muntazah	Ajman	6.11	58,500	23,099	43,582	2.0	0.7	1.5	0.05	224
9	Beginning of Comiche of Ajman	Ajman	5.63	74,500	29,585	55,821	0.0	0.0	0.0	0.06	419

Table 4. Physico-Chemical analysis of the concentration of inorganic substances and properties in several areas in Ajman and Sharjah, samples 1-9 (in mg/L).

Sample number	Beach	Location	Sulfate (SO4²⁻)	Bromine (Br2)	Nitrate (NO3⁻)	Nitrate-Nitrogen (NO₃⁻-N)
Standard Value			2700	65 - 80	< 5	< 3
1	Kanat Al Qasha	Sharjah	3,000	0.25	0.0	0.0
2	Tawan next to Mamzar	Sharjah	2,500	0.00	0.0	0.0
3	Next to Zalal water & electricity industry	Sharjah	2,500	0.00	0.0	0.0
4	Port Nayya	Sharjah	3,000	0.10	0.0	0.0
5	Ajman Port beside the creek of Ajman	Ajman	2,500	0.11	2.4	0.5
6	Comiche of Ajman	Ajman	2,500	0.03	0.0	0.0
7	Ajman Beach	Ajman	2,500	0.22	0.0	0.0
8	Fishing boat area at Al Muntazah	Ajman	2,000	0.03	12.5	2.8
9	Beginning of	Ajman	2,500	0.08	1.7	0.4

4. Conclusion

Human activity has a fairly important impact on the water quality as seen with a high range in phosphates in entertainment and industrial areas. Based on the analysis of the results, Port Nayya as well as water collected next to the industry in Sharjah and the beginning of the Corniche of Ajman had higher iron content due to the port activity and industrial activity leading to corrosion and other negative impacts. The electrical conductivity and the salinity were found to be within an acceptable range which is a good indication for low pollution levels except for the fishing boat area where salinity ranges were little bit lower than normal which may be a point to consider in a long term aspect. However, the high TDS levels do imply dominant human activities increasing the risk of future high pollution levels if practiced over a long period of time. Many other inorganic substances involving PO_4^{3-}, P, P_2O_5, Fe^{3+}, Cu^{2+}, SO_4^{2-}, Br^-, NO_3^-, and NO_3^- as nitrogen, were also analysed to determine the impact of human activity on water quality. Even though the seawaters are not highly polluted with the analyzed parameters, the pH is relatively low compared to the normal range. This implies that the acidity levels caused by dissolved carbon dioxide or acid generated salts are relatively high. This would be something interesting to look at in the near future because acidic water conditions may represent a threat to marine life. In order to draw a clear conclusion on how surroundings affect the water quality, further studies may be needed to look at the interactions between the different physico-chemical parameters and the biotic factors (marine organisms and plants).

Acknowledgment

The authors would like to acknowledge Ms. Zena Muhtaseb for her help and support throughout this study.

References

[1]   R.H. Friis "Essentials of Environmental Health" Sudbury, USA: Jones & Bartlett learning, 2012.

[2]   G.T. Miller and S. Spoolman "Environmental Science" Mississauga, Canada: Thomson, 2008.

[3]   M. Kwong, "You're looking at 140 litres of water" [Online] Available: http://www.thenational.ae/ news/uaenews/science/youre-looking-at-140-litres-of-water.

[4]   R. El Khatib, A. Falah, G. Tavakoli, C. D'cruz and J. Pereira "A Study of Water Quality Near to a Coral Reef Site in the Region of Dubai, United Arab Emirates." Canadian Journal on Chemical Engineering & Technology ,vol. 3, no.3, April 2012.

[5]   A. Falah, R. El Khatib, N. Yahfoufi "Water Quality Survey of Arabian Peninsula in Regions of Dubai in the United Arab Emirates." Canadian Journal on Chemical Engineering & Technology, vol. 3, no. 1, January 2012.

[6]   Aqualab Systems Phosphate Problemsin Pools [Online]. 2011 Available: http://www.askalan-aquestion.com/phosphate_pool_problems.htm.

[7]   J. D. Fitzgerald. "Safety guidelines for copper in water." The American Journal of Clinical Nutrition , 1998.

[8]   H. H. Hammud, "Quality and Pollution Studies of Water in Lebanon", Ultra Science - Dimension of Pollution, 1, 19, 2001.

[9]   G. Anderson "Seawater Composition." [Online] Available:http://www.marinebio.net/marine-science/02ocean/swcomposition.htm. 2008.

[10]   E. Wenner "Water Quality" [Online] . Available: http://www.nerrs.noaa.gov/doc/siteprofile/acebasin/html/envicond/watqual/wqintro.htm 2012.

[11]   R. Nave "Seawater" [Online] Available:http://hyperphysics.phy-astr.gsu.edu/hbase/chemical/seawater.html. 2000.

[12]   Apps Laboratories "Salinity - what do those figures mean?" [Online] Available: http://appslabs.com.au/salinity.htm. 2014.

[13]   Water Quality Association "Water Classifications" [Online]. Availble: http://www.pacificro.com/watercla.htm. 2012.

[14]   World Health Organization "Sulfate in Drinking-water" 2004.

[15]   A. Kabbani, H.H. Hammud, H. Itani "Spring Water of Lebanese Bekaa Valley". Ultra Science - Dimension of Pollution, 1, 48, 2001.

[16]   Daat Solutions of ICL Group, "Bromine" [Online]. Available :http://icl-ip.com/?products=bromine. 2013.

20

Estimation of Concentration of Dissolved Organic Matter from Sediment by using UV–Visible Spectrophotometer

Shamshad Khan[1], Wu Yaoguo[1], Zhang Xiaoyan[1], Liu Jingtao[2], Sun Jichao[2], Hu Sihai[1]
[1]Department of Applied Chemistry, School of Science,
Northwestern Polytechnical University, Xi'an, 710072, China
shamshadkhan768@yahoo.com;wuygal@nwpu.edu.cn
[2]The Institute of Hydrogeology and Environmental Geology, Shijiazhuang, 050803, China

Abstract–The objective of this research was to develop a spectrophotometric technique for estimation of concentration of dissolved organic matter (DOM) from sediment. The results showed that the absorbance 272nm is good surrogate to the concentration of the DOM. In addition, absorption of DOM confirmed to be independent of pH values ranging from 3.0 to 11. Compare to the chemical oxygen demand method and TOC–analyzer, the develop correlation technique using the UV–visible absorption is very straightforward, needs much smaller sample volume, and shows a good reproducibility. The absorbance (272nm) increased with increasing the concentrations of DOM with very high correlation coefficient ($R^2 = 0.98$). So it is suggested that this technique should be used as a surrogate for estimation of concentration of DOM from sediment.

Keywords: absorbance, surrogate, sediment, dissolved organic matter, concentration.

1. Introduction

The importance of dissolved organic matter (DOM) as a key parameter in environmental studies is well understood. Dissolved organic matter is involved in mobilization, complexation of trace metals and transport of acidity, colloids, nutrients, metals, and pollutants [1, 2]. Dissolved organic matter act as an originator of carcinogenic disinfection by-products (DBPs) such as trihalomethanes (THMs) and haloacetic acids (HAAs) increased disinfectant demands in the distribution system [3–7] has been extensively studied for the protection of public health. These various interactions with the environment result from the fact that it represents a complex mixture of substances with chemical structure and different size. Therefore, the significance of DOM in the environment is very much associated to its composition.

Even if many studies have been accomplished to reduce the effects of DOM in water treatment, not many literatures are available for the characterization of DOM [8–12]. Therefore, the composition of DOM has been studied by various techniques including spectroscopic measurements, and physical and chemical fractionation, often in combination [1]. Result of these studies is that humic acids are the major part of DOM which includes carbonyl, carboxylic, hydroxyl, methoxyl, and phenolic functional groups. Primary composition for humic acids is carbon (52-56%), hydrogen (4-5.5%), oxygen (33-39%), and small percentages of nitrogen, sulfur, and phosphorus [13]. Based on the degradation products after alkaline hydrolysis and permanganate oxidation, Christman et al. [11] were able to suggest hypothetical structures for humic acids. However, it is for now impossible to observe the complete specific structure of DOM due to its complex structure. At this time, two analytical indirect techniques are commonly available to estimate the concentration of DOM. Total organic carbon (TOC) and UV absorbance measured at 254nm are used as substitute to evaluate the DOM concentration. Total organic carbon (TOC) and UV absorbance (254nm) estimate organic carbon content in the water and are used as a proxy to represent the DOM concentration and in particular, of THMs precursor material [14,15]. It was reported that correlation between UV absorbance at 254nm and disinfection by-products (DBPs) formation potential give a better estimation of organic matter reactivity than TOC [13,16]. But the limitation of this technique is also found in literature [17]. The value of absorbance (254nm) particularly depends on the concentrations of humic acids in water. UV absorbance (254nm) value of humic acids reduces by chlorination due to formation of organic halogen. When the concentration of humic acids is low, the UV254 acquired may be too low and generate a comparatively high random error [3].

Usually, the DOM in water produces an uncharacteristic spectrum by absorbing the ultraviolet light 200nm to 700nm with no sharp peaks [8] and the absorbance decreases as the wavelength increases. There is no visible variation between the spectra for humic and fulvic acids. It has been shown that no maximum or minimum peaks are present on the ultraviolet spectra of the aqueous solution of humic acids [8] and the optical absorbance reduces as the wavelength increases. Although the DOM with different sources may have different configurations and their chemical properties, their spectra are similar. Aromatic moieties parts in DOM offer the possibility to guess the proportion of dissolved organic carbon in DOM using UV–Visible spectrophotometer because UV absorption of organic solutes is directly proportional to their concentration in aromatic compounds [18,19]. This correlation demonstrated that the direct measurement of the UV absorbance gives suitable and valuable information for the characterization of the DOM in water. Wang and Hsieh [3] established a correlation between the absorbance spectrum of different wavelengths (250–350nm) and the concentration in natural organic matter. It is seems probable that a significant relationship could be developed between dissolved organic matter and absorbance at any wavelengths between 200nm and middle of visible range.

Based on the above discussions, the absorbance of DOM solutions at various wavelengths should be able to represent the concentrations of some specific constituents of DOM sample. Thus the objectives of this study are to establish the most favorable wavelength range that provides optimum predictabilty of dissolved organic matter from absorbance and the particular wavelength that should be able to represent the

concentration of some particular components of the DOM. In addition we will analysis the effect of pH on the UV–visible spectra of DOM.

2. Material and Methods

A sediment sample was collected from the topsoil (0-15cm) of Weihe riverbed in Xi'an, China. Sediment sample was air-dried, grounded, sieved to < 2 mm and stored in a plastic bottle until used. Selected physiochemical properties of sediment sample are shown in Table 1. Cation exchange capacity (CEC) of sediment was measured at pH 7.0 using $1mol\ L^{-1}$ ammonium acetate. The natural pH of sediment was measured in a 1:10 material/water ratio by pH meter (PHS-3C, Leici, Shanghai, China). Specific surface area (SSA) and micro porosity were evaluated by N_2-BET method (Tristar II 3020, Micromeritics, USA). Elemental composition was measured by energy dispersive spectrometer (INCA X-Act, TESCAN, Cezch). Soil organic carbon (SOC) content was determined by the wet dichromate oxidation method.

Table 1. Basic physicochemical properties of tested sediment.

pH	CEC (cmol.kg⁻¹)	SSA m²g⁻¹	Pore Size (nm)	SOC	C	Ca	Al	Fe
						%		
8.0	10.54	7.23	92.32	4.79	2.80	3.98	5.64	3.57

This sediment was used as source of DOM in water. Extraction of DOM was carried out just before absorption experiments by adding 200ml deionized water to 6gram of sediment in a plastic bottle and shaking for 24 h on an orbital shaker at (20 ± 1) °C. After that, the suspension was centrifuged at 900×g for 30 min and filtered through 0.45-μm pore size polycarbonate filter. Total organic carbon was determined by total organic carbon analyzer (TOC–V-VSH, Shimadzu, Tokyo, Japan). The non-purgable organic carbon (NPOC) technique was used to measure TOC (organic carbon remaining in an acidified sample after purging the sample with gas). The result TOC was calculated as a mean of the three valid measurements. Stock solution of 70 mg L^{-1} of DOM was used to prepare six different concentrations of diluted solutions of DOM. UV–visible absorbance measurements were performed on a high precision, double-beam spectrophotometer (model 2550) between 220 and 600 nm with the reference of distilled water. A quartz cell with 1.0 cm path length was used. Samples were allowed to warm to room temperature before measurement. Duplicates and measurement of the distilled water were made every 10-12 samples to ensure instrument stability. Buffer solution to maintain pH were not required because solutions were prepared at different pH have showed that UV–visible spectra were not notably affected by the 3 to 11 range of pH as shown in Figure 1.

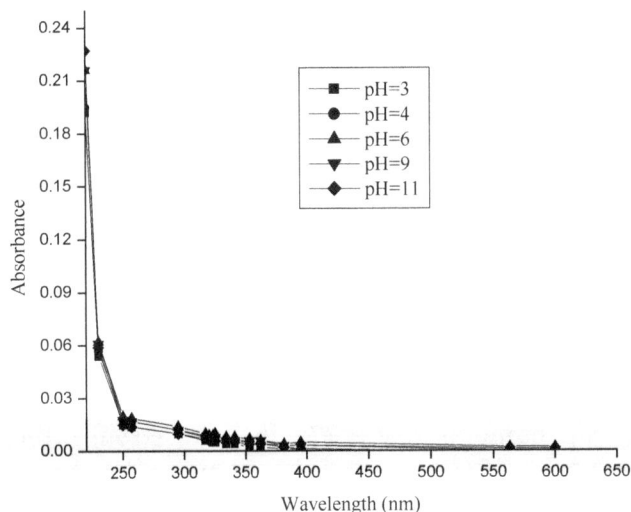

Figure 1. Effect of pH on UV-visible spectra of dissolved organic matter in water from sediment.

3. Results and Discussion

For selection of appropriate wavelength to the concentration of DOM and absorbance in water, various concentrations (9.1, 11.5, 13.9, 16.3, 18.7, 21.1, 23.5mg L^{-1}) of DOM. Figure 2 illustrates that the absorbance of DOM increases as the concentrations increase. The absorbance at wavelength length less than 250nm is much larger than of longer wavelength. As a result, a sharp slope is observed at the shorter wavelength (<275nm), this phenomenon is much more visible when the concentration of DOM is high. The absorbance measured at greater wavelength (>360nm) is relatively less than absorbance at 250-360nm. This result is approximately in agreement with the experimental results of Wang and Hsieh [3].

As can be seen in figure 2, absorbance values less than 250nm were not included in the study because of the strong absorbance by nitrates. Wavelength greater than 360nm is also not included due to the insensitivity of absorbance to changes in DOM at longer wavelength. As a consequence the correlation between absorbance and DOM data are restricted to the 250–360nm range.

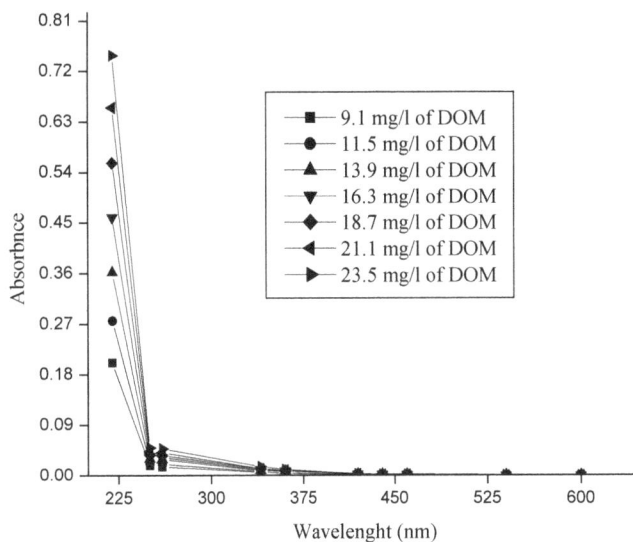

Figure 2. UV-visible spectra of different concentrations of dissolved organic matter in water from sediment.

For the estimation of the most favorable wavelength ranges between 250nm and 360nm, it is possible to add all the data to correlation regression analysis by considering the 10nm part length of restricted wavelength range 250–360nm. Correlation coefficient (R^2) was determined for each part length of10nm of the restricted wavelength range 250–360nm. The highest correlation coefficient (R^2= 0.97) was obtained between range of 270nm and 280nm. Figure 3 shows the correlation of DOM and selected wavelengths range (250–360nm) of 23.5mL solution of DOM. The maximum linear correlation coefficient (R^2= 0.97) obtained at wavelength range 270 and 280nm by analysis of all the data of between 250nm and 360nm.The most favorable correlation coefficient (R^2=0.97) was obtained between the 270 and 280nm. (The correlation coefficient lower than 270nm and higher 280nm are not shown here due to insignificant values). There is steady decline in the correlation coefficient below 270nm is probably due to growing amount of interferences by organic compounds of dissimilar classes. Drastic decline in correlation coefficient at higher wavelengths (>280nm) that is possibly due to declining the sensitivity of absorbance to the presence of organic compounds. It thus illustrates that DOM is mostly closely associated to absorbance in the wavelength range 270–280nm. The correlation coefficients propose that the ideal analytical wavelength is between 270 and 280nm.

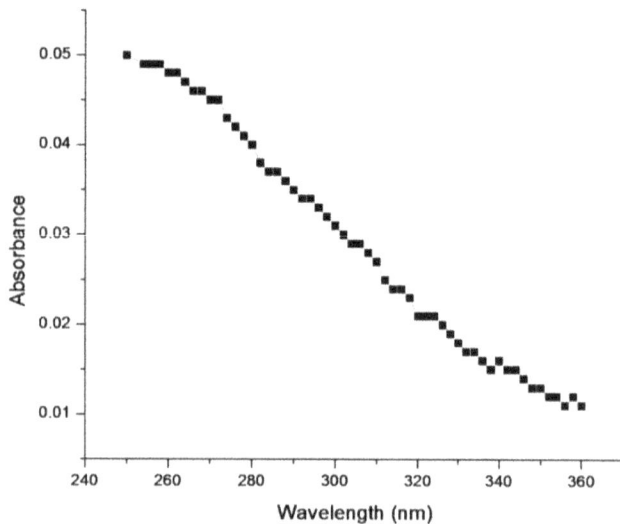

Figure 3. Relation between absorbance and concentration of dissolved organic matter as function of wavelengths.

For the selection of most favorable wavelength between the wavelength range 270–280nm, correlation coefficients were examined at interval of 2nm in the wavelength range 270–280nm while increasing the concentrations of DOM solutions. The optimum correlation coefficient (R^2=0.98) was obtained at the specific wavelength 272nm between the wavelength range 270–280nm (data not shown). For the verification of this technique, we developed a relationship between the absorbance (272nm) and the different concentrations of DOM determined by TOC carbon analyzer.

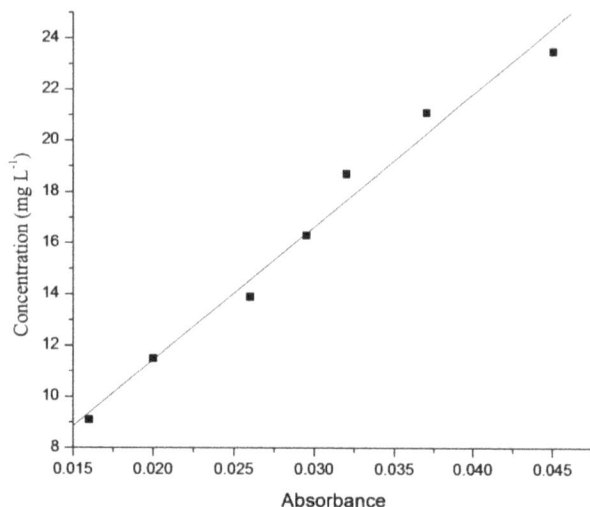

Figure 4. Correlation between absorbance at 272nm and concentration of dissolved organic matter in water from sediment determined by TOC carbon analyzer method.

Figure 4 shows the correlation between the absorbance (272nm) and the concentration of DOM of sediment in water. Absorbance at 272nm is highly related ($R^2> 0.98$; $P< 0.0001$) to the concentration of DOM determined by TOC carbon analyzer. It is recommended that the absorbance at 272nm should be used to estimate the concentration of DOM in sediments.

$$DOM\ (mg/l)= 518.93 \times Absorbance\ (272nm) + 1.065 \qquad (1)$$

In the above equation 1, 518.93 and 1.065 are the slope and y-intercept respectively. The intercept of the regression line illustrate the presence of 1.065mg L^{-1} of non–UV absorbing DOM, probably due to the organic matter containing no chromophores absorbing at 272nm. This DOM could contain carbohydrates, saturated aliphatic compounds. Data presented in Figure 5 compare the values of concentrations of DOM estimated by TOC carbon analyzer and UV–visible spectrophotometer at 272nm. A strong linear correlation (R^2=0.99) exists between these two methods.

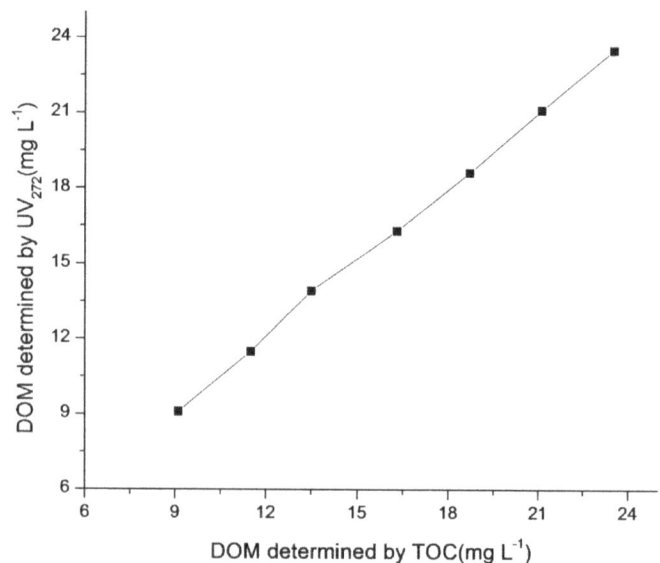

Figure 5. Correlation of the concentrations of DOM by UV_{272} method versus concentrations of DOM determined by the TOC carbon analyzer.

The higher correlation relationship of DOM with absorbance demonstrates that this technique should be used for assessment of concentration of DOM and it is very suitable and economical for estimation of DOM in water from sediment. Because other techniques like

Chemical oxygen method and TOC analyzer techniques are more time consuming, expensive, and need more sophisticated instruments, this method is interesting because it only requires a UV–visible spectrophotometer. Another benefit is that in normal working range of pH no need to use buffer solution because changes of pH within normal working ranges has no effect on this technique.

4. Conclusion

The most favorable correlation coefficient (R^2= 0.98) wavelength range is obtained among 270nm and 280nm by adding all the data to correlation regression analysis. There is high correlation coefficient (R^2= 0.98) at specific wavelength 272nm between the wavelength (270–280nm) and absorbance by using UV–visible spectrophotometer. The correlations of DOM with absorbance indicate that this method should be used for estimation of concentration of DOM in water form sediment. This method has benefit over general techniques for estimation of DOM concentrations with less time consumption. It is also valuable in normal working pH range (3–11).

Acknowledgements

This work was supported by the Land and Resources Scientific Research of China from special fund (20111020) in the public interest, the project titled "survey and assessment of groundwater pollution in main cities of Northwestern China (1212011220982)" and "NPU Foundation for Fundamental Research (NPU-FFR-JCR20130145)".

References

[1]   J. Dilling, K. Kaiser "Estimation of the hydrophobic fraction of dissolved organic matter in water samples using UV photometry" Water Research, 36(20), 2002, pp. 5037–5044.

[2]   K. Shamshad, W. Yaoguo, Z. Xiaoyan, H. Sihai, L. Tao, F. Yilin, L. Qiuge "Influence of dissolved organic matter from corn straw on Zn and Cu sorption to Chinese loess" Toxicological & Environmental Chemistry, 95(8), 2013, pp. 1318-1327.

[3]   G.S. Wang, S.T. Hsieh "Monitoring natural organic matter in water with scanning spectrophotometer" Environment International, 26(4), 2001, pp. 205–212.

[4]   H.O. Neung, A.P. Brian, A.M. Philip, J.H. Peter, M.B. Sandra, O. Noriaki, M.L. Kavvas, A.B. Brian, R.H.

William "The role of irrigation runoff and winter rainfall on dissolved organic carbon loads in an agricultural watershed" Agriculture, Ecosystems and Environment, 179(1), 2013, pp.1– 10.

[5]   E.R. Newall, F.D. Hulot, J.L. Janeau, A. Merroune "CDOM fluorescence as a proxy of DOC concentration in natural waters: a comparison of four contrasting tropical systems" Environmental Monitoring and Assessment, 186(1), 2014, pp.589–596.

[6]   G. Rachel, J.H. Peter, W. Naomi, F. Christopher "Dissolved organic carbon and trihalomethane precursor removal at a UK upland water treatment works" Science of the Total Environment. 468–469(1), 2014, pp.228–239.

[7]   A.L. Bonnie, M.C. Rose, S.W. Howard "Changes in dissolved organic matter fluorescence and disinfectionbyproduct formation from UV and subsequent chlorination/chloramination" Journal of Hazardous Materials, 264(1), 2014, pp. 411– 419.

[8]   G. Crozes, P. White, M. Marshall "Enhanced coagulation: its effect on NOM removal and chemical costs" Journal American Water Works Association, 87(1), 1995, pp. 78–89.

[9]   J.G. Jacangelo, J. Demarco, D.M. Owen, S.J. Randtke "Selected processes for removing NOM: an overview: Natural organic matter" Journal American Water Works Association, 87(1), 1995, pp. 64–77.

[10] S.J. Randtke "Organic contaminant removal by coagulation and related processes combinations" Journal American Water Works Association, 80(5), 1988, pp. 40–56.

[11] R.F. Christman, D.L. Norwood, Y. Seo, F.H. Frimmel "Humic substances: II. In search of structure" In: Hayes, M.H.B., Mac-Carty, P., Malcolm, R.L., Swift, R.S. (eds).Publishing Wiley, Chichester, England, 1985, pp. 451–463.

[12] G.G. Choudhry "Humic substances, structural, photophysical, photochemical and free radical aspects and interactions with environmental chemicals" Gordon & Breach, New York, 1984, pp. 98–106.

[13] D.A. Reckhow, P.C. Singer, R.L. Malcolm "Chlorination of humic materials: byproduct formation and chemical interpretations" Environmental Science & technology, 24(11), 1990, pp. 1655–1664.

[14] I.N. Najm, N.L. Patania, J.G. Jacangelo, S.W. Krasner "Evaluating surrogates for disinfection byprod-

ucts" Journal American Water Works Association, 86(6), 1994, pp.98–106.

[15] A. Eaton "Measuring UV-absorbing organics: a standard method" Journal American Water Works Association, 87(2), 1995, pp.86–90.

[16] P. Roccaro, F.G.A. Vagliasindi "Differential vs. absolute UV absorbance approaches in studying NOM reactivity in DBPs formation: Comparison and applicability" Water Research, 43(3), 2009, pp.744–750.

[17] N. Ates, U. Yetis, M. Kitis "Effects of Bromide Ion and Natural Organic Matter Fractions on the Formation and Speciation of Chlorination By-Products" Journal of Environmental Engineering, 133(10), 2007, pp.947–954.

[18] S.J. Traina, J. Novak, N.E. Smeck "An ultraviolet absorbance method of estimation of the percent aromatic carbon of humic acids" Journal of Environmental Quality, 19(1), 1990, pp. 151– 153.

[19] Y.P. Chin, G. Aiken, E. O'Loughlin "Molecular weight, polydispersity, and spectroscopic properties of aquatic humic substances" Envorimental Science & technology, 28(11), 1994, pp.1853–1858.

21

NOx De-Pollution Using Innovative Mortars and Concretes - From Garage Prototypes to Tunnel Pilot

E. Stora, M. Horgnies, I. Dubois-Brugger, L. Dao-Castellana

Lafarge Centre de Recherche, 95 rue du Montmurier,
Saint Quentin-Fallavier, F-38291 France

eric.stora@lafargeholcim.com; matthieu.horgnies@lafargeholcim.com; isabelle.dubois-brugger@lafargeholcim.com

Abstract - *Air pollution generated by transportation (cars, trucks, etc.) affects the health of millions of persons around the world. Some of the most toxic air pollutants are composed of nitrogen oxides (especially nitrogen dioxide, NO2) and volatile organic compounds (VOCs). These gas pollutants affect also the environment by promoting the formation of ground-level ozone and micro-particles harmful to human health and ecosystems. These compounds are responsible for the rise of acute respiratory diseases, asthma and allergies in urban areas.*

The patented technology of de-polluting concrete does not rely on photo-catalysis and can function perfectly well without sun light, which is especially suitable for use in confined areas prone to pollution peak (tunnels or parking garages). The experiments done in laboratory demonstrated that the addition of certain activated carbons into the mix improves the NO2 absorption properties without affecting the mechanical strength of concrete.

In order to demonstrate the interest of this innovation, two parking garages were built with the walls made of a de-polluting concrete or covered by a de-polluting mortar. The tests done using gasoline generator as a source of pollutants confirmed a significant abatement of NO2. The de-polluting effect still detected after one year of aging (carbonation) helped us in scheduling a field test in a tunnel located in the region of Lyon (France). This field-test consisted in spraying the de-polluting mortar on the walls of the ventilation plant located at the top of a chimney of a motorway tunnel. The results after 6 months of trials confirm a significant NO2 reduction rate in the air released by the chimney.

Keywords: Air pollution, concrete, mortar, tunnel, confined spaces, NOx, parking.

1. Introduction

Hazardous gaseous pollutants, such as nitrogen oxides (NOx), are mainly produced by car and truck traffic. They affect the health of millions of people by worsening allergic, respiratory and cardiovascular diseases [1-2]. Indeed, the development of new de-polluting materials that could reduce the NOx concentrations in confined spaces (such as in office or residential buildings, but also in basements, underground garages, industrial plants, warehouses and motorway tunnels) could improve the life quality in urban areas [3-4].

We have previously shown that ordinary Portland cement pastes are porous alkaline material that can absorb NO_2 at ambient temperature until the complete carbonation of the paste [5]. In addition, the activated carbons are known to be excellent adsorbents for NO_2 [6-8] and can be added to the mix of mortar and concrete [9]. The validity of the solution at a lab scale has been demonstrated previously [9-10] and the challenge is now to prove that it works in-situ in prototypes.

The objectives of this paper consist in presenting: (i) the NO_2 reduction rates in instrumented garage boxes built with de-polluting building materials (ii) the results of a field test undertaken in the ventilation plant of a chimney of a motorway tunnel.

2. Presentation of the de-polluting material

It is well established that NO_2 can react in alkaline aqueous solutions to give nitrite and nitrate ions [11]. The most strongly alkaline phases like $Ca(OH)_2$ of the hydrated cement pastes show a high absorption capacity but this effect is temporary. Indeed, the NO_2 reduction rate by the reference cement paste is affected by carbonation caused by atmospheric CO_2: this gas converts the highly alkaline hydrates to calcium carbonates, which are less reactive with NO_2. Laboratory-scale experiments show that the addition of small percentages of activated carbon (AC) powder to the cement paste increases the NO_2 reduction rate and reduces the influence of carbonation, by prolonging the de-polluting effect.

Two types of hardened cement pastes were tested in a lab reactor. The reference cement pastes were prepared by mixing 0.45 parts of water per 1 part of cement. Other cement pastes were manufactured with additions of a selected activated carbon, keeping the same water/cement (W/C) ratio. Previous work showed that the addition of AC powder did not affect significantly the microstructure and the mechanical strengths of the cement pastes [9]. Finally, the composition of the polluted gas was modified by diluting the NOx in ambient air in order to quantify the NO_2 abatement in presence of carbon dioxide (380 ppmv of CO_2). The addition of activated carbon in the cement paste improved significantly the NO_2 abatement. Figure 1 evidences a stable NO_2 abatement close to 70%, even in presence of CO_2, which reacts with the cement-based hydrates present at the surface of the hardened paste.

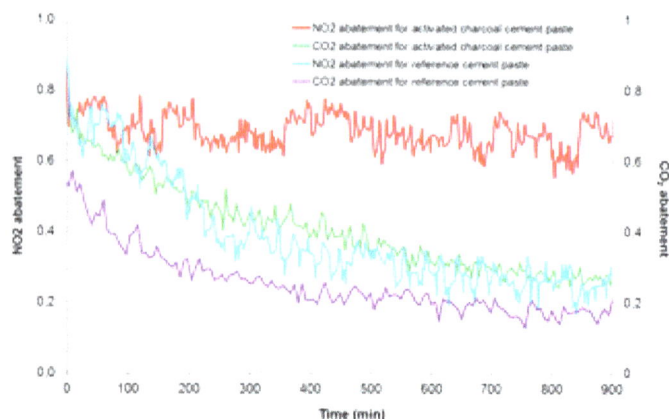

Figure 1. Variations of the NO_2 and CO_2 abatements using reference and activated carbon cement pastes.

NOx absorption experiments confirmed the predominant role of the activated carbon powder. The NO_2 absorption in the alkaline cement pastes is then governed by two consecutive phenomena:

- NO_2 physisorption on the activated carbon powder,

- Reaction of the adsorbed NO_2 with the alkaline pore solution of the surrounding cement paste, which depend on its carbonation.

3. Pilot tests done in parking boxes

3.1. Description of the instrumented parking boxes

Two garages (dimensions: 4.0x2.0x2.2 m, see Figure 2) were built using walls made of reference concrete and activated carbon concrete, respectively. The volume was 18 m^3 for an exposed surface area of 18 m^2 (the ceiling and floor were insulated by sticking a plastic liner). The both concretes were manufactured from Portland cement, limestone fillers, sand, gravels and admixtures (superplasticizer). Activated carbon was added to the mix of de-polluting. No special treatment was applied to the surfaces of these concretes after demoulding. For the measurements done in the garages, a petrol-engine generator was used, allowing the injection of multiple air pollutants, similar to the ones released by the road traffic. The concentrations of NOx were measured every 5 minutes using an automatic gas analyzer. The estimated residence time is 14 minutes.

Figure 2. Picture of the garages made of reference (left) and activated carbon concrete (right).

3.2. NO_2 abatement measured in the parking boxes

Table 1 compares the NO_2 abatements measured during the different tests, according to: (i) the NO_2 concentration injected at the input of the garages, and (ii) the conditions in terms of temperature and relative humidity. By comparing the concentrations measured at the input and output of the garage, no significant absorption of NO_2 was detected in the garage, while an abatement of 20-25% was observed in the garage made of activated carbon concrete, whatever the weather conditions and even in presence of other pollutants released by the generator.

Table 1. De-polluting tests undertaken using the garages made of reference concrete or activated carbon mortar and concrete.

Material type	T (°C)	RH (%)	Input NO2 concentration (µg/m3)	Output NO2 concentration (µg/m3)	NO2 absorption rate (%)
Reference concrete	24	39	421	407	< 3
Reference concrete	22	53	382	382	< 1
AC concrete	27	32	453	346	24
AC concrete	27	32	937	688	26
AC mortar	31	29	721	402	44
AC mortar	24	52	604	335	45
AC mortar	28	34	717	411	42
AC mortar	20	71	476	279	41

The absence of absorption for the reference concrete can be explained by the presence of CO_2 in the atmosphere leading to a carbonation of the surface of the walls, which tends to reduce the reactivity of NO_2 with the alkaline medium [12]. On the other hand, no effect of carbonation is observed on the NO2 reduction measured in the activated carbon concrete garage supporting the laboratory scale observations. The solutions leached from the walls of the garages were analyzed by ion chromatography after the gas absorption tests. The results showed almost no nitrate or nitrite in the leachate from the reference concrete and significant levels of nitrates and nitrites in the leachate coming from the activated carbon concrete yet below the water quality standards [11].

3.3. Improvement of the NO2 abatement with a de-polluting spray mortar

One of the garages (previously used to study the NO_2 absorption on the reference concrete) was refurbished by spraying a new de-polluting mortar on the walls. This de-polluting mortar was mixed according to the same AC/cement ratio as the one used for the activated carbon concrete. The W/C ratio and the type and amount of admixtures of this spray mortar were designed to stick well on the walls. The overall porosity (measured by mercury intrusion porosimetry) of this de-polluting mortar was close to 30%, which is significantly higher than the one measured for the activated carbon concrete (about 15%). Figure 3 compares the NO_2 concentrations alternately

measured at the inlet and at the outlet of the garage covered by the de-polluting spray mortar. The NO_2 concentrations were measured every 5 minutes and during a few hours using an automatic gas analyzer. The temperature and the relative humidity were about 28°C and 34%, respectively. NO_2 abatement close to 42% was detected by using the activated carbon mortar, which is a significant improvement compared with the previous tests done with a de-polluting concrete.

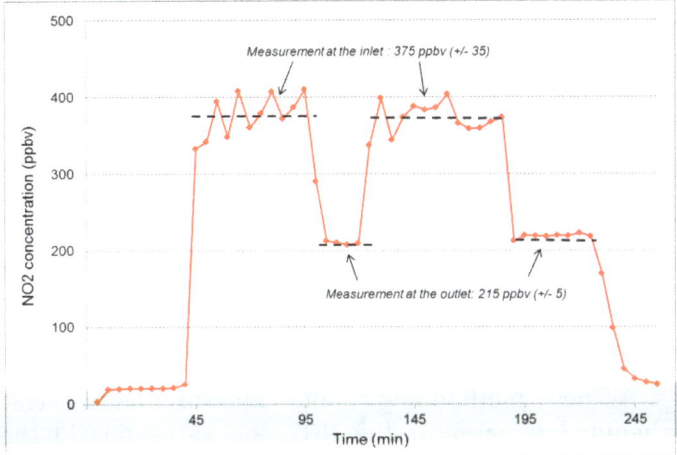

Figure 3. Examples of NO2 concentrations measured in the garage covered by a de-polluting spray mortar (1ppb = 1.912 µg/m3).

Table 1 summarizes the NO_2 abatements measured during the different tests carried out in the garage covered by the de-polluting spray mortar (using a residence time estimated to 14 min). A stable abatement of about 40-45% was always detected whatever the levels of NO_2 concentrations injected at the inlet of the garage. As previously demonstrated using the de-polluting concrete, the presence of other gas pollutants released the generator (CO, CO_2, VOCs) did not affect the NO_2 abatement. Final tests were also done by decreasing the residence time of the gas pollutants into the garage. As detailed in [12], the NO_2 abatement varies as a function of the residence time and decrease by 38% NO2 as the residence time is divided by a factor three. This last parameter was taken into account for planning bigger field-tests.

4. Pilot tests in a chimney of a motorway tunnel

4.1. Spray of the de-polluting mortar on the walls of the ventilation plant

The test done to assess the depolluting effect of the activated carbon mortar took place in 2014 at the top of a chimney of a motorway tunnel, which is located in the region of Lyon (one of the biggest cities of France). The length of this 4-ways tunnel is 1750 m and its traffic is about 50000 vehicles per day. The polluted air coming from the tunnel is extracted by five chimneys topped by a ventilation plant. Each ventilation plant can be separated in three distinct rooms: inlet room, fan room and outlet room (as detailed in Figure 4). The roof of the outlet room is partially open to release the polluted gas into the atmosphere.

The field test undertaken in September 2014 consisted in spraying more than 125 m² of de-polluting mortar on the walls of the inlet room. The final thickness of the layer of mortar was about 2 cm. As shown by the pictures of Figure 4, the activated carbon mortar was sprayed on the four walls of the inlet room (inside volume close to 200 m³). Note that the turbulent polluted gas coming from the chimney is supposed to be in contact with the walls of the inlet room covered by the depolluting mortar, before being ejected into the outlet room thanks to the three extractors located in the fan room.

The methodology of measurements was scheduled to monitor the NO2 concentrations in the inlet and outlet rooms, before and after the spray of the de-polluting mortar. During the tests, the NO_2 concentrations were measured by regularly positioning at least six absorbent sensors (Radiello® code 166) in each room at a height of at least 45 cm and at less than 1 m from the air ejection. The validity of the measurement by these sensors was checked by the following experiments:

1. The data given by this type of sensor were previously validated during the pilot tests done in the garages by using an automatic gas analyzer, where we observed a good agreement between the two measurement methods.

2. Another test was performed recently inside the inlet room of the tunnel chimney to compare the Radiello® sensor and an automatic gas analyzer located at the same position. The NO2 concentration given by the sensor (942 µg/m3) was in good agreement with the average measured with the automatic NOx analyzer (991 µg/m3).

3. During the different tests, 2 sensors were positioned side by side to check the variability of the measurement. The average difference observed between two sensors located side by side is 6 %.

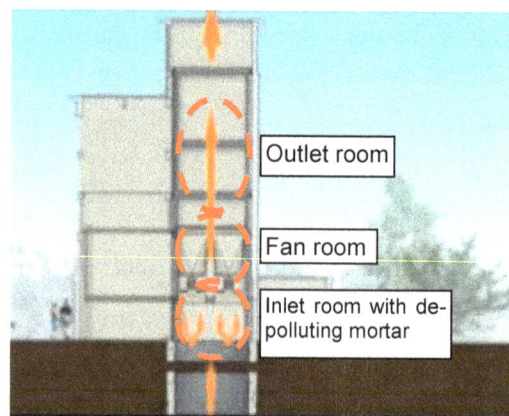

(a)

(b)

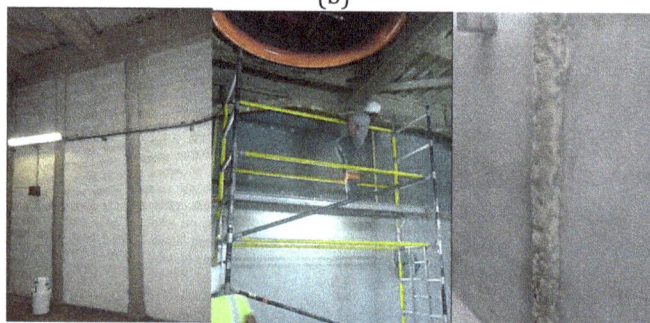

(c)

Figure 4. Field test done for a motorway tunnel: (a) schematic of the ventilation plant (the NO2 concentrations were measured in the inlet and outlet rooms at the top of one of the chimneys); (b) inlet room (6.3x6.3x5m) where the walls were covered by a de-polluting mortar; (c) photos before, during and after spray of the de-polluting mortar (from left to right).

A few series of tests detailed on Table 2 were carried out between 2014 and 2015 to measure the de-polluting effects obtained by the application of the mortar. Note that the tests were performed at a slow ventilation rate. One of the three fans was regulated in manual configuration at 300 rds/min, while the two others were closed. This allowed for an average air speed of 1.6 m/s with variations 1.4 to 1.8 m/s in the inlet room and a sufficient residence time for the air pollutants to react with the de-polluting walls.

Table 2. De-polluting tests undertaken using the garages made of reference concrete or activated carbon concrete.

Test series	T (°C)*	RH (%)*	NO_2 level ($\mu g/m3$)*	Traffic (vehicles per hour)	Sensor exposure time (in hours)
One month before spraying the mortar	28	No data	421	1700	18h
One month after spraying the mortar	21	64	450	Max 2000	6h
One month after spraying the mortar	20	62	600-653	Max 4000	18h and 53h
Five months after spraying the mortar	9	38	453	No data	6 h
Seven months after spraying the mortar	14	48	937	No data	3 h

* measured in the inlet room

4.2. Results one month before and after mortar application

Figure 5 compares the NO_2 concentrations measured in the inlet and outlet rooms, before and after the spray of the de-polluting mortar. All the absorbent sensors used to monitor the NO_2 concentrations were exposed for 18h to the pollutant gas flow. The levels of concentrations detected in the inlet room were similar to the one measured in the outlet room before the spray of the de-polluting mortar (according to the uncertainty of measurement), giving an average value close to 600 $\mu g/m^3$ of NO_2. However, the level of NO_2 concentrations measured in the outlet room decreased drastically after the spray of the de-polluting mortar in the inlet room, reaching an average value close to 200 $\mu g/m^3$ (for a NO_2 reduction rate of about 60%). Even if the comparison at two similar but different testing periods of time has to be taken cautiously, a significant effect of the material was observed on NO_2 concentration confirming the previous results obtained in lab and in the garage prototypes.

Figure 6 shows the NO_2 concentrations measured into the inlet and outlet rooms after the spray of the mortar in the inlet room. Several absorbent sensors were continuously exposed to the polluted gas flow for distinct periods (from 6h to 53h). In every case, the NO_2 concentrations detected into the outlet room were lower than the ones measured in the inlet room.

The average NO_2 reduction rate in 24h measured is of 48% with some variations observed depending on the exposure duration. Considering the concentrations measured during the longest periods of exposure (from 18h to 53h) characterized by a high traffic (peak close to 4000 vehicles per hour), the NO_2 reduction rate could reach over 60%, which is close to the rate calculated from the concentrations measured before and after the spray of the activated carbon mortar (see Figure 6). On the other hand, the NO_2 reduction goes down to 32 %, considering only the values measured for a period of 6h (9AM-3PM) characterized by a low traffic (maximum of about 2000 vehicles per hour), which could reduce the concentrations detected in the inlet room (and thus under-estimate the NO_2 reduction rate). Thus the higher the NO_2 concentration the more effective is the depolluting material. This may be explained by the fact a higher NO_2 concentration induces that more NO_2 gas penetrates faster into the material and react with it.

These first results about the NO2 reduction rate measured in this field test should be considered cautiously. Indeed, they were measured a few weeks after the spray of the activated carbon mortar. Other sets of measurements were then scheduled in order to confirm the robustness of the de-polluting effect.

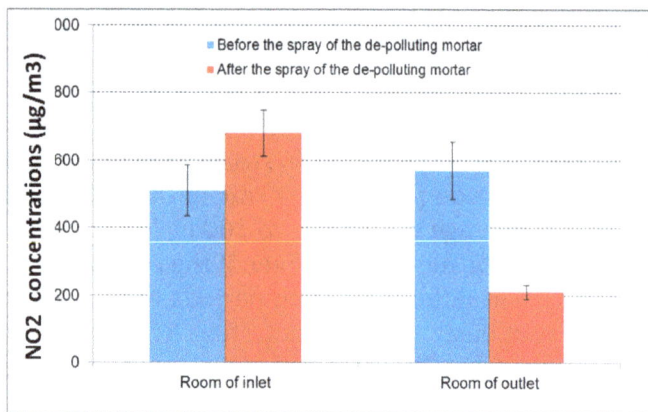

Figure 5. NO2 concentrations measured in the inlet and outlet rooms, before and after the spray of the de-polluting mortar (the absorbent sensors were exposed for 18h to the polluted gas flow).

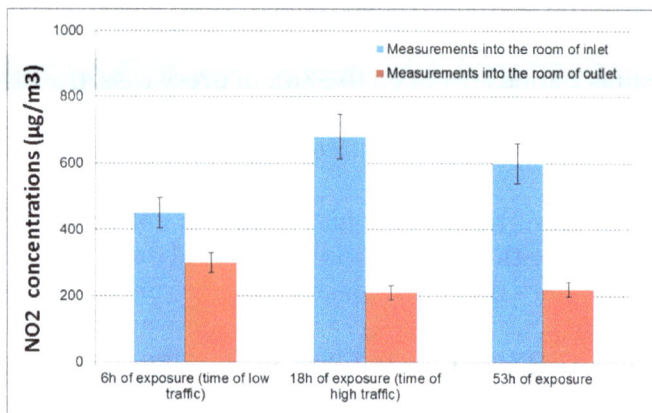

Figure 6. NO2 concentrations measured in the inlet and outlet rooms after the spray of the activated carbon mortar in the inlet room; the absorbent sensors were continuously exposed for different time durations (from 6h to 53 h).

4.3. Results measured five and seven months after the application of the mortar

Other tests done five and seven months after the application of the spray mortar, confirm the significant reduction rate of NO_2 due to the action of the de-polluting mortar containing activated carbon.

After 5 months, the level of pollution measured in the inlet room (Figure 7) was a bit higher than the ones measured previously but it may be due to a more intense traffic. However, the reduction rate of NO_2 deduced from the measurements done in the inlet and outlet rooms was similar to the ones measured one month after the application of the de-polluting mortar. The results after 5 months of test thus do not highlight any sign of saturation of the material as it was observed from the lab previous experiments (Figure 1).

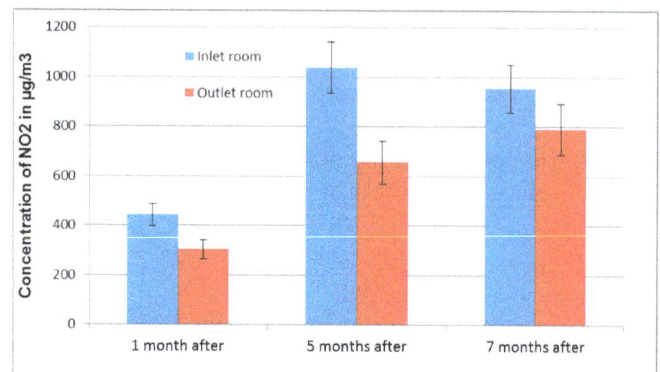

Figure 7. NO2 concentrations measured in the inlet and outlet rooms, 1, 5 and 7 months respectively, after the spray of the activated carbon mortar in the inlet room; the absorbent sensors were continuously exposed for 6h for the 1 and 5 months trial and 3h for the 7 months trial.

A decrease is detected after 7 months, but the NO_2 abatement of about 20 % is still significant. It is important to remind that the material is exposed to conditions where the pollution is particularly concentrated. It is thus possible to anticipate that the de-polluting effects should last for years in more common applications such as underground parking lot, where the NO_2 concentration is about 5 times lower. Tests after one year of exposure are being scheduled to confirm if the material gets saturated over time. In that case, a solution can be to wash the material with water and collect the nitrite and nitrate ions formed by reaction of gaseous NO_2 in the mortar [11]. This is a way to reactivate the de-polluting effect induced by the material.

4.4. Error measurement assessment

The fiability of the measurements with the Radiello® sensors was assessed by the following tests done during the last campaign:

1. Comparison with an automatic NOx analyzer: a test was performed inside the inlet room of the tunnel chimney to compare the Radiello® sensor and an automatic gas analyzer located at the same position. The NO_2 concentration given by the sensor (942 µg/m3) was in good agreement with the average recorded by the automatic NOx analyzer (991 µg/m3). Note that a peak over 1200 µg/m3 was observed in the middle of the rush hours, showing that the NO_2 concentration can strongly vary with traffic.

2. In four different locations of the inlet and oulet rooms, two sensors were positioned side by

side to check the variability of the measurements. The average differences observed between two sensors located side by side in the inlet room is 5.5 %, which is an acceptable error. The mean error in the outlet plenum is a bit higher (6.8%). This is probably due to the fact that the oulet room is in semi-open conditions where the air flux is more random.

3. According to the fabricants, the measurement with Radiello® sensor is difficult with air speeds higher than 10 m/s. The air speeds in the chimneys were also monitored using two anemometers located at the same position of two NO2 sensors. The average recorded is 1.6 m/s, all the values being comprised between 1.4 m/s and 1.8 m/s. It shows that the ventilation worked regularly during the tests.

These tests show some variability of the NO_2 measures, which remains acceptable for a full-scale pilot in real conditions.

5. Conclusion

The objective of this paper was to validate at a higher scale the laboratory-scale experiments showing that the addition of specific activated carbon powders to the cement paste increases durably the NO_2 reduction rate. Pilot-scale tests done with both garages built respectively in reference concrete and activated carbon concrete confirmed the general conclusions drawn from laboratory-scale tests: NO_2 reduction rates of 20-25% were measured in the activated carbon concrete garage whatever the weather conditions and after one year of carbonation while the reference garage gave almost no absorption.

Moreover, it was demonstrated that spraying a porous de-polluting mortar (containing activated carbon powder) can enhance the de-polluting effect (reaching a NO_2 reduction rate of 40-45% in certain conditions) even in presence of other gaseous pollutants (CO_2, CO, VOCs) released by a generator. The tests performed into the garages helped us to plan a test at a higher scale in a motorway tunnel.

This field test consisted in monitoring NO_2 before and after spraying about 125 m² of de-polluting mortar on the walls of the inlet room of the ventilation plant located at the top of a chimney of a tunnel located in the south-east of France. The first results confirm a significant NO_2 reduction rate by about a factor 2 by comparing the concentrations measured in the outlet room of the ventilation plant. The more recent results are also encouraging, since no decrease had been observed before 6 months of continuous exposure in the urban tunnel.

Future experiments are planned to model the long-term evolution of the de-polluting effect under various exposure conditions of ventilation in this tunnel where the traffic can reach about 50000 vehicles per day. Thanks to these encouraging results, other field tests are now on-going in different parking lots in Spain.

References

[1] P. Blondeau, V. Lordache, O. Poupard, D. Genin and F. Allard, "Relationship between outdoor and indoor air quality in eight French schools," *Indoor Air*, vol. 15, pp. 2-12, 2005.

[2] C. S. Mitchell, J. Zhang, T. Sigsgaard, M. Jantunen, P. J. Lioy, R. Samson and M. H. Karol, "Current state of the science: health effects and indoor environmental quality," *Environ. Health Perspect.*, vol. 115, pp. 958-964, 2007.

[3] S. Shen, M. Burton, B. Jobson and L. Haselbach, "Pervious Concrete with Titanium Dioxide as a Photocatalyst Compound for a Greener Urban Road," in *Environment TRB proceeding*, Washington, 2012.

[4] A. Challoner and L. Gill, "Indoor/outdoor air pollution relationships in ten commercial buildings: PM2.5 and NO2," *Build. Environ.*, vol. 80, pp. 159-173, 2014.

[5] N. J. Krou, I. Batonneau-Gener, T. Belin, S. Mignard, M. Horgnies and I. Dubois-Brugger, "Mechanisms of NOx entrapment into hydrated cement paste containing activated carbon – Influences of the temperature and carbonation," *Cem. Concr. Res.*, vol. 43, pp. 51-58, 2013.

[6] W. J. Zhang, A. Bagreev and F. Rasouli, "Reaction of NO2 with activated carbon at ambient temperature," *Indust. Eng. Chem. Res.*, vol. 47, pp. 4358-4362, 2008.

[7] X. Gao, S. Liu, Y. Zhang, Z. Luo, M. Ni and K. Cen, "Adsorption and reduction of NO2 over activated carbon at low temperature," *Fuel Process. Technol.*, vol. 92, p. 139–146, 2011.

[8] N. Shirahama, S. H. Moon, K.-H. Choi, T. Enjoji, S. Kawano, Y. Korai, M. Tanoura and I. Mochida, "Mechanistic study on adsorption and reduction of NO2 over activated carbon fibers," *Carbon*, vol. 40, p. 2605–2611, 2002.

[9] M. Horgnies, I. Dubois-Brugger and E. Gartner, "NOx de-pollution by hardened concrete and the influence of activated carbon additions," *Cem. Concr. Res.*, vol. 42, p. 1348–1355, 2012.

[10] M. Horgnies, I. Dubois-Brugger, F. Serre and E. Gartner, "NOx de-pollution using activated charcoal concrete - From laboratory experiments to tests with garage prototypes," in *4th International Conference on Environmental Pollution and Remediation*, Prague, 2014.

[11] L. J. Ignarro, J. M. Fukuto, J. M. Griscavage, N. E. Rogers and R. Byrns, "Oxidation of nitric oxide in aqueous solution to nitrite but not nitrate: Comparison with enzymatically formed nitric oxide from L-arginine," *Proc. Nat. Acad. Sci.*, vol. 90, pp. 8103-8107, 1993.

[12] M. Horgnies, I. Dubois-Brugger and E. Stora, "An innovative depolluting concrete doped with **activated carbon to enhance air quality**," in *10th Int. Concr. Sus. Conf.*, Miami, 2015.

22

Hybrid Clay Nanomaterials with Improved Affinity for Carbon Dioxide through Chemical Grafting of Amino Groups

Saadia Nousir, Andrei-Sergiu Sergentu[1], Tze Chieh Shiao, René Roy and Abdelkrim Azzouz*

Nanoqam, Department of Chemistry, University of Quebec in Montreal,
Montreal, Qc, Canada H3C3P8
[1]Laboratory of Catalysis and Microporous Materials (LCMM),
University of Bacau, Romania
azzouz.a@uqam.ca

Abstract – Chemical grafting of (3-aminopropyl) triethoxysilane on Na-montmorillonite in ethanol-water mixture or ethylene glycol as solvent resulted in two organoclays (NaMt-S-EW and NaMt-S-EG, respectively). The latter were characterized through X-ray diffraction, Fourier transform infrared spectroscopy, thermal analysis, differential scanning calorimetry and ^{29}Si solid-state nuclear magnetic resonance (NMR). Nitrogen adsorption isotherms measurements revealed lower specific surface area as compared to the starting clay mineral. This was explained by the formation of compact lamella stacks bound by silylation at the edges of the clay sheets. Thermal programmed desorption analyses (TPD) revealed an improved affinity towards carbon dioxide (CO_2) as compared to the starting clay mineral. NaMt-S-EW displayed higher affinity towards CO_2 than NaMt-S-EG, with retention efficiency factor exceeding 16 $\mu mol.m^{-2}$ for high amine content. Differential scanning calorimetry gave desorption enthalpy ranging from 148 to 467 $kcal.mol^{-1}$, suggesting that only chemical interactions are involved between the amino groups grafted and CO_2. CO_2 retention capacity exceeding 1.0 $mmol.g^{-1}$ with efficiency factor higher than 16 $micromol.m^{-2}$ can be obtained for higher amine content, in optimum content, when no CO_2 removal through forced convection takes place.

Keywords: Silylated montmorillonite, (3-aminopropyl)triethoxysilane, chemical grafting, TPD, CO_2 retention.

1. Introduction

A growing interest is now devoted to clay-based nanocomposites obtained through chemical grafting of organic silane moieties [1-4] and to the features of the silylating agents to be grafted [5,6]. Chemical grafting of amines has been intensively studied [7], but the use of clay minerals as inorganic supports for preparing CO_2 adsorbents has been scarcely tackled. The solvent used during silylation is expected to play a significant role, because it should strongly influence the swelling capacity and dispersion grade of both the clay mineral lamellae and the chemical species to be grafted [8,9]. For instance, ethylene glycol (EG) displays a surface energy close to that of montmorillonite [10], and favor interaction between ethoxy-silanes and OH groups of the clay mineral.

Attempts to modify montmorillonite resulted in the insertion of a single molecular silane layer within the interlamellar space [7]. Ethanol/water mixtures were found to enhance hydrolysis and polymerization of the silane molecules into siloxane bonds [7,11,12]. These two processes can occcur even before clay pillaring [13], leading to the formation of a wide variety of polymer sizes and shapes. Reportedly, the amount of water in such reaction mixtures was found to play a key role [14]. After previous polymerization, if any, only those molecules having appropriate size and spatial geometry can be incorporated in the interlayer space. At the edges and the structural defects of montmorillonite layers, the terminal OH groups are expected to act as active sites for the silylation process [7,11]. Consequently, the grafting process of silane species should strongly depend on the surface density of the terminal OH groups. These sites should be relatively more accessible for silylation than those arising from structural defects, if any, or clay sheet edges sandwiched within the interlayer space. The accessibility to these potential interlayer silylation sites could be more or less accentuated by using different solvents that promote clay exfoliation.

For this purpose, the present study was achieved in order to clarify the role of the two solvents used, namely ethylene glycol and a water-ethanol mixture. The exfoliation grade of the solvent will be discussed in terms of specific surface area, porosity and CO_2 retention capacity of the corresponding organoclay. The results obtained will certainly be very useful for designing effective silylation processes on lamellar crystalline matrices.

2. Experimental
2.1. Adsorbent Preparation

A montmorillonite-rich material ion-exchanged into the sodium form (NaMt) was used as inorganic support [15]. Further, NaMt-S samples were obtained through NaMt silylation with (3-aminopropyl)triethoxysilane (γ-APTES) according to two different grafting methods. The first one was carried out by impregnating 10 g NaMt under stirring at 80°C for 5h in 1000 mL of a 25:75 vol. water/ethanol mixture containing 3g of γ-APTES. The resulting powder (NaMt-S-EW-1, designated as NaMt-S-EW) was filtered, washed and dried at 80°C for 24 hours. Two other samples (NaMt-S-EW-2 and NaMt-S-EW-3) were prepared with [APTES:Montmorillonite] ratios of 0.8 and 1.0 g.g^{-1}), respectively in the impregnating solution. The second method involved the preparation of a mixture containing on one hand a dispersion of 1g of NaMt in 100 mL of ethylene glycol (EG) under stirring, and a solution of 1g γ-APTES in 100 mL of ethylene glycol (EG). This mixture was stirred at room temperature for 30 min. The resulting product (**NaMt-S-EG**) was filtered and dried at 80 °C.

2.2. Characterization and Thermal Desorption Measurements

X-Ray powder diffraction (XRD) was carried out by means of a Siemens D5000 instrument (Co-Kα at 1.7890

Å). BET measurements of the specific surface area and BJH assessment of porosity and pore size distribution were performed through nitrogen adsorption-desorption isotherm, using a Quantachrome device, with an Autosorb automated gas sorption system control. Thus, samples of 80-100 mg were previously dried at 80°C for 24 h, degassed at 80°C for 16 h under a 10^{-4} Torr vacuum and then the nitrogen adsorption was made at -195.7 °C. Thermal gravimetric analysis (TGA) was performed by means of a TG/TDA6200 thermal analyzer (Seiko Instrument Inc.) under a 120 mL.min^{-1} nitrogen stream and a 5°C.min^{-1} heating rate. The amount of the grafted APTES was assessed on the basis of the mass loss between 200 and 600°C. Besides, infrared spectra were recorded using a KBr IR cell and Fourier Transform Infrared spectroscopy equipment (Model IR 550, Magna Nicolet). 29

^{29}Si nuclear magnetic-resonance (^{29}Si NMR) analysis was achieved by means of a Brüker DSX-300 spectrometer operating at 12.5 KHz, at a contact time of 5 ms, a recycle delay of 1 s and a spinning rate of 12.5 KHz. Tetramethylsilane (TMS) was used as the external reference.

The CO_2 retention capacity (*CRC*) of NaMt and NaMT-S samples was assessed by thermal programmed desorption of carbon dioxide (CO_2-TPD) according to a procedure fully described elsewhere [16]. TPD measurements were performed between 20°C and 200°C under different dry nitrogen streams (1, 2, 5 and 15 mL.min^{-1}) at a 5 °C/min heating rate, using a tubular glass reactor having a 10 mm internal diameter. The latter was incorporated in a tubular ceramic oven and coupled to a CO_2-detector (Li-840A CO_2/H$_2$O Gas Analyzer). Prior to TPD investigations, CO_2 was injected in excess at room temperature through the reactor under nitrogen flow rate (15 mL.min^{-1}). Full saturation of the adsorbent and impregnation were achieved during fixed times according to the experiments, followed by a purge in order to release the excess CO_2 not adsorbed at 20°C. Here, special measures were taken to avoid forced convection under strong nitrogen stream, which may affect CRC measurement accuracy [15,17].

3. Results and Discussion
3.1. Changes in the Surface Structure

FT-IR analyses revealed a shift of the Si-O stretching band from 1032 cm^{-1} for NaMt to 1038 cm^{-1} for both NaMt-S-EW-1 (designated as NaMt-S-EW) and NaMt-S-EG (Figure 1). This provides evidence of the strengthening of the Si-O bond belonging to the inorganic support, presumably due to the silylation process.

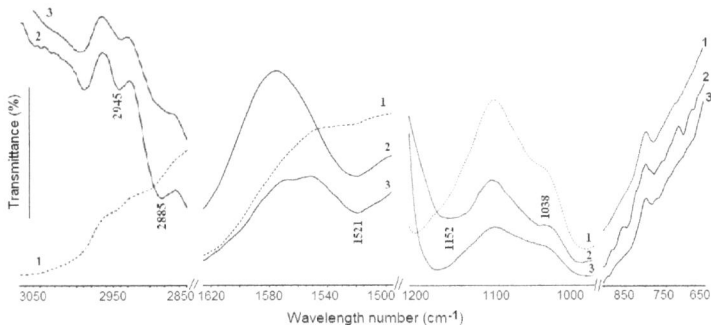

Figure 1. FTIR spectra of NaMt (1), NaMt-S-EG (2) and NaMt-S-EW (3). The FTIR spectra were recorded between 500 and 4000 cm-1, using KBr as support for the different adsorbents in powder form, but closeups were focused on the main characteristics bands of the grafted form of APTES.

The presence of APTES on the clay mineral surface was supported by new bands observed at 1521, 2885 and 2945 cm^{-1}, and assigned to the deformation of -CH$_2$ and to both the asymmetric and symmetric stretching of the -CH belonging to the -CH$_2$ groups respectively. Besides, other bands were also registered between 900-950 cm-1, and were attributed to the –NH$_2$ bending.

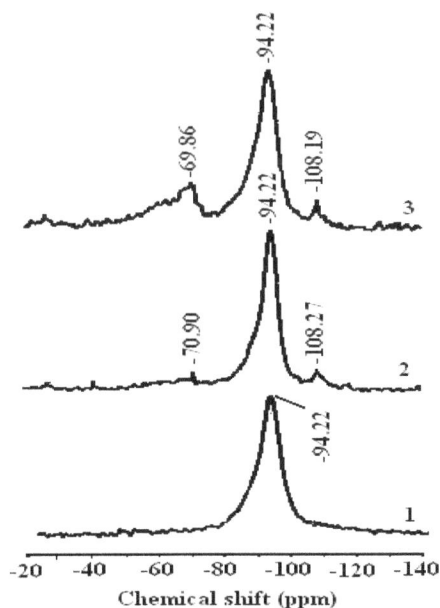

Figure 2. Si29 NMR spectra of NaMt (1), NaMt-S-EG (2) and NaMt-S-EW (3).

Amine grafting produced two additional [29]Si signals at -70.9 and -108.27 ppm for NaMt-S-EG and at -69.86 and -108.19 ppm for NaMt-S-EW. These two signals, attributed to the hydrolyzed tridentate $T^3[Si(OSi)_3R]$ (R

= $CH_2CH_2CH_2NH_2$) and $T^4[Si(OSi)_4]$, respectively indicate the formation of Clay-O-Si-R bonds, i.e. the chemical grafting of propylamine on the surface of montmorillonite.

3.2. Interlayer Structure and Changes Upon Silylation

NaMt-S-EW and NaMt-S-EG showed higher basal spacing of 1.67 and 1.89 nm respectively, as compared to NaMt (1.13 nm) (**Figure 3**). This indicates an increase in the interlayer space as a result of montmorillonite modification. The discrepancy between both basal spacing indicates a significant influence of the grafting procedure. Surprisingly, amine grafting seems to have a negative effect on the specific surface area, which dramatically decreased from 59 m^2.g-1 for NaMt to 28 m^2.g-1 (for NaMt-S-EG) and 21 m^2.g-1 (for NaMt-S-EW) (Table 1). A possible explanation is that amine grafting through silylation on montmorillonite seems to promote mesoporosity (38 Å ≤ pore size ≤ 73 Å) at the expense of micropores. The latter almost totally disappeared by NaMt chemical modification. A deep analysis of the pore size distribution and XRD patterns provides arguments in this regard, inasmuch as NaMt-S-EG possesses a lower 001 spacing (1,89 nm) as compared to NaMt-S-EW (1,67 nm) (Figure 3) and a slightly higher pore volume (Figure 4).

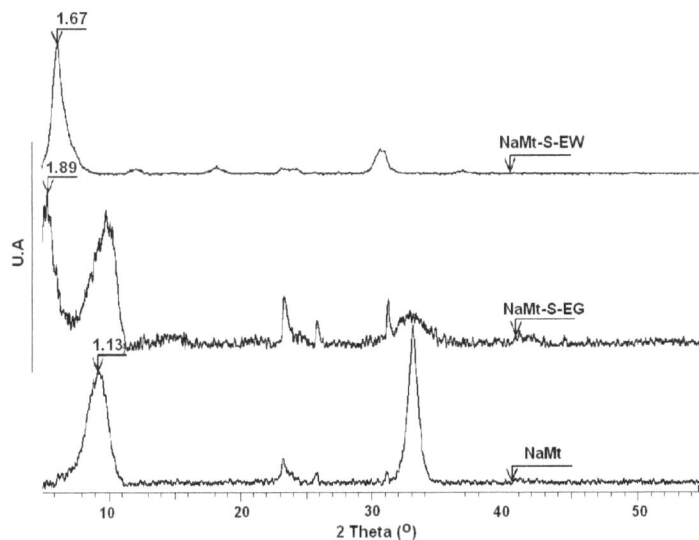

Figure 3. Powder XRD patterns of NaMt and NaMt-S samples (Co-Kα at 1.7890 Å).

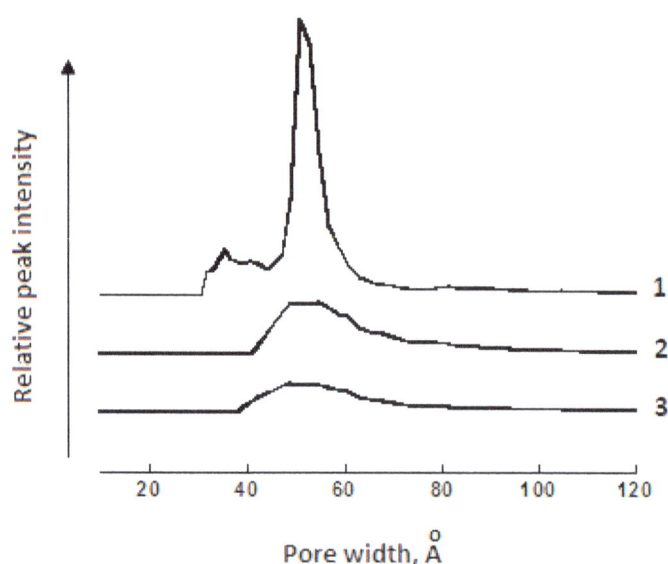

Figure 4. Changes in pore size distribution of NaMt (1), NaMt-S-EG (2) and NaMt-S-EW (3).The analyses were carried out through nitrogen adsorption-desorption isotherm at -195,7°C.

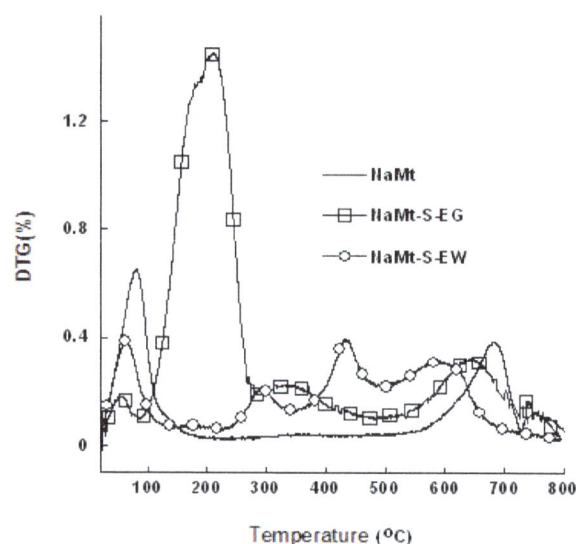

Figure 5. DTG patterns of NaMt and NaMT-S samples between 25 and 800°C in nitrogen stream at a 5 °C.min−1 heating rate.

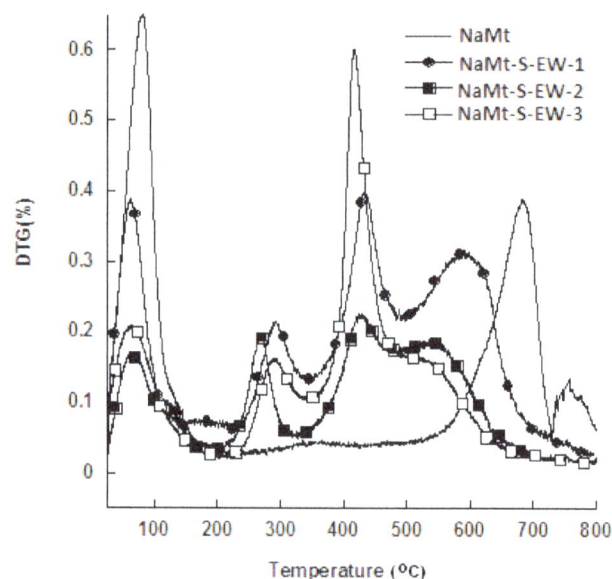

Figure 6. DTG patterns of NaMt and NaMT-S-EW samples between 25 and 800°C in nitrogen stream at a 5 °C.min−1 heating rate.

Nonetheless, the occurrence of two basal spacing for NaMt-S-EG suggests that a significant part of NaMt did not undergo full exfoliation. Here, amine grafting through silylation on the terminal OH groups located at the clay lamella edges is supposed to produce compact clay sheet stacks with low porosity and accessibility to the interlayer space [9]. Clay compaction, if any, should be more pronounced on NaMt-S-EG, presumably due to the strong sandwiching effect of ethylene glycol, well known to hinder clay delamination. In this case, the remaining part of mesopores accessible to CO_2 molecules should arise only from the interparticle void volume. Subsequently, NaMt-S-EG is expected to retain less CO_2 than NaMt-S-EW.

3.3. Thermal Stability

TG and DTG investigations of NaMt chemical modification showed the appearance of new mass loss peaks between 100 and 600°C, beside that of dehydration below 100°C and of dehydroxylation beyond 550°C belonging to montmorillonite (Figure 5 and Figure 6).

NaMt-S-EW samples were found to display higher thermal stability up to 250°C as compared to NaMt-S-EG, which starts decomposing even at 100°C. The mass losses registered beyond 250°C are common features of NaMt-S-EW samples. They must be due to the thermal decomposition of physically adsorbed APTES (330°C and 430°C) and chemically grafted amines (540°C). The latter showed increasing amount with increasing [APTES:Montmorillonite] ratio in the starting reaction mixture (0.3, 0.8 and 1.0 g.g[-1]) (Table 1).

Table 1. Some features of the as synthesized montmorillonite based materials.

ample	APTES contacted (g.g^{-1})[a]	Mass Loss[b] (wt. %)		-NH$_2$ amount for different number of silyl bridges (mmol.g^{-1})[c]			CRC[d] (μmol.g^{-1})	S$_{BET}$ (m^2.g^{-1})	EF[e] (μmol.m^{-2})
		200-600°C	490-600°C	1	2	3			
NaMt	0	0.7	0.7	-	-	-	1.25	59	0.021
NaMt-S-EG	0.3	11.26	1.65	0.16	0.11	0.08	115	28	4.11
NaMt-S-EW-1[f]	0.3	8.12	2.92	0.28	0.19	0.15	250	21	11.9
NaMt-S-EW-2	0.8	10.50	3.65	0.35	0.24	0.19	495	31.6	15.7
NaMt-S-EW-3	1.0	12.89	3.07	0.29	0.20	0.16	> 1000	31.2	> 32

[a] Amount of APTES in the impregnating solution contacted per gram of dry montmorillonite.
[b] Mass loss values measured through thermogravimetry between 200 and 600°C in air stream. This accounts for the amount of removed ethanol and propylamine, taking into account that the total decomposition of each APTES molecule results in one residual Si(OH)$_4$ molecule deposited on the clay mineral surface.
[c] The number of amino groups grafted on montmorillonite was calculated for different number of silyl bridges by dividing the mass loss (g.g^{-1}) by the sum of the molecular weights of propylamine (C$_2$H$_7$NH$_2$, 59 g.mol^{-1}) and of one (46 g.mol^{-1}), two (2x46 g.mol^{-1}) or three ethanol molecules (3x46 g.mol^{-1}). These calculations take into account that one amino group may be grafted by APTES silylation with one, two or three clay silanol groups, resulting in the formation of one, two or three silyl bridges, respectively.
[d] The CRC value was expressed in terms of desorbed amount of CO$_2$ between 20 and 80°C. These values were assessed by TPD (1.0 mL.min^{-1} nitrogen flow rate) after CO$_2$ injection and purge at 15 mL.min^{-1} nitrogen throughput.
[e] The efficiency factor (EF) was defined as being the [CRC : specific surface area] ratio (Azzouz et al., 2013-a).
[f] Throughout the entire manuscript text, NaMt-S-EW-1 was designated as NaMt-S-EW.

The amount of amino groups incorporated assessed on the basis of the overall mass loss between 200 and 600°C gave [CO$_2$:-NH$_2$] mole ratios much lower than the expected 1:1 stoichiometric value, which is a special feature of the formation of a carbamate group from one amino group with one CO$_2$ molecule. Any discrepancy from this value suggests that a part of amino groups do not contribute to CO$_2$ retention, not being accessible, as previously stated.

In contrast, similar calculations on the basis of the third decomposition peak at 490-600°C provided values of the [CO$_2$:-NH$_2$] mole ratios close to unity, more particularly when assuming the formation of one silyl group via the loss of one ethanol molecule. Increasing number of hypothetical number of silyl groups up to three induced an increase of the [CO$_2$:-NH$_2$] mole ratio. This indicates that an excess of CO$_2$ with respect to stoichiometry adsorbs also via physical interactions, presumably due to the base character induced in the adsorbent bulk.

3.4. Affinity towards CO$_2$ and Heat of CO$_2$ Retention

NaMt-S-EW showed higher amount of thermally desorbed CO$_2$ as compared to NaMt-S-EG and the starting NaMt material (Figure 7). The thermal stability of both adsorbents was confirmed by reproducible CO$_2$-TPD triplicate performed between 20 and 200°C for NaMt-S-EW and between 20 and 80°C for NaMt-S-EG

under constant moisture content (Figure 8 and 9). Dehydration after repetitive adsorption-desorption cycles was found to reduce unavoidably the amount of desorbed CO$_2$. This was already explained by the contribution of water in CO$_2$ retention [17,18].

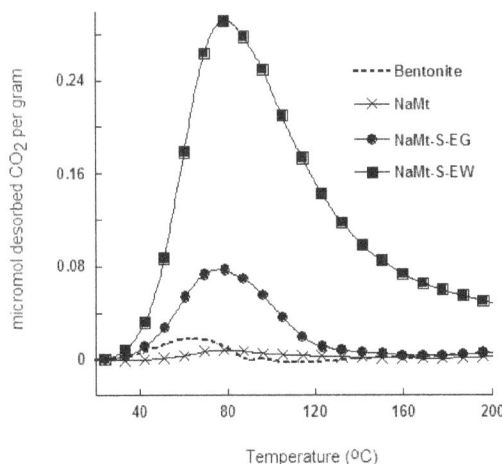

Figure 7. CO2 TPD patterns between 20 and 200 °C. Here, TPD measurements were carried out after by a dynamic impregnation with 0.016 mmol of CO2/g at 20°C in a 15 mL.min-1 of dry nitrogen stream, followed by a purge under similar conditions for 40 min. Accurate CRC assessments between 20 °C and 200 °C required a progressive heating at 5°C.min-1 up to 200°C, and to maintain temperature at 200°C for at least 20 min.

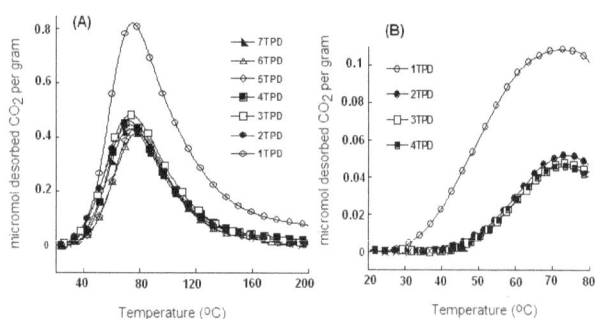

Figure 8. Repetitive TPD cycle of desorbed CO2 for NaMt-S-EW (A) and NaMt-S-EG (B). The temperature range was established according to the thermal stability of each sample. TPD measurements were carried out after dynamic impregnation with 0.016 mmol of CO2/g at 20°C in a 15 mL.min-1 of dry nitrogen stream, followed by a purge under similar conditions for 40 min. Accurate CRC assessment between 20 °C and the final temperature requires a progressive heating at 5 °C.min-1, and to maintain temperature at the respective upper limit for at least 20 min.

By increasing the nitrogen flow rates from 1 to 15 mL.min^{-1}, the amount of desorbed CO_2 decreased dramatically from ca. 70 to ca. 12-15 µmol.g^{-1} for NaMt-S-EW, and from 25-35 to less than 5 µmol.g^{-1} for NaMt-S-EG (Figure 10). Here also, strong nitrogen stream are supposed to remove CO_2 even at room temperature through forced convection, as already reported for OH-enriched montmorillonites [15,17]. This result is of great importance, because it demonstrates that CO_2 removal without heating is possible even when chemical adsorption is involved.

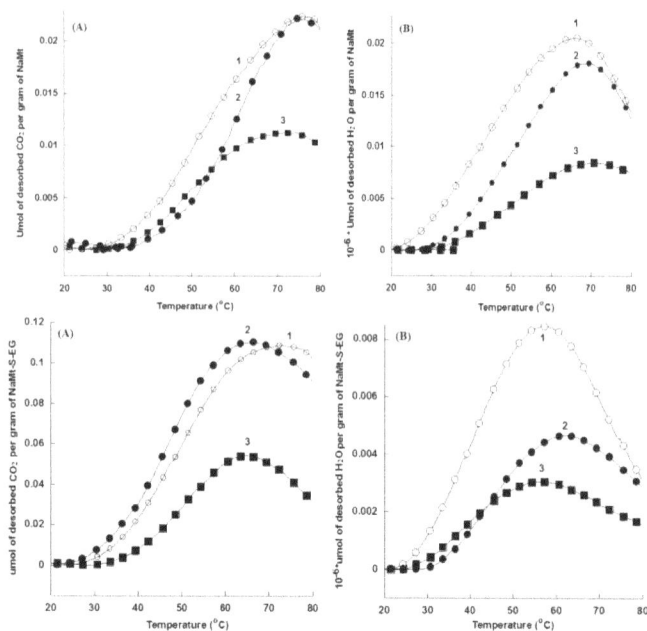

Figure 9. Repetitive TPD cycles of desorbed CO2 (A) and water (B) for NaMt, NaMt-S-EW and NaMt-S-EG. Previous dynamic impregnation was achieved with 0.016 mmol CO2/g at 20°C in a 15 mL.min-1 dry nitrogen stream, followed by a purge under similar conditions. Accurate CRC assessment was achieved through TPD measurements under a dry nitrogen stream of 15 mL.min-1 at 5 °C/min heating rate from 20 °C to 80°C, which was then maintained constant for 20 min. Legend: 1. is the first TPD performed for each fresh sample; 2. is the second TPD obtained after cooling down to 20°C and impregnation with wet air stream for 12h; 3. is the third TPD performed directly after the second TPD without refreshing with air.

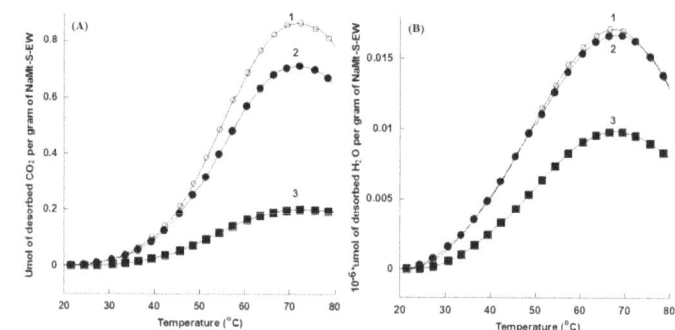

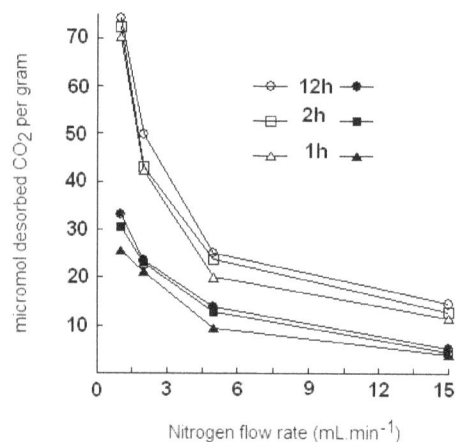

Figure 10. CRC values (umol.g-1) between 20°C and 80°C versus N2flow rate at different saturation times. NaMt-S-EG: (dark symbols); NaMt-S-EW (white symbols). Saturation of both samples with CO2 was performed in static mode (without nitrogen stream), and was followed by a purge at 20°C in a 15 mL.min-1 dry nitrogen stream till no CO2 is detected. TPD experiments were performed at a heating rate of 5 °C/min up to the upper temperature value, which was then maintained constant for 20 min.

DSC measurements revealed that CO_2 desorption requires heat amount of 264-467 kcal.mol^{-1} for NaMt-S-EW between 26 and 153°C, and of 148-229 kcal.mol-1 for NaMt-S-EG between 26 and 115°C. This indicates that CO_2 adsorption involves stronger interaction with

NaMt-S-EW as compared to NaMt-S-EG, and provides clear evidence that only chemical interaction is involved between CO_2 and the amino groups incorporated. Under optimum conditions, i.e. when CO_2 removal by forced convection at room temperature is avoided or minimized, the CRC value may exceed by far 1.0 mmol.g^{-1} (Table 1).

3.5. Effect of Grafting Grade

These values explain the relatively high affinity of NaMt-S-EW towards CO_2 as compared to NaMt-S-EG, in agreement with its higher amine content. Deeper insights in this regard showed that the amount of CO_2 thermally desorbed by TPD increases with increasing amount of injected CO_2 (Figure 11).

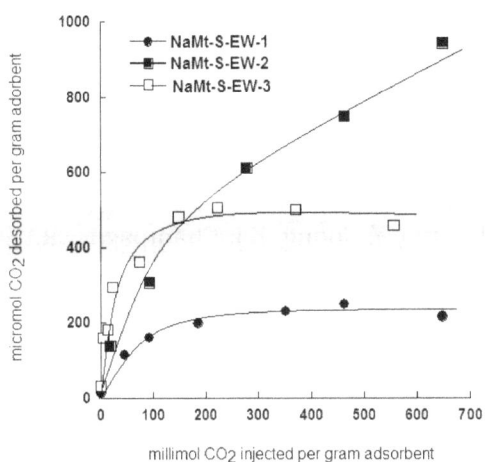

Figure 11. CO2 retention capacity between 20°C and 80°C versus injected CO2 amount for NaMt-S-EW with different loading amounts of OH groups. Dynamic impregnation of the samples was achieved with different CO2 amounts (1.5 - 700 mL) at 20°C in a 15 mL.min-1 dry nitrogen stream, followed by a purge under similar conditions till no CO2 is detected. TPD measurements were performed at 1.0 mL.mn-1 of dry nitrogen stream and a heating rate of 5 °C/min up to 80 °C, which was then maintained constant for 20 min.

Saturation was attained at 350 mmol.g^{-1} for NaMt-S-EW-1 and 200 mmol.g^{-1} for NaMt-S-EW-2, providing CO_2 retention capacity (CRC) of 240-250 mmol.g^{-1} and 490-500 mmol.g^{-1}, respectively. This suggests an almost linear proportionality with respect to the amount of incorporated amine. No saturation took place on NaMt-S-EW-3, even up to 700 mmol.g^{-1} of injected CO_2. This indicates CRC values much higher 1000 mmol.g^{-1}, in agreement with other data [19].

Indeed, in spite of its lower specific surface area, NaMt-S-EW showed an efficiency factor (11.9) approximately three times higher than that of NaMt-S-

EG (4.11). As compared to OH-enriched montmorillonite [15,17], NaMt-S-EW showed appreciable efficiency factor that may exceed 16 micromol CO_2.m^{-2} for higher amine content.

4. Conclusion

Grafting montmorillonite with (3-aminopropyl)triethoxysilane induced a significant improvement of the CO_2 retention capacity as compared to the starting clay mineral. The latter was found to be influenced by the grafting procedure. It was found that the use of ethanol/water mixture in the preparation leads to materials with high thermal stability and affinity towards CO_2. NaMt-S-EW displayed higher affinity towards CO_2 than NaMt-S-EG, with retention efficiency factor exceeding 16 μmol.m^{-2} for high amine content. Differential scanning calorimetry demonstrated that only chemical interactions are involved between the amino groups grafted and CO_2. However, CO_2 removal without heating is possible upon mere forced convection at high nitrogen throughput. **This innovative concept opens new prospects for** providing new inorganic–organic materials that act as respiratory matrices for the reversible capture of polluting gases and other environmental purposes.

Acknowledgements
This work was supported by a grant from MDEIE-FQRNT to R.R. and A.A.

References
[1] K.W. Park, S.Y. Jeong, O.Y. Kwon "Interlamellar silylation of H-kenyaite with 3-aminopropyltriethoxysilane" Applied clay science, 27(1-2), 2004, pp. 21-27.

[2] N.N. Herrera, J.M. Letoffe, J.L. Putaux "Aqueous dispersions of silane-functionalized laponite clay platelets: A first step toward the elaboration of water-based polymer/clay nanocomposites", Langmuir, 20(5), 2004, pp. 1564-1571.

[3] M. Park, I.K. Shim, E.Y. Jung "Modification of external surface of laponite by silane grafting", Journal of Physics and Chemistry Solids, 65(2-3), 2004, pp. 499-501.

[4] K. Isoda, K. Kuroda "Interlamellar grafting of γ-meththacryloxypro- pylsilyl groups on magadiite and copolymerization with methyl methacrylate", Chemistry of Materials, 12, 2000, pp. 1702-1707.

[5] J.X. Zhu, H.P. He, J.G. Guo, D. Yang, X.D. Xie "Arrangement models of alkylammonium cations in the interlayer of HDTMA+ pillared montmorillonites", Chinese Science Bulletin, 48(4), 2003, pp. 368-372.

[6] K.A. Carrado, L.Q. Xu, R. Csencsits, J.V. Muntean "Use of organo- and alkoxysilanes in the synthesis of grafted and pristine clays" Chemistry of Materials, 13(10), 2001, pp. 3766-3773.

[7] H.P. He, J. Duchet, J. Galy, J.F. Gerard "Grafting of swelling clay materials with 3-aminopropyltriethoxysilane" Journal of Colloid and Interface Science, 288(1), 2005, pp. 171-176.

[8] S. Linna, Q. Tao, H. He, J. Zhu, P. Yuan, R. Zhu "Silylation of montmorillonite surfaces: Dependence on solvent nature" Journal of Colloid and Interface Science, 377(1), 2012, pp. 328-333.

[9] A.M. Shanmugharaj, K.Y. Rhee, S.H. Ryu "Influence of dispersing medium on grafting of aminopropyltriethoxysilane in swelling clay materials" Journal of Colloid and Interface Science, 298(2), 2006, pp. 854-859.

[10] J.E. Jordan "Organophilic bentonites I. Swelling in organic liquids" Journal of Physical and Colloid Chemistry, 53(2), 1949, pp. 294-306.

[11] S. Ek, E.I. Iiskola, L. Niinistö "Atomic layer deposition of aminofunctionalized silica surfaces using N-(2-Aminoethyl)-3-aminopropyltrimethoxysilane as a silylating agent" The Journal of Physical Chemistry B, 108(28), 2004, pp. 9650-9655.

[12] C.J. Brinker, G.W. Scherer "Sol-Gel Science, the Physics and Chemistry of Sol-Gel Processing" Academic Press, London, 1990 pp. 97.

[13] J. Ahenach, P. Cool, E. Vansant, O. Lebedev, J.V. Landuyt "Influence of water on the pillaring of montmorillonite with aminopropyltriethoxysilane" Physical Chemistry Chemical Physics, 1(15), 1999, pp. 3703-3708.

[14] M.M. Sprung, F.O. Guenther "The partial hydrolysis of methyltriethoxysilane" Journal of the American. Chemistry Society, 77(15), 1955, pp. 3990-3996.

[15] A. Azzouz, N. Platon, S. Nousir, K. Ghomari, D. Nistor, T.C. Shiao, R. Roy "OH-enriched organo-montmorillonites for potential applications in carbon dioxide separation and concentration" Separation and Purification Technology, 108, 2013, pp. 181-188.

[16] A. Azzouz, D. Nistor, D. Miron, A.V. Ursu, T. Sajin, F. Monette, P. Niquette, R. Hausler "Assessment of acid–base strength distribution of ion-exchanged montmorillonites through NH3 and CO2-TPD measurements" Thermochimica Acta, 449(1-2), 2006, pp. 27-34.

[17] S. Nousir, N. Platon, K. Ghomari, A.S. Sergentu, T.C. Shiao, G. Hersant, J.Y. Bergeron, R. Roy, A. Azzouz "Correlation between the hydrophilic character and affinity towards carbon dioxide of montmorillonite-supported polyalcohols" Journal of Colloid and Interface Science, 402, 2013, PP. 215-222.

[18] A. Azzouz, S. Nousir, N. Platon, K. Ghomari, T.C. Shiao, G. Hersant, J.Y. Bergeron, R. Roy "Truly reversible capture of CO2 by montmorillonite intercalated with soya oil-derived polyglycerols" International Journal of Greenhouse Gas Control, 17, 2013, pp. 140–147.

[19] M.L. Gray, Y. Soong, K.J. Champagne, R.W. Stevens Jr., P. Toochinda, S.S.C. Chuang "Solid amine CO2 capture sorbents" Fuel Chemistry Division Preprints, 47(1), (2002). pp. 63-64.

23

Phorbol Esters Degradation and Enzyme Production by *Bacillus* using *Jatropha* Seed Cake as Substrate

Chin-Feng Chang[1], Jen-Hsien Weng[1], Kao-Yung Lin[2], Li-Yun Liu[3], Shang-Shyng Yang[4*]

[1]Department of Food Science, China University of Science and Technology,
Nankang, Taipei 11581, Taiwan
afengb_b@yahoo.com.tw; toshibasl650@gmail.com

[2]Department of Living Science, National Open University,
Luzhou, New Taipei City 24701, Taiwan
lgu@mail.nou.edu.tw

[3]Department of Food Science, Nutrition and Nutraceutical Biotechnology,
Shih Chien University, Taipei 10464, Taiwan
liyun@mail.usc.edu.tw

[4]Department of Food Science, China University of Science and Technology,
Nankang, Taipei 11581, Taiwan

[4]Department of Biochemical Science and Technology, National Taiwan University,
Taipei 10617, Taiwan
ssyang@ntu.edu.tw

Abstract- *The purposes of this research were to evaluate phorbol esters (PEs) degradation rate and enzyme production yield using submerged fermentation (SMF) as screening method and further using solid-state fermentation (SSF) as pilot scale-up study. SMF was carried out with 20 g seed cake in 100 ml minimal salt medium for 7 days incubation, while SSF was done with 20 g seed cake at 50% moisture content for 9 days incubation. Bacillus strains grew well on J. curcas seed cake with 10^8-10^{11} CFU/ ml in SMF for 3 days incubation, while they were 10^8-10^{10} CFU/ g in SSF. PEs reduced 76.5%, 77.1%, 78.4%, 85.5%, and 92.0% in SMF with B. smithii G16, B. sonorensis D12, B. licheniformis A3, B. subtilis H8 and B. coagulans C45 for 3 days incubation, respectively, and PEs completed degraded by these five strains for 7 days incubation. Maximum amylase, cellulase, lipase, pectinase, protease and xylanase productions in SMF were observed in B. sonorensis D12 (5.49 ± 0.49 U/ ml; day 7), B. subtilis H8 (17.03 ± 4.90 U/ ml; day 2), B. licheniformis A3 (59.03 ± 0.26 U/ ml; day 7), B. sonorensis D12 (1.70 ± 0.04 U/ ml; day 3), B. coagulans C45 (15.95 ± 0.35 U/ ml; day 7) and B. smithii G16 (1.40 ± 0.01 U/ ml; day 3), respectively. For SSF, PEs were reduced 86.0%, 83.2%, and 93.0% with B. sonorensis D12, B. subtilis H8 and B. smithii G16 for 3 days incubation, respectively. Maximum amylase, cellulase, lipase, pectinase, protease and xylanase productions in SSF were observed in B. smithii G16 (16.08 ± 0.36 U/ g; day 4), B. sonorensis D12 (2.94 ± 0.06 U/ g; day 2), B. smithii G16 (3.87 ± 0.64 U/ g; day 4), B. sonorensis D12 (8.13 ± 1.06 U/ g; day 2), B. smithii G16 (14.13 ± 0.30 U/ g; day 4) and B. smithii G16 (9.72 ± 0.97 U/ g; day 3), respectively. J. curcas seed cake could be detoxified by Bacillus and the high-protein seed cake could be potentially used for enzyme production in industry.*

Keywords: *Jatropha curcas* seed cake, phorbol esters, degradation, submerged and solid state fermentation, enzyme production.

1. Introduction

Due to the demand of local energy production, *Jatropha curcas* has high adaption capacity for land fertilization and can grow in marginalized land. It does not competitive with other vegetation crops for agricultural lands. The oil of *J. curcas* seed can be used as the raw materials for biodiesel production. However, biodiesel production from *J. curcas* seeds generate large quantum of residual de-oiled seed cake with an average of 700 g cake per kg of seed used [23]. The de-oiled *J. curcas* seed cakes contain high protein and other different nutrients like minerals, and amino acids etc. [1]. They can be used as the nutrient for animal feeds or production of valuable products. However, the de-oiled *J. curcas* seed cake contains anti-nutritional substances ex. trypsin inhibitors, curcins, tannins, saponins, phytates and toxic factors - phorbol esters (PEs) that restricts the uses of the seed cake. PEs have been identified as the main toxicants in cake which cannot be destroyed even by heating at 160°C for 30 min [23].

Physical methods, chemical treatments and microbial fermentation have been used to detoxify the toxic compounds in the seed cake ([4], [12], [17], [20], [22], [29]). However, the biological method would be more advantageous than the others for environmental friendly with safety and energy concerns. Especially solid state fermentation (SSF) could detoxify PEs and use for production of valuable products [17].

SSF and SMF techniques are common and conventional biotechnology processes for production of value-added products such as enzymes, biopharmaceuticals, organic acids, biosurfactants, vitamins, flavoring compounds, biofuels, biopesticides etc. [10]. SSF has economical and practical advantages over SMF, which includes use of raw materials as substrates, low capital cost, low energy expenditure, and less expensive downstream processing [6]. Therefore, SSF is adapted as commercial production and SMF is adapted as screening method for basic research.

In comparison with the previous works, *Bacillus* species were often subjected to the enzyme production (30) and most of *Bacillus* species are considered as Generally Regarded as Safe (GRAS) by FDA of USA. Additionally, *Bacillus* spp. was often using in production of food additives and probiotic products ([14], [32]). Therefore, *Bacillus* strains were applied to evaluate the degradation efficiency of PEs toxins factors in *J. curcas* seed cake, and *J. curcas* seed cake was used as the substrate for enzyme production in this paper.

2. Materials and Methods
2. 1. *J. curcas* Seed Cake and Tested Microbes
De-oiled *J. curcas* seed cake was kindly provided by Shin-Feng Energy Technology Co., LTD (Pingtung, Taiwan), then ground and pressed the defatted *J. curcas*

seed cake. Five *Bacillus* spp. (*B. coagulans* C45, *B. licheniformis* A3, *B. smithii* G16, *B. sonorensis* D12, and *B. subtilis* H8) isolated from different compost plants and biofertilizers were used for PEs degradation and enzyme production [7].

2. 2. Biodegradation of PEs and Enzyme Production of *J. curcas* Seed Cake by *Bacillus*
For SMF, one ml of inoculums (about 10^8 CFU/ ml) was inoculated to 100 ml of mineral salts medium (MSM). Each litter of MSM medium contained seed cake 200 g, $Na_2HPO_4 \cdot 7H_2O$ 6.7 g, KH_2PO_4 1.5 g, $(NH_4)_2SO_4$ 1 g, $MgSO_4 \cdot 7H_2O$ 0.2 g, ferrous ammonium citrate 0.06 g, $CaCl_2 \cdot 2H_2O$ 0.01 g and trace-element solution 1 ml. The trace element solution contained H_3BO_3 0.3 g, $CoCl_2 \cdot 6H_2O$ 0.2 g, $ZnSO_4 \cdot 7H_2O$ 0.1 g, $MnCl_2 \cdot 4H_2O$ 0.03 g, $NaMoO_4 \cdot 2H_2O$ 0.03 g, $NiCl_2 \cdot 6H_2O$ 0.02 g and $CuSO_4 \cdot 5H_2O$ 0.01 g in 1 litter of 0.1N HCl [28]. The cultures were incubated in rotary shaker at 25°C, 150 rpm for 0, 1, 2, 3 and 7 days. PEs degradation and enzyme production were determined.

For SSF, 1ml of freshly prepared bacterial cells (about 10^9 CFU/ ml) was transferred to each of 250-ml Erlenmeyer flasks containing 20 g sterile seed cake at 50% initial moisture content. The flasks were then manually shaken well and incubated at 25°C for 0, 1, 2, 3, 5, 7 and 9 days. PEs degradation and enzyme production were determined.

2. 3. Phorbol Ester Extraction and Analysis
PEs were extracted and determined by the modified method of Hass and Mittelbach [13]. The mixture of *J. curcas* crushed seed cake 5 g and 95% ethanol 20 ml was shaken at 200 rpm for 5 min, then centrifuged at 14,000 *g* for 5 min. The residue was extracted two additional times with 95% ethanol. The extract fractions were combined and dried under vacuum at 50°C. The dried extract was dissolved in 1 ml 95% ethanol and passed through a 0.2-μm membrane filter. Phorbol esters were analyzed by HPLC system (Thermo Separation Products, U.S.) consisted of an AS1000 autosampler, P2000 pump and UV1000 detector. The solvents were water and acetonitrile: start with 60% water and 40% acetonitrile for 15 min, then 25% water and 75% acetonitrile for the next 20 min, and finally 100% acetonitrile for the next 20 min. Separation was performed at room temperature (25°C) with flow rate 1.3 ml/ min. The detector wavelength was set at 280 nm. The results were expressed as

equivalent to phorbol-12-myristate-13-acetate (PMA) (Sigma, U.K.) used as an external standard.

2. 4. Enzyme Extraction and Assay

For enzyme assay in SMF, the sample was centrifuged at 12,000 g for 30 min and the supernatant was used for enzyme activities determination. While in SSF, 5 g fermented substrate was extracted with 20 ml 0.1 M Tris-HCl buffer pH 8.0 at 200 rpm orbital shaking for 30 min. The suspension was then centrifuged at 12,000 g for 30 min and the supernatant was used for enzyme assay.

Amylase, cellulase, lipase, pectinase, protease, and xylanase activities were determined by the methods of Bernfeld [5], Hu et al. [15], Kilcawley et al. [19], Janani et al. [16], Shimogaki et al. (31) and Joshi and Khare [18], respectively.

2.5. Chemical Analysis

Moisture, pH, protein, fat, and ash contents were analyzed by the standard methods of Association of Official Analytical Chemists [3]. Total organic carbon (TOC) was determined by TOC-5000A total organic carbon analyzer (Shimadzu, Japan).

3. Results and Discussion

3.1. Chemical Composition and Phorbol Ester Content in *J. curcas* Seed Cake

The chemical compositions of *J. curcas* seed cakes from Pingtung County, Taiwan are shown in Table 1. *J. curcas* seed cakes had pH 6.58 ± 0.02, moisture 5.92 ± 0.10%, ash 6.15 ± 0.20%, protein 24.24 ± 2.13%, fat 7.72 ± 0.26%, TOC 44.41 ± 0.01%, C/N ratio 11.02 ± 1.23 and PEs 0.60 ± 0.06 mg/ g dry sample.

PEs concentrations in seed cakes were lower than those of unshelled seed cakes from Nicaragua (1.78 mg/ g dry sample) [2], and Zimbabwe (0.70 mg/ g dry sample) [8], but they were higher than seed cakes from four provinces of Thailand (0.21-0.47 mg/ g dry sample) [11] This is possibly caused by various amounts of residual oil left in the samples and the variations of *J. curcas* in cultivation areas, soils, and climatic conditions ([4], [23], [26]).

Table 1. Chemical composition of *J. curcas* seed cakes.

Composition	Amount
pH	6.58 ± 0.02
Moisture (%)	5.92 ± 0.10
Ash (%)	6.15 ± 0.20
Protein (%)	24.24 ± 2.13
Fat (%)	7.72 ± 0.26
TOC (%)	44.41 ± 0.01
C/N ratio	11.02 ± 1.23
Phorbol ester (mg g^{-1} dry sample)	0.60 ± 0.06

Means ± S. D. (n=3)

3.2. Degradation of Phorbol Esters and Enzyme Production by Submerged Fermentation

The *J. curcas* seed cake 20 g in 100 ml MSM medium was fermented with 5 *Bacillus* strains at 25°C. The *Bacillus* strains had initial cell numbers 10^4-10^6 CFU/ ml and then increased to 10^8-10^{11} CFU/ ml after 7 days incubation. While the PEs contents reduced 76.5%, 77.1%, 78.4%, 85.5%, and 92.0% with *B. smithii* G16, *B. sonorensis* D12, *B. licheniformis* A3, *B. subtilis* H8, and *B. coagulans* C45 for 3 days incubation, respectively (Figure 1).

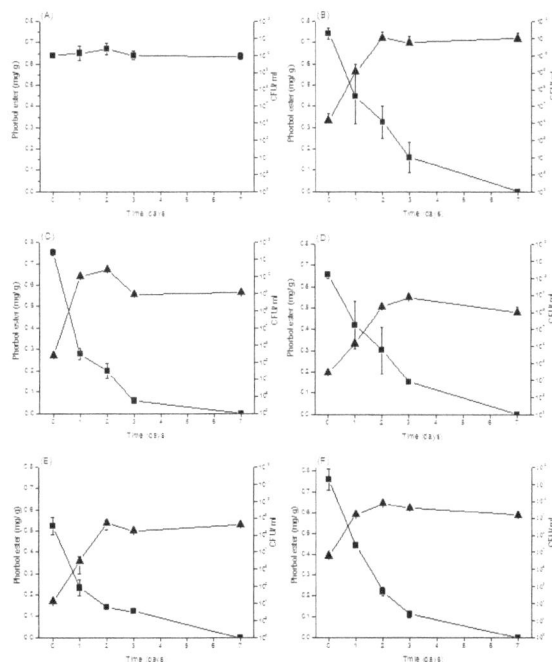

Figure 1. Phorbol esters contents (■) and *Bacillus* populations (▲) in *J. curcas* seed cake submerged fermentation. (A) Blank; (B) *B. licheniformis* A3; (C) *B. coagulans* C45; (D) *B. sonorensis* D12; (E) *B. smithii* G16; and (F) *B. subtilis* H8.

PEs contents in deoiled *J. curcas* seed cake were extracted by ethanol and analyzed by HPLC. The chromatogram (Figure 2) shows four major peaks of phorbol esters at retention times of 29.09-32.74 min, closely related to the reported by Makkar et al. [23]. PEs could be completely degraded by these five strains for seven days cultivation (Figures 1 and 2).

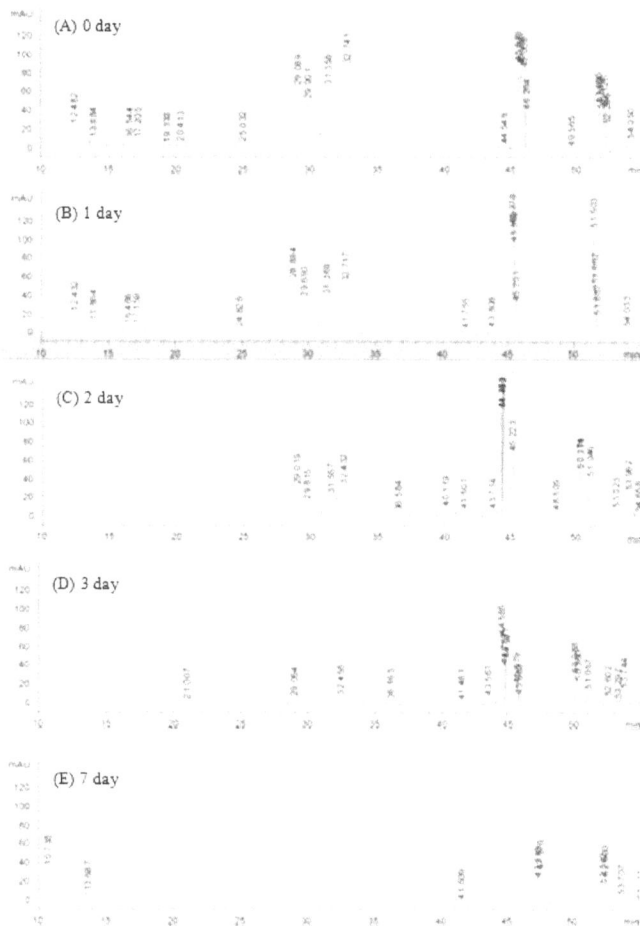

Figure 2. HPLC chromatograms of phorbol esters (PEs) biodegradation by *B. coagulans* C45 for 0-7 days (A–E).

Enzyme productions of *Bacillus* strains in SMF at 25°C were shown in Figure 3. *B. sonorensis* D12 had the maximum amylase production 5.49 ± 0.49 U/ ml for 7 days cultivation. *B. subtilis* H8 had the maximum cellulase production 17.03 ± 4.90 U/ ml for 2 days cultivation. *B. licheniformis* A3 had the maximum lipase production 59.03 ± 0.26 U/ ml for 7 days cultivation. *B. sonorensis* D12 had the maximum pectinase production 1.70 ± 0.04 U/ ml for 3 days cultivation. *B. coagulans* C45 had the maximum protease production 15.95 ± 0.35 U/ ml for 7 days cultivation. *B. smithii* G16 had the

maximum xylanase production 1.40 ± 0.0 U/ ml for 3 days cultivation.

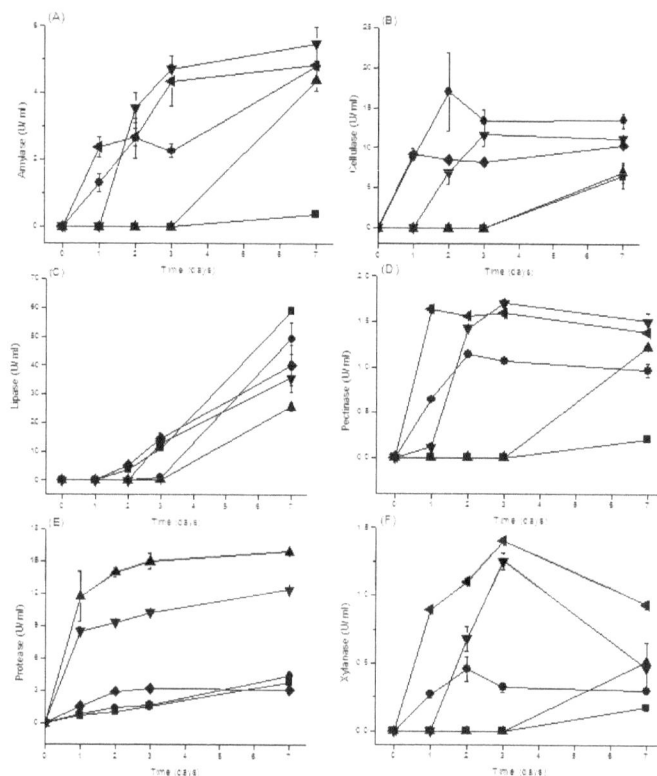

Figure 3. Enzyme production of *J. curcas* seed cake with submerged fermentation by *Bacillus* spp. (A) Amylase, (B) Cellulase, (C) Lipase, (D) Pectinase, (E) Protease, (F) Xylanase. A3 (■), H8 (●), C45 (▲), D12 (▼), and G16 (◆).

3.3 Degradation of Phorbol Esters and Enzyme Production by Solid-state Fermentation

For SSF, 20 g seed cake with 50% initial moisture content at 25°C was fermented with 3 *Bacillus* spp. (H8, D12 and G16) had high enzyme production in SMF. The *Bacillus* strains had initial cell numbers 10^4-10^6 CFU/ g and then increased to 10^8-10^{11} CFU/ g after 7 days incubation. The PEs contents in seed cake reduced 86.0%, 83.2% and 93.0% with *B. sonorensis* D12, *B. subtilis* H8, and *B. smithii* G16 for 3 days incubation at 25°C, respectively (Figure 4).

The enzyme productions of tested microbes increased with cultivation and had the maximal production for 2-4 days cultivation (Figure 5). *B. smithii* G16 had the maximum amylase production 16.08 ± 0.36 U/ g for 4 days cultivation. *B. sonorensis* D12 had the maximum cellulase production 2.94 ± 0.06 U/ g for 2 days cultivation. *B. smithii* G16 had the maximum lipase production 3.87 ± 0.64 U/ g for 4 days cultivation. *B. sonorensis* D12 had the maximum pectinase production

8.13 ± 1.06 U/ g for 2 days cultivation. *B. smithii* G16 had the maximum protease production 14.13 ± 0.30 U/ g for 4 days cultivation. *B. smithii* G16 had the maximum xylanase production 9.72 ± 0.97 U/ g for 3 days cultivation.

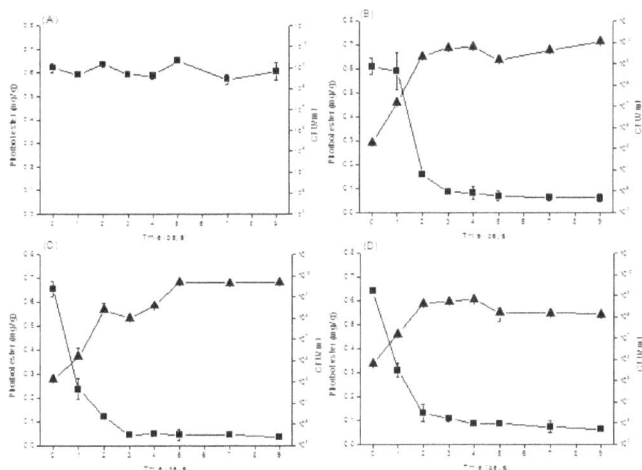

Figure 4. Phorbol esters contents (■) and *Bacillus* populations (▲) in *J. curcas* seed cake by solid state fermentation. (A) Blank; (B) *B. sonorensis* D12; (C) *B. smithii* G16; (D) *B. subtilis* H8.

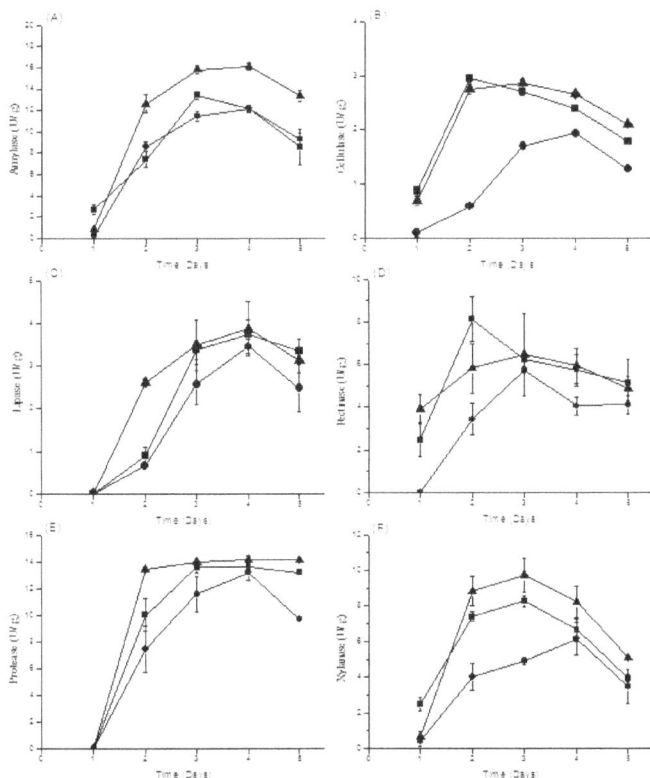

Figure 5. Enzyme production of *J. curcas* seed cake with solid-state fermentation by *Bacillus* spp. (A) Amylase; (B) Cellulase; (C) Lipase; (D) Pectinase; (E) Protease; (F) Xylanase. D12 (■), H8 (●), and G16 (▲).

The world market of industrial enzymes is estimated to be 1.6 billion \$US, including food enzymes (29%), feed enzymes (15%), and general technical enzymes (56%). *Bacillus* can secrete a variety of soluble extracellular enzymes [27] and most of *Bacillus* species are considered as GRAS. It is estimated that *Bacillus* spp. enzymes make up about 50% of the total enzyme market around the world [25]. Therefore, *Bacillus* is a good candidate microbe for enzyme production of *J. curcas* seed cake. Each *Bacillus* strain in this study has different capacities of enzyme production, and can be applied by SMF or SSF for enzyme production in the future.

Phengnuam and Suntornsuk [26] reported that *B. licheniformis* and *B. subtilis* could degrade PEs 60% and 40% for 5 days incubation with SMF. In this study, 5 *Bacillus* strains can degradation PEs 76.5-92.0% for 3 days incubation and 100% for 7 days incubation by SMF. Joshi et al. [17] reported that *Pseudomonas aeruginosa* degradated 60-73% phorbol esters in the seed cake for 6 days incubation with SSF, and Belewu and Sam [4] also reported that *Aspergillus niger* degraded 77% phorbol esters in the seed cake for 7 days incubation by SSF. In this study, 3 *Bacillus* strains can degrade PEs 83.2-93.0% for 3 days incubation by SSF. Both of SMF and SSF can be used for PEs degradation [26].

Jatropha seed oil has been used extensively for biodiesel production. The residue of *J. curcas* seed cake can be used as substrate in SMF or SSF for production of useful substances. In SMF, Choudhury et al. [9] used *J. curcas* seed cake as nutrients for pullulan production by an osmotolerant yeast *Aureobasidium pullulans*. Phengnuam and Suntornsuk [26] used *J. curcas* seed cake as nutrients for protease, phytase, and esterase production by *Bacillus licheniformis*. In SSF, different kinds of microbes like *Aspergillus niger*, *Aspergillus oryzae*, *Pseudomonas aeruginosa*, and *Scytalidium thermophilum* were used for xylanase, cellulase, protease, and lipase production ([18], [21], [24], [33]). From this study, it showed that SMF and SSF could degrade PEs efficiency, and the high-protein seed cake could be used for enzyme production or other valuable products production in the future.

References

[1]   P.C. Abhilash, P. Srivasatav, S. Jamil "Revisited *Jatropha curcas* as an oil plant of multiple benefits: critical research needs and prospects for the future"

Environmental Science and Pollution Research, 18, 2011, 127-131.

[2] E.M. Aregheore, K. Becker, H.P.S. Makkar "Detoxification of a toxic variety of *Jatropha curcas* using heat and chemical treatments, and preliminary nutritional evaluation with rats" The South Pacific Journal of Natural Science, 21, 2003, 50-56.

[3] Association of Official Analytical Chemists (AOAC) "Official Methods of Analysis of AOAC International, 16th edition. AOAC International" 1998, Gaithersburg, MD, USA.

[4] M.A. Belewu, R. Sam "Solid state fermentation of Jatropha curcas kernel cake: proximate composition and antinutritional components" Journal of Yeast and Fungal Research, 1, 2010, 44-46.

[5] P. Bernfeld "Amylase α and β. In Methods in Enzymology" Academic Press Inc., New York, 1955, pp. 149-158.

[6] L.R. Castilho, C.M.S. Polato, E.A. Baruque, G.L. Sant'Anna Jr, D.M.G. Freire "Economic analysis of lipase production by *Penicillium restrictum* in solid-state and submerged fermentations" Biochemical Engineering Journal, 4, 2000, 239-247.

[7] C.H. Chang, S.S. Yang "Thermo-tolerant phosphate-solubilizing microbes for multi-functional biofertilizer preparation" Bioresource Technology Journal, 100, 2009, 1648-1658.

[8] E. Chivandi, J.P. Mtimuni, J.S. Read, S.M. Makuza "Effect of processing method on phorbol esters concentration, total phenolics, trypsin inhibitor activity and the proximate composition of the Zimbabwean Jatropha curcas provenance: A potential livestock feed" Pakistan Journal of Biological Sciences, 7, 2004, 1001-1005.

[9] A.R. Choudhury, N. Sharma, G.S. Prasad "Deoiled jatropha seed cake is a useful nutrient for pullulan production" Microbial Cell Factories, 11, 2012, 39-47.

[10] S.R. Couto, M.A. Sanromán "Application of solid-state fermentation to ligninolytic enzyme production" Biochemical Engineering Journal, 22, 2005, 211-219.

[11] S. Donlaporn, W. Suntornsuk "Toxic compound, anti-nutritional factors and functional properties of protein isolated from detoxified *Jatropha curcas* seed cake" International Journal of Molecular Science, 12, 2011, 66-77.

[12] G. Goel, H.P.S. Makkar, G. Francis, K. Becker "Phorbol esters: structure, biological activity, and toxicity in animals" International Journal of Toxicology, 26, 2007, 279-288.

[13] W. Hass, M. Mittelbach "Detoxification experiments with the seed oil from *Jatropha curcas* L" Industrial Crops and Products,12, 2000, 111-118.

[14] H.A. Hong, L.H. Duc, S.M. Cutting "The use of bacterial spore formers as probiotics" FEMS Microbiology Reviews, 29, 2005, 813-835.

[15] C.C. Hu, L.Y. Liu, S.S. Yang "Protein enrichment, cellulase production and *in vitro* digestion improvement of pangolagrass with solid state fermentation" Journal of Microbiology, Immunology, and Infection, 45, 2012, 7-14.

[16] L.K. Janani, G. Kumar, K.V. Bhaskara Rao "Screening of pectinase producing microorganisms from agricultural waste dump soil" Asian Journal of Biochemical and Pharmaceutical Research, 1, 2011, 329-336.

[17] C. Joshi, P. Mathur, S.K. Khare "Degradation of phorbol esters by *Pseudomonas aeruginosa* PseA during solid-state fermentation of deoiled Jatropha curcas seed cake" Bioresource Technology, 102, 2011, 4815-4819.

[18] C. Joshi, S.K. Khare "Utilization of deoiled Jatropha curcas seed cake for production of xylanase from thermophilic *Scytalidium thermophilum*" Bioresource Technology, 102, 2011, 1722-1726.

[19] K.N. Kilcawley, M.G. Wilkinson, P.F. Fox "Determination of key enzyme activities in commercial peptidase and lipase preparations from microbial or animal sources" Enzyme and Microbial Technology, 31, 2002, 310-320.

[20] A. Kumar, S. Sharma "An evaluation of multipurpose oil seed crop for industrial uses (*Jatropha curcas* L.): a review" Industrial Crops and Products, 28, 2008, 1-10.

[21] N. Mahanta, A. Gupta, S.K. Khare "Production of protease and lipase by solvent tolerant *Pseudomonas aeruginosa* PseA in solid-state fermentation using *Jatropha curcas* seed cake as substrate" Bioresource Technology, 99, 2008, 1729-1735.

[22] H.P.S. Makkar, A.O. Aderibigbe, K. Becker "Comparative evaluation of nontoxic and toxic varieties of *Jatropha curcas* for chemical composition, digestibility, protein degradability and toxic factors" Food Chemistry, 62, 1998, 207-215.

[23] H.P.S. Makkar, K. Becker, F. Sporer, M. Wink "Studies on nutritive potential and toxic constituents of different provenances of *Jatropha curcas*" Journal of Agricultural and Food Chemistry, 45, 1997, 3152-3157.

[24] T. Ncube, R.L. Howard, E.K. Abotsi, E.L.J. van Rensburg, I. Ncube "*Jatropha curcas* seed cake as substrate for production of xylanase and cellulase by *Aspergillus niger* FGSCA733 in solid-state fermentation" Industrial Crops and Products, 37, 2012, 118-123.

[25] H. Outtrup, S.T. Jorgensen "The importance of *Bacillus* species in the production of industrial enzymes" In Applications and systems of *Bacillus* and relatives. Blackwell Science Inc., 2002, Malden, Mass. pp. 206–218.

[26] T. Phengnuam, W. Suntornsuk "Detoxification and anti-nutrients reduction of *Jatropha curcas* seed cake by *Bacillus* fermentation" Journal of Bioscience and Bioengineering, 115, 2012, 168-172.

[27] F.G. Priest "Extracellular enzyme synthesis in the genus *Bacillus*" Bacteriology Review, 41, 1977, 711-753.

[28] B.A. Ramsay, K. Lomaliza, C. Chavarie, B. Dubé, P. Bataille, J.A. Ramsay "Production of poly-(β-hydroxybutyric-co-β-hydroxyvaleric) acids" Applied and Environmental Microbiology, 56, 1990, 2093-2098.

[29] D. Saetae, W. Suntornsuk "Antifungal activities of ethanolic extraction from *Jatropha curcas* seed cake" Journal of Microbiology and Biotechnology, 20, 2010, 319-324.

[30] M. Schallmey, A. Singh, O.P. Ward "Developments in the use of *Bacillus* species for industrial production" Canadian Journal of Microbiology, 50, 2004, 1-17.

[31] H. Shimogaki, K. Takeuchi, T. Nishino, M. Ohdera, T. Kudo, K. Ohba, M. Iwama, M. Irie "Purification and properties of a novel surface active agent and alkaline-resistant protease from *Bacillus* sp. Y" Agricultural and Biological Chemistry, 55, 1991, 2251-2258.

[32] M.L. Teixeira, F. Cladera-Olivera, J. dos Santos, A. Brandelli "Purification and characterization of a peptide from *Bacillus licheniformis* showing dual antimicrobial and emulsifying activities" Food Research International, 42, 2009, 63-68.

[33] A. Thanapimmetha, A. Luadsongkrama, B. Titapiwatanakunc, P. Srinophakuna "Value added waste of *Jatropha curcas* residue: Optimization of protease production in solid state fermentation by Taguchi DOE methodology" Industrial Crops and Products, 37, 2012, 1-5.

24

Leaching and Mechanical Behaviour of Solidified/Stabilized Nickel Contaminated Soil with Cement and Geosta

Nicholas Hytiris[1], Panagiotis Fotis[1], Theodora-Dafni Stavraka[1, 2], Abdelkrim Bennabi[2], Rabah Hamzaoui[2]
[1]Glasgow Caledonian University, School of Engineering & Built Environment (SEBE), Glasgow G4 0BA, United Kingdom.
N.Hytiris@gcu.ac.uk; Panagiotis.Fotis@gcu.uk
[2]Université Paris-Est, Institut de Recherche en Constructibilité, ESTP, 28 avenue du Président Wilson, 94234 Cachan, France.
tstavraka@adm.estp.fr; abennabi@adm.estp.fr; rhamzaoui@adm.estp.fr

Abstract- In the present work, solidification/stabilization (S/S) of nickel contaminated soil using Ordinary Portland Cement-OPC and commercial was carried out. Effects of different binder combinations of OPC and commercial stabilizer wt% in the S/S mix designs and physical and chemical characteristics of the treated samples were investigated. The mechanical property studied was unconfined compressive strength-UCS while chemical characterization of the samples was focused on the leachability of nickel. Results indicated that the optimum mix design, in terms of mechanical efficiency, was 10% OPC wt% and 4.2 wt% commercial stabilizer while in terms of chemical efficiency 10% OPC wt% and 1.4 wt% commercial stabilizer.

Keywords: Cement, Graded gravelly sand, Nickel, UCS, Leaching, S/S, Geosta® stabilizer.

1. Introduction

Stabilization/solidification (S/S) is one of the major methods in treating hazardous wastes prior to land disposal and also an effective technique for reducing the leachability of contaminants in soils like, heavy metals [1, 2]. Entrapment of wastes that expresses hazardous characteristics within a cementitious matrix (solidification) and binding of contaminants (organic or inorganic) of a hazardous stream into a stable insoluble form (stabilization) are the mechanisms that best describe the principle behind solidification and stabilization (S/S) treatment. Solidification and stabilization (S/S) related processes such as chemical and physical stabilization of contaminants, dangerous to natural and build environment, have been identified as Best Demonstrated Available Technology-BDAT for 57 different hazardous wastes under the Resource Conservation and Recovery Act-RCRA [3, 4].

Heavy metals are well known to be toxic to most organisms and harmful to the environment when present in excessive concentrations [5, 6]. Nickel has recently become a serious pollutant which is mainly released, from metal processing operations and from increased combustion of coal and oil [7]. Nickel is considered to be one of the most dangerous chemical elements, which may cause permanent soil contamination due to its specific physicochemical properties and mechanisms of action [8, 9].

Portland cements is the most commonly used primary binder for S/S matrix because it can restrict the mobility of heavy metals due to high pH and due to its capability to precipitate the metals in insoluble forms [1, 2]. Yin et al., 2008 [10] studied S/S of nickel hydroxide sludge using OPC and oil palm ash-OPA. They investigated the possibility to reduce the availability of Ni by increasing the amount of OPA and reducing the amount of OPC and found that the optimum mix design is 15 wt% OPA, 35 wt% OPC and 50 wt% sludge. Grega

& Domen, 2011 [11] examined the effectiveness of OPC, calcium aluminate cement-CAC, pozzolanic cement-PC and different additives in immobilizing Cd, Pb, Zn, Cu, Ni and As, in contaminated soil. The effectiveness was evaluated using leaching experiments, mechanical strength and geochemical modeling. Based on the model calculation, the most efficient S/S formulation was CAC + Akrimal® (a cement repair mortar modified with aqueous acrylic polymer dispersion), which reduced soil leachability of Ni up to 4.7 times. Eisa et al., 2011 [12] investigated the immobilization of Ni(II) in various cement matrices (neat Portland cement in absence and presence of water reducing- and water repelling-admixtures as well as blended cement with kaolin) using the S/S technique. The degree of immobilization was assessed by using static mode and semi static mode of leaching and it was found to be very high (99%).

Secondary binders could be described as materials that are not very effective on their own, when used in S/S methods, and are useful only when used in conjunction with lime or cement. Secondary additives are used mostly as stabilizers comprised of fly ash, zeolites, calcium or sodium or ammonium chlorides, enzymes, polymers, and potassium compounds. Several researches have shown that zeolites may be more suitable than other additives for the decontamination of soils polluted by heavy metals because they adjust soil pH value and as cation exchangers [13, 14]. Shanableh & Kharabsheh, 1996 [15] used a natural zeolite as additive to reduce the leaching of Pb^{2+}, Cd^{2+} and Ni^{2+} from a contaminated soil and found that using up to 50% additive, nickel leaching was reduced by a maximum of approximately 50%. Belviso et al., 2010 [16] in their study, used an artificially Ni contaminated soil, treated with coal fly ash for synthesizing zeolite at low temperatures and they found that newly-formed zeolites reduce the toxicity of the element in the polluted soil. Table 1 shows all the research works reported until now in this section.

Table 1. All research works reported in introduction section.

Research work	Binders		Optimum contents	Water/Binder Ratio	Additives	Tests
	Primary	Secondary				
Yin et al., 2007	OPC	OPA	50 wt% sludge, 15 wt% OPA, 35 wt% OPC	0,45		*UCS *Leaching method (TCLP) *Atomic Absorption Spectrometer (AAS) *XRD
Grega & Domen, 2011	OPC PC CAC		67% soil, 10% CAC, 1.39% Akrimal®	0.29	*Cement plasticizers *Fibrous material polypropylene fibers *Aqueous acrylic polymer dispersion Akrimal-E®	*UCS *Leaching method (TCLP) * PTMs mass transfer (modeling) *Statixtical analysis
Eisa et al., 2011	OPC	Kaolinite	The degree of immobilization of Ni(II) in the various used cement pastes was very high	*0.28 (without superplasticizer) *0.24 (with supeplasticizer)	*Superplasticizer *Calcium stearate	*UCS *Tank leaching
Shanableh & Kharabsheh, 1996			Nickel leaching was reduced by a maximum of approximately 50% zeolite		*Zeolite	*Leaching method (TCLP)
Belviso et al., 2010	*Coal fly ash		Newly-formed zeolites reduce the toxicity of the element in the polluted soil		*NaOH	*Bench-scale experiments *XRD *BCR three-step sequential extraction

The aim of this work was to use OPC and Geosta® at different binder combinations in order to study the strength development as well the leachability aspects of OPC -treated Ni contaminated soil.

2. Materials and Methods

The experiments were carried out using an artificially restored and polluted soil. For soil, the particle size distribution (ASTM D6913) and the British Standard light compaction test (BS 1377) were applied. After particle size analysis, the soil consisting of 60% of sand (ranging between fine, medium and coarse sand) and 40% of fine gravel. Therefore, the soil could be described as a well graded gravelly sand. Regarding the compaction test, the values of optimum water content and maximum dry density were 8.1% and 1724 kg/m³, respectively. The nickel used was Nickel (II) sulfate hexahydrate ($NiSO_4 6H_2O$) which had a solubility of 625 g/l at 20°C and a final concentration of 2300 mg/kg by weight of soil. The experimental program consisted of one primary binder (Medium strength, type I 35/A OPC) at different quantities (5%, 7.5% and 10% by dry weight of the soil) and Geosta® at 1.4% (100gr), 2.8% (200gr) and 4.2% (300gr) (by dry weight of the soil). A water to soil ratio of 0.1 was used. Geosta® was a secondary stabilization agent consisting of artificial zeolites A4, chlorides and alkalis. Table2 reflects the mix design symbol used for each mixture. Geosta® chemical composition is shown in Table 3.

Table 2. Mix design symbol for each mixture.

Mix design symbol	Mixed
5% OPC	
5% OPC	5% cement 0%. Geosta®
5%OPC-1.4% G	5% cement, 1.4% Geosta®
5%OPC-2.8% G	5% cement, 2.8% Geosta®
5%OPC-4.2% G	5% cement, 4.2% Geosta®
7.5% OPC	
7.5% OPC	7.5% cement 0%. Geosta®
7.5% OPC-1.4% G	7.5% cement, 1.4% Geosta®
7.5% OPC-2.8% G	7.5% cement, 2.8% Geosta®
7.5% OPC-4.2% G	7.5% cement, 4.2% Geosta®
10% OPC	
10% OPC	10% cement 0%. Geosta®
10% OPC-1.4% G	10% cement, 1.4% Geosta®
10% OPC-2.8% G	10% cement, 2.8% Geosta®
10% OPC-4.2% G	10% cement, 4.2% Geosta®

OPC = Ordinary Portland Cement, G = Geosta® stabilizer

Table 3. Chemical composition of Geosta®.

Component	Quantity (%)	Component	Quantity (%)
$MgCl_2.6H_2O$ (tech. pure)	14.00	$FeCl_2$	3.00
NaCl (tech. pure)	13.00	$KHCO_3$	2.80
KCl (tech. pure)	11.60	Amorphous SiO_2 (5 - 40 μm)	2.55
$CaCl_2.2H_2O$ (tech. pure)	10.00	Na_2SO_4	2.00
Synthetic Zeolite A4	9.17	$FeSO_4$	1.02
K_2CO_3 (tech. pure)	5.10	$Al_2(SO_4)_3$	0.31
MgO (tech. pure)	5.00	Cobalt	0.31
$Na_2S_2O_3$ (Thiosulphate)	3.40	Confidential component(s)	16.75

Prior to mixing, soil sample was allowed to dry in a pre-heated oven for 24 hours and at 105 °C. The dried soil was then placed in the mechanical mixer which thoroughly mixed the uncontaminated soil in order to homogenize particle distribution. During the mixing process of the uncontaminated soil and after 2 minutes, OPC was added and the admixture was allowed to mix for another 2 minutes. Having ensured by now an even distribution among soil and cement particles, contaminant solution was poured over the admixture and the mixing process was allowed to run for another 10 minutes. While the mixing procedure was still active, the weight of cubical moulds was determined and recorded. With the completion of the mixing process, the moulds were filled with the admixture, which was compacted in three layers. For the mixtures containing Geosta®, Geosta® powder (different quantity for each mixture) was added 5 minutes prior the end of the mixing procedure. The mass of the moulds with the compacted contaminated soil was again determined and recorded. The moulds were then covered with a plastic film and allowed to solidify for 24 hours. After 28 days of curing time, S/S samples were subjected to the standard protocol of the Leaching Characteristics Of Moulded Or

Monolithic Building And Waste Materials, "The Tank Test" (17) (Environmental Agency UK, 2004). The leaching experiments were performed at room temperature, (~24.8° C). The specimen tank (plastic container, with Width=240mm, Height=180 mm and Length=290 mm) was filled with distilled water to achieve a liquid to solid ratio (L/S) of 5:1. The leachate was removed and replaced after 0.25, 1.0, 2.25, 4.0, 9.0, 16.0, 32.0 and 64.0 days (giving a total leaching time of ~128 days). Then the samples were collected and centrifuged at a speed of 10,000 rpm for 10 min. The supernatant was then analyzed, through a Flame Atomic Absorption Spectrometer (FAAS) and according to DIN 38406 [18].

Cube specimens of 100x100x100 mm were used for UCS test of the pastes. UCS was determined at 156 days in order to obtain the same time of curing as the leaching samples (28 days of air curing at 25°C and then a leaching period of 128 days) according to ASTM 2166. UCS was the average value of three samples. The test was carried out with a MTS machine of 100kN for small values and an MTS machine of 300 kN for high values. The constant rate applied on both MTS machines is 1 mm/min.

3. Results and Discussion

Table 4 lists the S/S waste acceptance criteria which are utilized to assess the effectiveness of the treatment. The leachability limits are extracted from the Interdepartmental Committee on Redevelopment of Contaminated Land (ICRCL) while the UCS limits are extracted from regulatory waste limit at a disposal site in the United Kingdom.

Table 4. Stabilized/solidified waste acceptance criteria for Ni concerning UCS and leachability.

Indicators	Regulatory (acceptance)	Level
UCS at 28-day of curing	Landfill disposal limit	0.34 (MPa)
Ni Leachability	Residential	130 (mg/kg)
	Allotment	230 (mg/kg)
	Commercial	1800 (mg/kg)

3.1. Leachability

The cumulative leach values of 5, 7.5 and 10% OPC-Geosta® mix designs, in mg/l, are plotted against time,

in days, in Fig.1 (a, b, c) respectively. The poor environmental performance of only OPC mixtures is evident when compared to the mixtures containing Geosta®, however, all mixtures seem to have a downward tendency towards Ni release.

This poor environmental performance in the absence of Geosta® relies on its chemical composition (see Table 3). Geosta® is mainly composed of chlorides ($MgCl$, $NaCl$, KCl, and $CaCl_2$) and zeolites. As it has already mentioned these chlorides are good stabilizers. The key mechanism involved in producing stabilization is ion exchange between soil-cement constituents and chlorides of Geosta®. More specifically, when clay particles (usually negatively charged) are covered with like-charged particles they repel each other, but if some particles have unlike charges, they attract. Then a displacement of sorbed heavy metals occurs and leads to a formation of heavy metal-Cl complexes [19, 20].

The second essential compound of Geosta® is the zeolite. Zeolites are a class of alkaline porous aluminosilicates [13] with permanent negative charges on their surfaces. They have a high cation exchange capacity because their structure is made of a framework of SiO_4 and AlO_4 tetrahedra with a replacement of Si^{4+} by Al^{3+} and as a result they are natural cation exchangers and appropriate to remove toxic cations [13, 14]. In fact this negative charge is balanced by exchangeable cations like calcium, potassium or sodium. On their turn these cations are exchangeable with the heavy metal cations [21]. As a consequence, heavy metals can be trapped inside the zeolitic structure [22]. Due to these effects the environmental performance with the addition of Geosta® is more effective and this becomes clearer in Fig. 2, where the cumulative measured leaching for Ni, for all mix designs, is presented.

For every cube specimen was used 2.33 kg of soil, thus 136 mg of Ni were leached from every cube. The degree of contamination was fixed to 2300 mg/kg by weight of soil; as a consequence every cube contains 5359 mg:

2300 mg/kg X 2.33 kg = 5359 mg.

The difference between 5359 mg and 136 mg is the amount of Ni that entrapped into the sample's structure:

5359 mg − 136 mg = 5223 mg.

In order to find the quantity of Ni entrapped into the structure in mg/kg, the 5223 mg had been divided by the weight of soil per cube:

5223 mg / 2.33 kg = 2241.6 mg/kg.

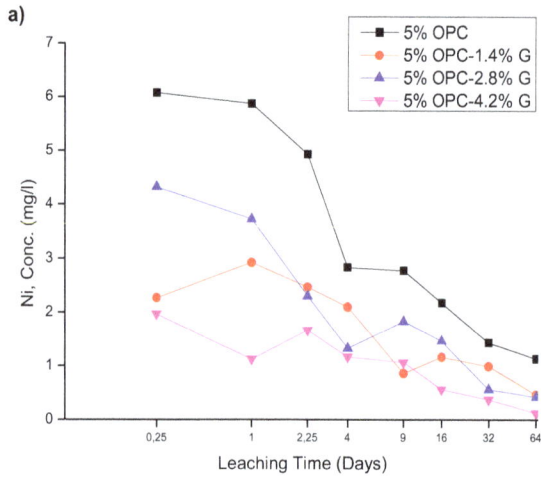

(a)

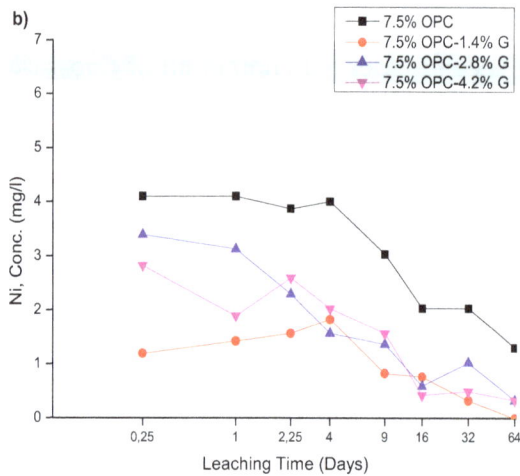

(b)

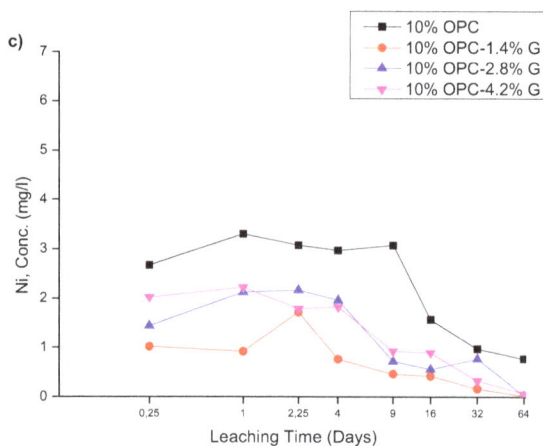

(c)

Figure 1. Ni concentration (mg/l) and leaching time (days), for a) 5% OPC, b) 7.5% OPC, c) 10% OPC.

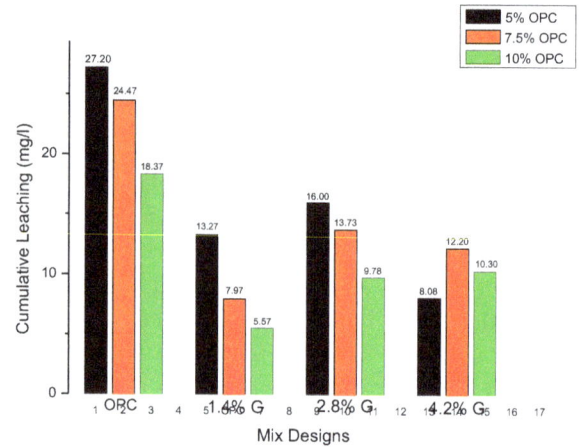

Figure 2. Cumulative Leaching of Ni mix designs.

Hence, 2241.6 mg/kg of contaminant were not leached while 58.4 mg/kg were leached (Ni **contamination below the trigger values proposed by** ICRCL). Finally, the 5%OPC-4.2%G mixture was proved more efficient in Ni confinement within its solidified matrix, since it was the only mixture that managed to maintain Ni leakage quite low 8.08mg/l (17.34 mg/kg).

Table 5 summarizes the leachability values in mg/l and mg/kg, in order to be easier to compare leaching results to waste acceptance criteria of Table 4. Moreover, Table 5 shows the ratio of stabilization for all mixtures.

Table 5. Leachability in mg/kg, mg/l and ratio of stabilization.

Mix Design	Leachability		Ratio of Stabilization (%)
	(mg/l)	(mg/kg)	
5% OPC	27.20	58.38	97.46
5% OPC-1.4% G	13.27	28.48	98.76
5% OPC-2.8% G	16.00	34.33	98.51
5% OPC-4.2% G	8.08	17.34	99.25
7.5% OPC	24.47	52.51	97.71
7.5% OPC-1.4% G	7.97	17.10	99.48
7.5% OPC-2.8% G	13.73	29.46	98.71
7.5% OPC-4.2% G	12.20	26.18	98.86
10% OPC	18.37	39.42	98.28
10% OPC-1.4% G	5.57	11.95	99.50
10% OPC-2.8% G	9.78	21.00	99.08
10% OPC-4.2% G	10.30	22.10	99.03

In mix design containing 7.5% OPC (see Fig 1b), a comparison of the environmental performance of all mixtures, strengthens the earlier observation over the poor ability of cement-only mixture (7.5%OPC) to restrain Ni release. Furthermore, Geosta® mixtures 7.5%OPC-1.4%G and 7.5%OPC-4.2%G, although are proved capable in Ni immobilisation when compared to 7.5%OPC, their behaviour seems to be reversed when compared to their environmental performance in the previous mix design (5% OPC). Mixture, 7.5%OPC-1.4%G shows optimum Ni restrainment in its cementitious matrix while, 7.5%OPC-2.8%G and 7.5%OPC-4.2%G exhibit similar behaviour.

In mix design containing 10% OPC (see Fig 1c); the difference in Ni release is also obvious between the OPC and the Geosta® containing mixtures. Ni release for all mixtures is lower when compared to 5% OPC and 7.5%OPC mix designs, indicating that the increase in cement (10% w/w) has a critical role in Ni release from solidified material. The cumulative release of 10%OPC mix design (Fig.2), verifies the previous statement, since all mixtures achieve better environmental performance when compared to 5% OPC mix design and 7.5% OPC mix design. However, as in previous mix design (7.5% OPC), mixture 10%OPC-1.4%G demonstrates optimum Ni retention ability (5.57 mg/l or 11.95 mg/kg) while, 10%OPC-2.8%G and 10%OPC-4.2%G performed in an antagonistic, although similar manner, by reaching cumulative concentrations of 9.78 mg/l (21.00 mg/kg) and 10.3 mg/l (22.10 mg/kg), respectively.

3.2. Mechanical Properties

The long-term viability of S/S waste was further assessed by analysing the mechanical performance of the S/S material. Fig.3 (a, b, c) shows the UCS-strain relation for the samples studied and Table 5 the values of UCS and elastic modulus (elastic modulus is defined as the slope of its stress–strain curve in the elastic deformation region).

It is evident that higher compressive strength values were obtained when higher amount of cement (OPC) was used for the solidification process. The mix design containing 10% of OPC has presented values almost three times higher than the mix design containing 7.5% OPC and four times higher when compared to the mix design containing 5% OPC. This effect could be attributed to the fact that by increasing the cement quantity, the amount of C_3S (tricalcium silicate) and C_2S (dicalcium silicate) increased in the stabilized soil

enabling more production of calcium–silicate–hydrate (C-S-H) [3].

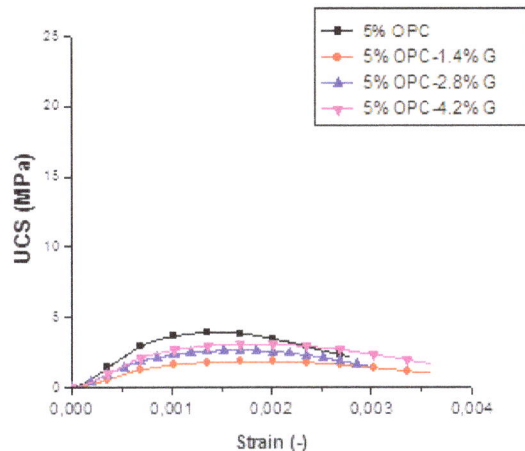

(a)

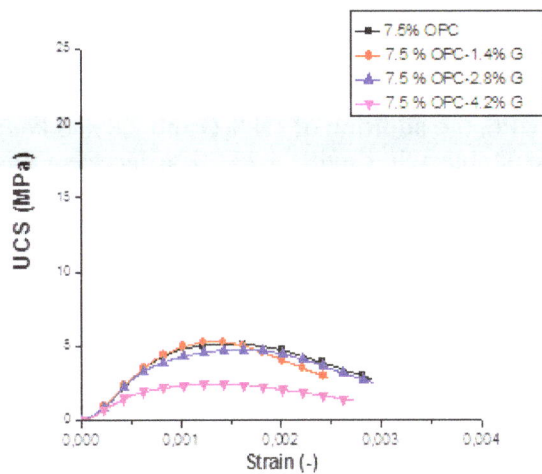

(b)

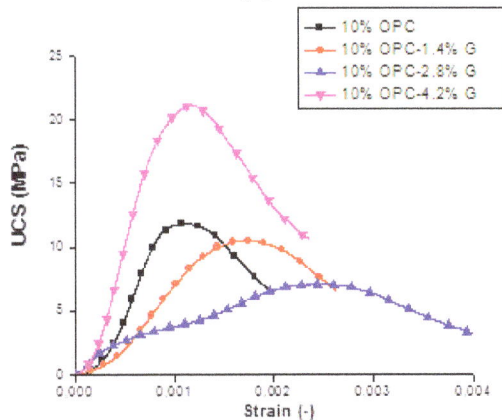

(c)

Figure 3. UCS-strain relation of Ni contaminated soil, treated with OPC and Geosta®: (a) 5% , (b) 7.5%, (c): 10%.

Table 6. Elastic Modulus and UCS of mix designs.

UCS (MPa) ± 1.00	Mix Design	Elastic modulus (GPa) ± 1.00
3.90	5% OPC	4.20
1.90	5% OPC-1.4% G	1.75
2.70	5% OPC-2.8% G	2.75
	5% OPC-4.2% G	2.84
5.10	7.5% OPC	5.46
5.30	7.5% OPC-1.4% G	5.70
4.80	7.5% OPC-2.8% G	5.28
2.50	7.5% OPC-4.2% G	3.45
11.84	10% OPC	11.80
10.51	10% OPC-1.4% G	6.50
7.11	10% OPC-2.8% G	6.20
21.11	10% OPC-4.2% G	22.75

Figure 4 shows the effect of Geosta® at OPC constant level. In the case of 5% OPC, it is observed that increasing the amount of Geosta®, UCS also increases but never exceeds the values of only cement mixture. Concerning 7.5% OPC, the addition of 1.4% G and 2.8% G keeps USC almost stable while with 4.2% G a decrease has been observed. Las but not least, with much more OPC and more specifically with 10% OPC, there is a strong increase with 4.2% G.

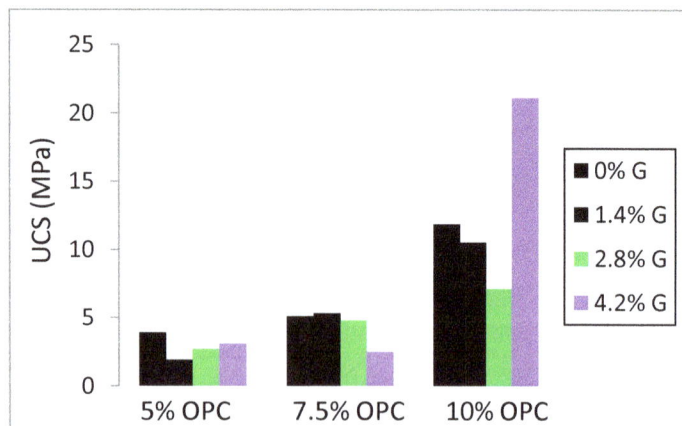

Figure 4. Effect of Geosta®, for the 3 OPC levels considered, on UCS.

4. Conclusion

Based on the environmental and physical performance of the OPC/ Geosta®-treated Ni contaminated soil, the following conclusions can be drawn:
• All mix designs managed to retain Ni contamination below the trigger values proposed by ICRCL. Higher value is 58.4 mg/kg while lower value is 11.95 mg/kg.

• The introduction of Geosta® into the S/S treatment increased the ability of the binder system over Ni retention by more than two times, when compared to only cement mixtures (it is the case for all mixtures).
• This difference in retention ability between the plain cement mixtures and the ones containing Geosta® could be attributed to the cation exchange ability of chlorides (formation of heavy metal-Cl complexes) and zeolites (heavy metals are trapped inside the zeolitic structure) in Geosta® powder.
• The optimum mixture, in terms of chemical efficiency was 10% OPC wt% and 1.4 wt% of Geosta® with 5.57 mg/l (11.95 mg/kg) and 99.50% ratio of stabilization.
• In relation to the mechanical performance of S/S Ni, higher UCS values were obtained when higher amount of cement (10% OPC) was used for the solidification process
• The optimum mixture, in terms of mechanical efficiency, was 10% OPC wt% and 4.2 wt% Geosta® with 21.11 MPa.
• The optimum mixture in terms of reuse, leaching, UCS and cost is 5%OPC-4.2%G and has tremendous potential in construction material applications such as engineering fil, pavement blocks and bricks among others.

References

[1] S. K. Gupta and M. T. Surwade. "Immobilization of Heavy Metals from Steel Plating Industry Sludge Using Cement as Binder at Different pH," in *Conf. on Moving Forward Wastewater Biosolids Sustainability: Technical, Managerial, and Public Synergy*, New Brunswick, Jun. 2007, pp. 773 -777.

[2] R. B. Kogbara and A-T. Abir, "Mechanical and Leaching Behaviour of Slag-cement and Lime-activated Slag Stabilised/Solidified Contaminated Soil," *Sci. of the Total Environment*, vol. 409, no. 11, pp. 2325-2335, May 2011.

[3] Y. C. Yin, G. M. Shaaban and H. B. Mahmud, "Chemical Stabilization of Scrap Metal Yard Contaminated Soil Using Ordinary Portland Cement: Strength and Leachability Aspects," *Building and Environment*, vol. 42, no. 2, pp. 794-802, Feb. 2007.

[4] S. Paria and P. K. Yuet, "Solidification/Stabilization of Organic and Inorganic Contaminants using Portland Cement: A literature review," *Environmental Rev.*, vol. 14, no. 4, pp. 217-255, Oct. 2006.

[5] K. E. Giller, E. Witter, and S. P. McGrath, "Toxicity of Heavy Metals to Microorganisms and Microbial Processes in Agricultural Soils: A Review," *Soil Biology and Biochemistry*, vol. 30, no. 10-11, pp. 1389-1414, Sep. 1998.

[6] P-C. Hsiau and S-L Lo, "Extractabilities of Heavy Metals in Chemically-Fixed Sewage Sludges," *J. Hazardous Materials*, vol. 58, no. 1-3, pp. 73-82, Feb. 1998.

[7] A. Kabata-Pendias and A. Mukherjee, *Trace elements from soil to human*, Pulawy, Poland: Springer-Verlag, 2007.

[8] K. Janicka and M. Cempel, "Effect of Nickel (II) Chloride Oral Exposure on Urinary Nickel Excretion and Some Other Element," *Polish J. of Environmental Stud.*, vol. 12, no. 5, pp. 563-566, Feb. 2003.

[9] C. Aydinalp and S. Marinova, "Distribution and Forms of Heavy Metals in Some Agricultural Soils," *Polish J. of Environmental Stud.*, vol. 12, no. 5, pp. 629-633, Mar. 2003.

[10] Y. C. Yin, A. Wan and Y. P. Lim, "Oil Palm Ash as Partial Replacement of Cement for Solidification/Stabilization of Nickel Hydroxide Sludge," *J. Hazardous Materials*, vol. 150, no. 2, pp. 413-418, Jan. 2008.

[11] G. E. Voglar and D. Lestan, "Efficiency Modeling of Solidification/Stabilization of Multi-metal Contaminated Industrial Soil Using Cement and Additives," *J. Hazardous Materials*, vol. 192, no. 2, pp. 753-762, Aug. 2011.

[12] E. E. Hekal, W. S. Hegazi, E. A. Kishar, and M. R. Mohamed, "Solidification/Stabilization of Ni(II) by Various Cement Pastes," *Construction and Building Materials*, vol. 25, no. 1, pp. 109-114, Jan. 2011.

[13] W. Shi, H. Shao, H. Li, M-A. Shaoa, and S. Du, "Progress in the Remediation of Hazardous Heavy Metal-Polluted Soils by Natural Zeolite," *J. Hazardous Materials*, vol. 170, no. 1, pp. 1-6, Oct. 2009.

[14] A. A. Mahabadi, M. A. Hajabbasi, H. Khademi, and H. Kazemian, "Soil Cadmium Stabilization Using an Iranian Natural Zeolite," *Geoderma*, vol. 137, no. 3-4, pp. 388-393, Jan. 2007.

[15] A. Shanableh and A. Kharabsheh, "Stabilization of Cd, Ni and Pb in Soil Using Natural Zeolite," *J. Hazardous Materials*, vol. 45, no. 2-3, pp. 207-217, Feb. 1996.

[16] C. Belviso, F. Cavalcante, P. Ragone, and S. Fiore, "Immobilization of Ni by Synthesising Zeolite at Low Temperatures in a Polluted Soil," *Chemosphere*, vol. 78, no. 9, pp. 1172-1176, Feb. 2010.

[17] NEN 7345 "Determination of the release of inorganic constituents from construction materials and stabilised waste products", in *NNI (Formerly Draft NEN 5432)*, Delft, Netherlands, 1993.

[18] H. H. Rump, *Laboratory Manual for the Examination of Water, Waste Water and Soil*. Dunfermline, UK: Wiley-Vch, 2000.

[19] D. N. Little and S. Nair, (2009, August). Recommended Practice for Stabilization of Subgrade Soils and Base Materials, TRB Publication. Washington, D.C., USA. [Online]. Available: http://onlinepubs.trb.org/onlinepubs/nchrp/nchrp_w144.pdf .

[20] M. F. Bertos, S. J. R. Simons, C. D. Hills, and P. J. Carey, "A Review of Accelerated Carbonation Technology in the Treatment of Cement-based Materials and Sequestration of CO2," *J. Hazardous Materials*, vol. 112, no. 3, pp. 193-205, Aug. 2004.

[21] Z. Li, S. J. Roy, Y. Zou, and R. Bowman, "Long-Term Chemical and Biological stability of Surfactant-Modified Zeolite," *Environmental Sci. & Technol*, vol. 32, no. 17, pp. 2628-2632, Jul. 1998.

[22] R. Terzavo, M. Spagnuolo, L. Medici, B. Vekemans, L. Vincze, K. Janssens, and P. Ruggiero, "Copper Stabilization by Zeolite Synthesis in Polluted Soils Treated with Coal Fly Ash," *Environmental Sci. & Technol.*, vol. 39, no. 16, pp. 6280-6287, Jul. 2005.

[23] E. Kontori, T. Perraki, S. Tsivilis, and G. Kakali, "Zeolite Blended Cements: Evaluation of their Hydration Rate by Means of Thermal Analysis," *J. Thermal Anal. Calorimetry*, vol. 96, no. 3, pp. 993-998, Jun. 2009.

25

Chitosan – A Natural Sorbent for Copper Ions

Simona Schwarz[1], Mandy Mende[1], Dana Schwarz[2]

[1] Leibniz-Institut für Polymerforschung Dresden e.V., Dept. of Polyelectrolytes and Dispersion
Hohe Straße 6, 01069 Dresden, Germany
simsch@ipfdd.de; mende@ipfdd.de
[2]Charles University in Prague, Faculty of Science, Department of Organic Chemistry
Hlavova 2030/8, 128 43 Prague 2, Czech Republic
schwarzda@natur.cuni.cz

Abstract - As a result of industrial activities and technological changes, a high and continuously increasing amount of heavy metals and heavy metal containing effluents are released into the environment by different industrial nations. These metals cannot be degraded. Furthermore, because of their toxicity, they are highly detrimental to the environment and human health. Heavy metals accumulate in the food chain and become permanent pollutants in the environment. In the human body they accumulate in different organs causing serious damage.

To overcome this problem, the sorption behaviour of heavy metal ions, in particular copper ions was investigated by apply chitosan flakes, powder, and beads as a natural adsorbent. Metal removal was studied using adsorbance measurements, SEM-EDX, and size measurements. The sorption capacity of chitosan was determined at different concentration and times. The received sorption capacities for copper ions were very promising, with a maximum value of 150 mg/g on chitosan powder.

Keywords: Chitosan, Removal, Heavy metal ion, Sorption, Waste water treatment.

1. Introduction

Waste water treatment, sludge dewatering, pulp and paper production are typical examples for solid-liquid separation through coagulation and flocculation. The overwhelming majority of these processes uses polyelectrolytes to regulate the stability and flocculation properties of the disperse systems. This resulted in a great variety of synthetic and natural flocculants like chitosan and starch as natural products which are now commercially available to meet the specific demands of industrial fields, where an efficient solid-liquid separation is required [1-7]. The removal of inorganic components/particles has been intensively investigated during the last years. Hence, it is much more difficult to remove soluble components like dye [8, 9], surfactants, or metal ions. This work will focus on the separation of heavy metal ions especially copper ions with chitosan as a natural polymer.

Heavy metal processing industry has always been a major of concern which affect surface water, drinking waters, ground waters, and rivers contamination. With the agreement to comply limiting values (according to the precautionary principle) [10] as they can be found in the European drinking water directive, the non-damaging of human health at lifelong consumption will be ensured. Actual limiting value of the German drinking water directive of copper ions is 2 mg/L [11].

Besides the endangerment for the environment and human health, heavy metal containing effluents are a source of potentially regenerative materials. The separation and recuperation of heavy metals from the material flow is ecologically and economically required. Thus, the implication is to design efficient and effective strategies to recycle and recover heavy metal ions from industrial effluents [12, 13]. The intention of this study is the development of a process, which can be used to separate heavy metals in a selective manner from the effluents, especially at low concentrations. The separation is supposed to be accomplished with the help of polyelectrolytes. Since several decades, researches carried out a method of separating heavy metals from effluents by using polyelectrolytes. In 1996, it was shown within an ultra filtration experiment that polyelectrolytes are promising substances to remove traces of heavy metals from industrial effluents [14]. Furthermore, it was proved that remaining amount of heavy metal in decontaminated water could significantly be reduced by application of polymeric flocculants, in comparison to inorganic flocculants [15].

Chitosan is a natural polymer and prepared by the deacetylation of chitin. The degree of deacetylation reclines between 50 and 100%. The good sorption properties of chitosan on heavy metal ions has long been known and is mainly attributed to the presence of the so formed secondary amino groups which occur both charged or uncharged depending on the pH of the solution. So far, various articles have been published focussing on the removal of watersoluble impurities especially heavy metal ions using polyelectrolytes like poly(acrylic acid) and poly(ethylene imine), or chitosan as a soluble flocculation or chelating agent [15]. In this work, we focused on the removal of copper ions by using defined chitosan flakes, powder, and beads as efficient adsorber material.

2. Experimental

Chitosan was purchased from BioLog GmbH. The degree of deacetylation (DA) reclines between 85 and 90% for the investigated samples displayed in Table 1:

Table 1. Properties of the investigated chitosan samples, DA – degree of deacetylation.

Sample Name	DA	Molar Mass	Ash Content
Ch90/200/A1	90 %	200,000 g/mol	< 1%
Ch85/400/A2	85 %	400,000 g/mol	< 2%

The investigated Chitosan materials differed in their degree of deacetylation, particle size, and morphology (i.e. flakes, powder, or beads). With laser diffraction we determined a particle size distribution of the powder with a D 50 value of 100 μm. The flakes have a wide particle size distribution with a D 50 value of 635 μm. But few of the flakes have a size in the mm range. The beads had a diameter of about 3 mm in the swollen state. The beads are on basis of alginate coated with chitosan.

The sorption capacities were determined in dependence on the initial concentration of copper ions in solution, as well as on the time of sorption (contact time between adsorbent and adsorbate). Copper sulfate pentahydrate was utilized for all experiments. The sorption investigations were carried out as batch experiments. Sorption capacities were calculated by adsorbance measurements. For every experiment, the concentration of copper ions was measured before and after the sorption procedure at a defined time.

3. Results and Discussion

Contact time is a very important parameter for sorption experiments, the sorption capacity calculated later on at a certain time, and for a useful application. Figure 1 shows the increasing amount of copper ions adsorbed on chitosan as a function of time. The maximum sorption uptake has been reached after 24 h. Therefore, for calculating the sorption capacities all batch experiments were carried out after 24 hours sorption when equilibrium was reached. Almost 100 % of the copper ions were removed from water after 24 hours for an initial copper concentration of 180 mg/L. Furthermore, after 1 hour sorption time 60 % of the copper ions were adsorbed in chitosan already.

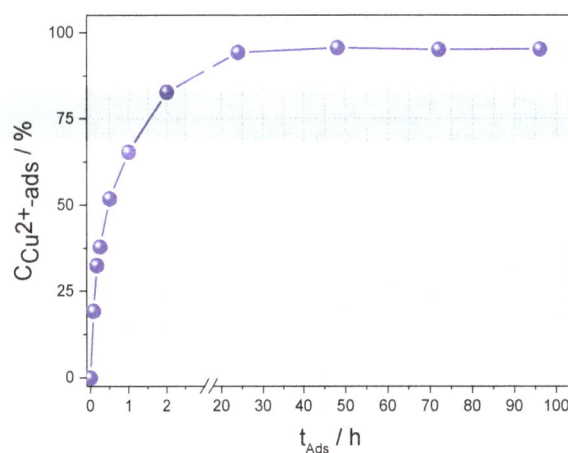

Figure 1. Adsorbed copper ions on chitosan flakes Ch90/200/A1 in dependence on time with an initial copper concentration (c_{Cu2+}) of 180 mg/L.

Chitosan is a white material which changes its colour during the sorption of heavy metal ions in aqueous solution towards the colour of the adsorbed metal salt (see Figure 2). The colour intensity depends on the initial concentration of the heavy metal salt as well as on the sorption time. (Copper sulfate is a grey white solid which appears in a beautiful blue colour as shown in Figure 2 in its hydrated form.) The colour change of the material gives evidence of the sorption of copper ions. Furthermore, the adsorbents were characterized be SEM-EDX measurements before and after sorption process (here not shown). The SEM-EDX measurements indicated an sorption of copper ions as well as sulfate ions.

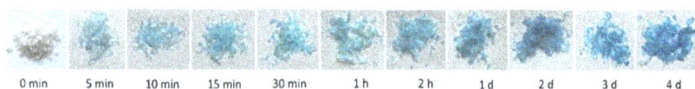

Figure 2. Images of chitosan flakes Ch90/200/A1 in dependence of the sorption time and as a result the increasing colour intensity of chitosan due to the increasing sorption of copper sulfate.

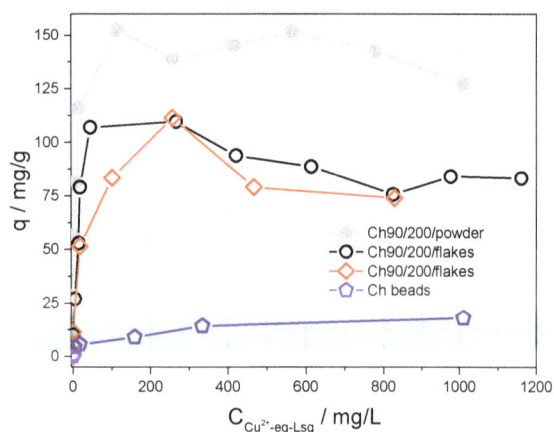

Figure 3. Sorption capacity of different types of chitosan materials in dependence of the initial copper ion concentration, sorption time of 24 h.

In Figure 3 the sorption capacity of the different types of chitosan in dependence of the initial Cu(II) concentration is shown. The sorption capacity is the difference between the initial concentration of the metal ion and the metal ion concentration after sorption multiplied by the volume used, and divided by the mass of the adsorbent. The pH of the solutions was not adjusted. The sorption capacities differ significantly from each other in the range powder >flakes>beads. In general, three different sorption capacities could be obtained depending on the morphology of the material. The highest sorption capacity with a maximum value of 150 mg/g was achieved with the chitosan powder. As the powder had the lowest particle size of all the examined materials, the surface area will be the highest. Hence, the higher the surface area, the higher will be the theoretical sorption capacity. Therefore, the observed values in Figure 3 show a strong dependence on the surface area, thus with the type of chitosan material used. The sorption capacities of the chitosan flakes with 85 and 90 % degree of deacetylation do not differ significantly. The reason could be the small difference in the degree of deacetylation. Hence, the sorption capacity does also depend on the accessibility of the

available amino groups. Furthermore, the analised chitosan beads are coated alginate beads with a small layer of chitosan on top of it. Therefore, the beads are difficult for comparison with the other three chitosan materials due to the possible different chemical moieties on the surface, as the amino groups might stick to the alginate surface. This could be another reason for the low sorption capacities on the beads. Investigation on studying the dependence of DA and molar mass for the sorption properties are in progress.

The achieved sorption capacities for copper sorption on chitosan as a natural sorbent are very high in comparison to other adsorbents, independent on the morphology of chitosan. Activated carbon functionalized with secondary amino groups exhibited an sorption capacity of 140 mg/g for copper ions [16]. Other non natural polymers (e.g. resins and carbamates) possessed sorption capacities between 80 and 70 mg/g [17, 18, 19]. Much higher sorption capacity were found with ion exchange resin. The maximum sorption capacity of Cu ions on a special resin was estimated to 324 mg/g [20]. The sorption capacity on a new type of sorbents on basis of $CaCO_3$ crystals is up to 1041.5 mg Cu(II)/g [21].

Hence, one should note that the sorption capacities of different types of materials were compared without any additional information to it. In this publication the achieved sorption capacities of one natural material with different morphologies was examined and compared. The morphology is one of a number of parameters that are dependent for the sorption capacities.

In further investigations of our work the sorption of other heavy metal ions were described [22-24].

4. Conclusion

In principle chitosan is a suitable adsorbent for copper ions in aqueous solution. The highest sorption capacity maximum value of 150 mg/g was achieved for chitosan powder because of the high accessible surface for copper ions. Chitosan beads have much lower sorption capacities than the chitosan flakes and powder probably due to the small amount of chitosan that was coated on the surface of the alginate beads.

5. Acknowledgement

This work was supported by the Central Innovation Programme (ZIM) of the Federal Ministry of Economy and Energy (BMWi) (KF 2022812RH1). The authors thank Heppe Biolog GmbH from Germany for the support of the materials and discussion and cooperativeness.

References

[1] S. Genest, G. Petzold, and S. Schwarz, " Removal of micro-stickies from model wastewaters of the paper industry by amphiphilic starch derivatives," *Colloids and Surfaces A: Physicochemical and Engineering Aspects*, vol. 484, pp. 231-241, 2015.

[2] S. Bratskaya, S. Genest, K. Petzold-Welcke, T. Heinze, and S. Schwarz, "Flocculation Efficiency of Novel Amphiphilic Starch Derivatives: A Comparative Study," *Macromolecular Materials and Engineering*, vol. 299, pp. 722-728, 2014.

[3] S. Schwarz and G. Petzold, "Polyelectrolyte Complexes in Flocculation Applications," *Advances in Polymer Science*, vol. 256, pp. 25-65, 2014.

[4] R. Rojas, S. Schwarz, G. Heinrich, G. Petzold, S. Schütze, and J. Bohrisch, "Flocculation efficiency of modified water soluble chitosan versus commonly used commercial polyelectrolytes," *Carbohydrates Polymers*, vol. 81, pp. 317-322, 2010.

[5] S. Bratskaya, S. Schwarz, G. Petzold, T. Liebert, and T. Heinze, "Cationic Starches of High Degree of Functionalization: 12. Modification of Cellulose Fibers toward High Filler Technology in Papermaking," *Industrial & Engineering Chemistry Research*, vol. 45, pp. 7374-7379, 2006.

[6] S. Schwarz and G. Petzold, "Polyelectrolyte Interactions with Inorganic Particles," in *Encyclopedia of Surface and Colloid Science*, Third Edition, vol. 6, CRC Press, 2006, pp. 4735-4754.

[7] S. Bratskaya, V. Avramenko, S. Schwarz, and I. Philippova, "Enhanced flocculation of oil-in-water emulsions by hydrophobically modified chitosan derivatives," *Colloids and Surfaces A: Physicochemical and Engineering Aspects*, vol. 275, pp. 168-176, 2006.

[8] G. Petzold, S. Schwarz, M. Mende, and W. Jaeger, "Dye flocculation using polyampholytes and polyelectrolyte-surfactant nanoparticles," *Journal of Applied Polymer Science*, vol. 104, pp. 1342-1349, 2007.

[9] G. Petzold and S. Schwarz, "Dye removal from solutions and sludges by using polyelectrolytes and polyelectrolyte–surfactant complexes,"

Separation and Purification Technology, vol. 51, pp. 318-324, 2006.

[10] Deutscher Bundestag, "Bundesrat, Verordnung zur Novellierung der Trinkwasserverordnung," 2000. [Online]. Available: http://dipbt.bundestag.de/doc/brd/2000/D721+00.pdf.

[11] World Health Organization, *Guidelines for drinking-water quality*. World Health Organization, Geneva, 2011.

[12] BGBl1, "Verordnung zur Novellierung der Trinkwasserverordnung (TrinkwV).," *Bundesanzeiger Verlag*. Anlage 2 (zu § 6 Abs. 2 Teil I und II TrinkwV + Novellierung Nov. 2011), vol. 24 .zu Bonn, p. 959 ff., 2001.

[13] B. Rivas, E. Pereira, M. Palencia, and J. Sánchez, "Water-soluble functional polymers in conjunction with membranes to remove pollutant ions from aqueous solutions," *Progress in Polymer Science*, vol. 36, no. 2, pp. 294-322, 2011.

[14] R. S. Juang and M. N. Chen, " Retention of copper(II)—EDTA chelates from dilute aqueous solutions by a polyelectrolyte-enhanced ultrafiltration process," *Journal of Membrane Science*, vol. 119, no. 1, pp. 25-37, 1996.

[15] S. Bratskayaa, A. Pestovb, Y. Yatlukb, and V. Avramenkoa, "Heavy metals removal by flocculation/precipitation using N-(2-carboxyethyl)chitosans," *Colloids and Surfaces A: Physicochemical and Engineering Aspects*, vol. 339, no. 1-3 , pp. 140-144, 2009.

[16] M. Mahaninia, P. Rahimina, and T. Kaghazchi, "Modified activated carbons with amino groups and their copper adsorption properties in aqueous solution," *Chinese Journal of Chemical Engineering*, vol. 23, p. 50, 2015.

[17] A. Atia, A. Donia, S. Abou-El-Enein, and A. Yousif, "Studies on uptake behaviour of copper(II) and lead(II) by amine chelating resins with different textural properties," *Separation and Purification Technology*, vol. 33, no. 3, p. 295, 2003.

[18] Y. Yang, Z. Wei, C. Wang, and Z. Tong, "Lignin-based Pickering HIPEs for macroporous foams and their enhanced adsorption of copper(II) ions," *Chemical Communications*, vol. 49, pp. 7144-7146, 2013.

[19] X. Jing, F. Liu, X. Yang, P. Ling, L. Li, C. Long, and A. Li, "Adsorption performances and mechanisms of the newly synthesized N,N′-di (carboxymethyl) dithiocarbamate chelating resin toward divalent

heavy metal ions from aqueous media," *Journal of Hazardous Materials*, vol. 167, p. 589 , 2009.

[20] C. Xiong and C. Yao, "Adsorption Behavior of Cu(II) in Aqueous Solutions," *Iran. J. Chem. Chem. Eng.*, vol. 32, no. 2, p. 57, 2013.

[21] M. Mihai, I. Bunia, F. Doroftei, C. Varganici, and, B. Simionescu, "Highly Efficient Copper(II) Ion Sorbents Obtained by Calcium Carbonate Mineralization on Functionalized Cross-Linked Copolymers," *Chemistry A European Journal*, vol. 21, no. 13, p. 5220–5230, 2015.

[22] S. Schwarz, C. Steinbach, D. Schwarz, M. Mende, and R. Boldt, "Chitosan – the Application of a Natural Polymer against Iron Hydroxide Deposition," *American Journal of Analytical Chemistry*, vol. 7, pp. 623-632, 2016.

[23] M. Mende, D. Schwarz, C. Steinbach, R. Boldt, and S. Schwarz, "Simultaneous adsorption of heavy metal ions and anions from aqueous solutions on chitosan—Investigated by spectrophotometry and SEM-EDX analysis," *Colloids and Surfaces A: Physicochemical and Engineering Aspects*, 2016.

[24] M. Mende, D. Schwarz, C. Steinbach, R. Boldt, and S. Schwarz, "The Influence of Salt Anions on Heavy Metal Ion Adsorption on the Example of Nickel," *Journal of Materials Chemistry*.

26

A Study of Adaptive Cruise Control System to Improve Fuel Efficiency

Changwoo Park[1], Hyeongcheol Lee[2]

[1]Hanyang University, Department of Electrical Engineering

222 Whangsimni-ro, Seongdong-gu, Seoul, Republic of Korea, 133-791

changwoo@hanyang.ac.kr

[2]Hanyang University, Department of Electrical and Biomedical Engineering

222 Whangsimni-ro, Seongdong-gu, Seoul, Republic of Korea, 133-791

hclee@hanyang.ac.kr

Abstract - This paper presents modelling and control strategy of adaptive cruise control system. The algorithm utilize the information of subject and target vehicles to against waste energy. The Control strategy operates to follow the preceding vehicle with the optimal fuel consumption speed trajectory while the target vehicle is in the control distance range. This algorithm focuses on reducing unnecessary acceleration and deceleration. The simulation is conducted by Matlab/Simulink® and CarSim® simulator to verify its effectiveness.

Keywords: Adaptive Cruise Control, Eco Drive, Fuel Efficiency, Energy Saving.

1. Introduction

As globally increasing interest in environmental and energy issues, research in many field is being achieved a resolution of energy saving and environmental problems simultaneously improving fuel efficiency of automotive. Preceding studies have showed that high emission rates connected to the using fossil fuels and road traffic strongly affect the air quality in urban areas [1], [2]. In particular, for dealing with the problems, it is researched in its field to improve fuel economy of existing gasoline and diesel vehicles. There are important parts to develop a new material for reduction weight of chassis and to design vehicles shape for lower air resistance. [3], [4] Also industries try to find other systems for efficiency such as CVT (Continuously Variable Transmission) to be able to control engine speed on the optimal operation line. [5] In addition, hydrogen and electric vehicles have been developed and become wide spread as alternative conventional one with emissions of carbon dioxide (CO_2) which causes greenhouse gas. [6], [7]

Despite these efforts, however, these studies are hard to apply in most vehicles because of high cost to develop new material or limit of structural problems. It is effective method improving fuel efficiency of vehicle not to only develop new system, but also controlling vehicle speed, acceleration and braking to against waste energy on the road. In this study, new vehicle speed control strategy is suggested based on the adaptive cruise control system. This system is automatically adjusting the subject vehicle speed to maintain safe distance from preceding vehicle.

An adaptive cruise control system is an extension of a cruise control system in conventional vehicles. [8] This system, generally, focus on following performance and keep the distance between subject and target vehicle while existing a vehicle ahead. The control algorithm in this paper suggests to change the reactivity depending on current circumstance of subject and target vehicle with flexible distance to maintain of safety.

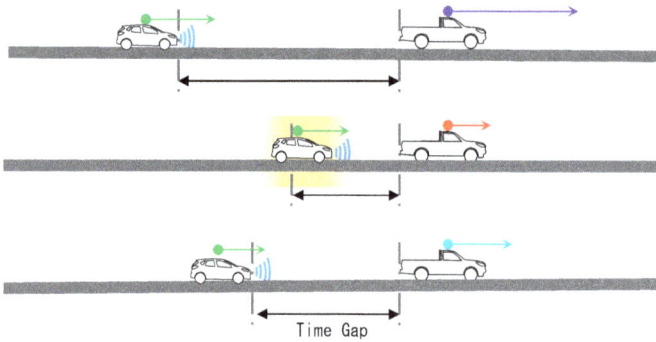

Figure 1. Example of Flexible Distance.

2. Control Algorithm

The control algorithm in this paper is based on optimal control algorithm, linear quadratic regulation (LQR), so both errors and control input can be considered. One of errors is different of desired and actual relative distance and the other is relative velocity between subject and target vehicle. The control input is an acceleration of subject vehicle. This algorithm determines the optimal fuel efficiency acceleration as control input and guarantees safety.

While a vehicle accelerates rapidly, it causes more fuel consumption [9], [10]. It means that the vehicle should smoothly accelerate to improve fuel efficiency. If subject vehicle is controlled to follows target too slowly, however, it can give a room to collide when the target vehicle has emergency stop. To solve this problem, this control algorithm has variable acceleration weighting with two considered influences, relative distance and sign of target vehicle's acceleration.

2.1. Active Linear Quadratic Regulation

Required acceleration is designed with LQR and the state equation can be written as follows:

$$\dot{x} = Ax + Bu = \begin{bmatrix} 0 & -1 \\ 0 & 0 \end{bmatrix} x + \begin{bmatrix} 1 \\ -1 \end{bmatrix} u \tag{1}$$

$$x^T = \begin{bmatrix} x_1 & x_2 \end{bmatrix} = \begin{bmatrix} d_d - d & v_t - v_s \end{bmatrix} \tag{2}$$

The x is states matrix as error, u is control input as an acceleration of subject vehicle in Eqs. (1). The d_d is a desired distance between target and subject vehicle and it is set by Time-gap. d is an actual relative distance, v_t and v_s are each velocity of target and subject vehicle.

The goal of this optimal controller is to minimize a cost function of a state equation. The cost function contains error of relative distance and velocity as shown in Eqs. (3) below.

$$J = \int_0^\infty \left(x^T Q x + u^T R u \right) dt \tag{3}$$

The weighting matrixes Q and R are defined as below.

$$Q = \begin{bmatrix} \rho_1 & 0 \\ 0 & \rho_2 \end{bmatrix}, \quad R = [r] \tag{4}$$

In Eqs. (4), ρ_1 and ρ_2 are weighting factors of distance and velocity states, r is for control input acceleration of subject vehicle.

As the feedback input $u = -kx$ is calculated to minimize the cost function, input acceleration can be decided by solving Eqs. (3). By Lyapunov's second method and Riccati equation, a coefficient matrix K can be written as follows:

$$A^T P + PA - PBR^{-1}B^T P + Q = 0 \tag{5}$$

$$K = R^{-1} B^T P \tag{6}$$

The P is the value of Riccati equation on the steady state condition, so required acceleration a is shown as Eqs. (7).

$$a = u = -Kx = -k_1 \cdot (d_d - d) - k_2 (v_t - v_s) \tag{7}$$

K is gain matrix of controller, and weighting matrix Q and R affect following performance, fuel efficiency and ride-quality. Therefore, it is important to choose proper factors ρ_1, ρ_2 and r.

2.2. Active Weighting Factor

First, the weighting factor can be set by a sign of acceleration to against collision. While preceding vehicle is decelerating with small gap, subject vehicle has to decelerate fast to avoid crush. Another influence is a relative distance. When preceding vehicle accelerates with small relative distance, the subject vehicle should not follow too fast because the target vehicle has possibility to decelerate soon. On the other hand, the subject vehicle has more weighting factor not to miss the target with big relative distance.

The active weighting factor map has two input, acceleration and relative distance, and output as weighting factor as Fig. 2. Fundamentally, the factor is high. If it is high enough, controller can against big acceleration. Therefore, the fuel efficiency is expected higher and following performance for preceding vehicle is lower. Using this map, however, the controller maintains safety with low weighting factor while minus acceleration of target vehicle and small distance between subject and target.

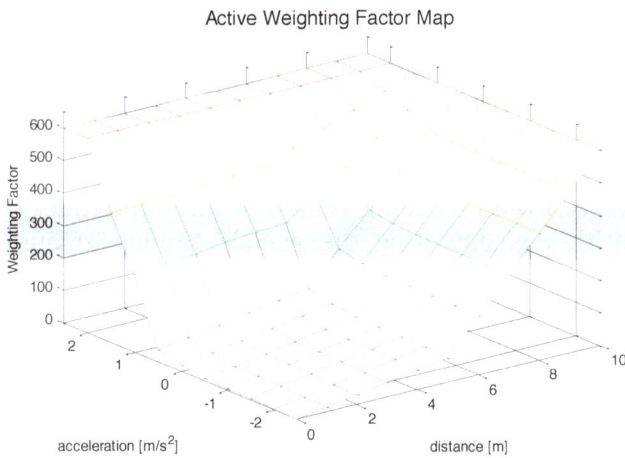

Figure 2. Map of Active Weighting Factor.

3. Simulation and Result

The optimal control with active weighting factor algorithm is simulated using Simulink® for control algorithm and CarSim® for vehicle model. The Simulink® is a simulation tool for model-based design and CarSim® is a co-simulator with vehicle dynamics. The simulation environment has set with real vehicle model and legislative driving cycle. This simulation is focused on fuel efficiency and following performance as safety side both.

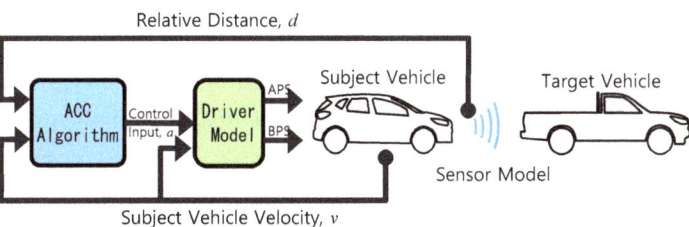

Figure 3. Structure Map of Adaptive Cruise Control System.

3.1. Simulation Environment

A conventional SUV (Sport Utility Vehicle) using 2.2 diesel engine is modelled for controlled vehicle and FTP-75 (EPA Federal Test Procedure), commonly known as city driving cycle, is used to verify effectiveness of the algorithm. This driving cycle is not only legislative, but also frequent acceleration and deceleration are suitable for the system.

Compared controllers are simple PI controller (Proportion-Integration controller), basic LQR method with fixed weighting factor and LQR with active weighting factor. The simple PI controller is used control parameters from a preceding study by Rajamani [8]. LQR controllers with fixed factor are each designed by two different weighting factor. One of the controllers is used weighting factor for following performance and the other factor is selected for fuel efficiency because the two control goals are trade-off. Active LQR controller is adapted the map of active weighting factor as figure 2.

3.2. Simulation Result

The summary of the simulation result is showed as Table 1 and Figure 4, 5, 6. The table contains each amount of fuel used and the figures represent time-gap distance errors.

Fuel efficiency of simple PI controller is lowest as 9.95 km/L because it controls subject vehicle only to consider time-gap distance. Even the PI controller has only goal, it makes more overshoot. Thus, fuel efficiency is lower and following performance is not better than others.

Two compared LQR controllers have different character by a parameter setting. They bring about different results; the algorithms are same though. When the weighting factor R is a low value and focused on following performance as LQR#1 in Table 1 and compared in figure 5, relative distance trajectory is almost same as time-gap distance. For this setting, however, affects low fuel efficiency. If R factor is set a high value, the controller has higher fuel efficiency than the previous setting, but it has big distance errors.

Finally, the Active LQR controller has higher fuel efficiency than other controllers. At the same time, distance errors are much lower than LQR for fuel efficiency and similar to the others.

Table 1. Summary of the Simulation Result.

	Simple PI	Basic LQR#1	Basic LQR#2	Active LQR
Fuel Consumption (kg)	1.017	0.787	0.775	0.757
Fuel Efficiency (km/L)	9.95	12.86	13.07	13.38

Fuel efficiency of Active LQR is improved 34.5%, 4.0%, and 2.4% from others.

· Simple PI: PI controller for normal
  with parameters set: K_p=1, K_i=0.3
· Basic LQR#1: LQR for following performance
  with parameters set: ρ_1=1, ρ_2=50, r=1
· Basic LQR#2: LQR for Fuel Efficiency
  with parameters set: ρ_1=1, ρ_2=50, r=500
· Active LQR: LQR for Optimal basic LQR#1 and #2
  with parameters set: ρ_1= 1, ρ_2= 50, r=Map as figure 2.

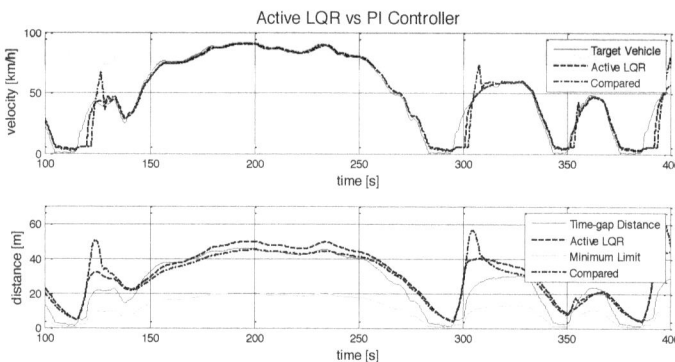

Figure 4. Comparison between Active LQR and PI Controller result.

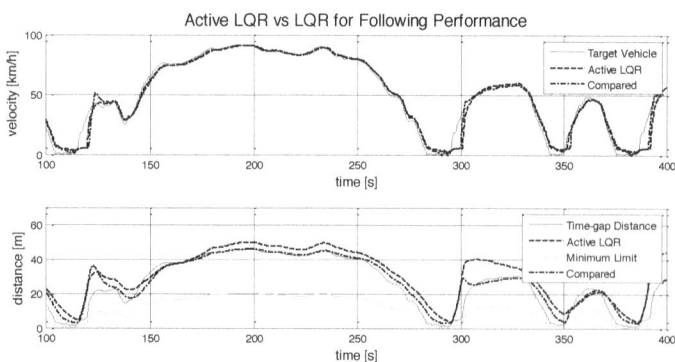

Figure 5. Comparison between Active LQR and LQR#1 for Following Performance result.

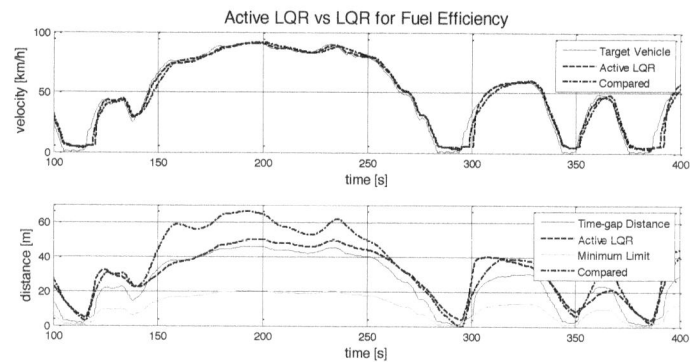

Figure 6. Comparison between Active LQR and LQR#2 for Fuel Efficiency result.

4. Conclusion

This study suggests the control algorithm for improving fuel efficiency on adaptive cruise control system. The algorithm is extended using variable weighting factor from preceding study which is linear quadratic regulation. Therefore, it controls the subject vehicle to improve fuel efficiency and to keep time-gap distance from preceding vehicle at the same time. This algorithm is operating the vehicle similar to human driver being.

However, this paper arrived at the result by a limited simulation of deduction, also the driving cycle is not for adaptive cruise system. The future study will develop and verify with more exclusive driving cycle for the system. Furthermore, the real-world test is demanded for the more accuracy.

Acknowledgements

This work was supported by the Industry Core Technology Development Project (grant no. : 10052501) from the Ministry of Trade, Industry and Energy, Republic of Korea. And specific project name is development of design technology of a device visualizing the virtual driving environment and synchronizing with the vehicle actual driving conditions to test and evaluate ADAS.

References

[1]   C. Marino, A. Nucara, M. Pietrafesa, and A. Pudano, "The Assessment of Road Traffic Air Pollution by Means of an Average Emission Parameter," (in English), *Environmental Modeling & Assessment*, vol. 21, no. 1, pp. 53-69, 2016.

[2]   F. Nejadkoorki, K. Nicholson, I. Lake, and T. Davies, "An approach for modelling CO2 emissions

from road traffic in urban areas," *Sci Total Environ,* vol. 406, no. 1-2, pp. 269-278, 2008.

[3]   Y. Zhang, Choyu Lee, Hua Zhao, Tom Ma, Jing Feng, Zhiqiang Lin, Jie Shen, "Improvement of Fuel Economy and Vehicle Performance Through Pneumatic Regenerative Engine Braking Device (Reneged)," in *Proceedings of the FISITA 2012 World Automotive Congress,* 2013, pp. 55-66, Springer.

[4]   C. Brace, M. Deacon, N. Vaughan, R. Horrocks, and C. Burrows, "The compromise in reducing exhaust emissions and fuel consumption from a Diesel CVT powertrain over typical usage cycles," in *Proceedings of CVT'99 Congress,* Eindhoven, The Netherlands, 1999, pp. 27-33.

[5]   J. Ino, T. Ishizu, H. Sudou, and A. Hino, "Adaptive cruise control system using CVT gear ratio control," SAE Technical Paper 0148-7191, 2001.

[6]   H. Peng, Zheng Li, Bin Chen, Jieyu Wu, Zhenglan Zhao, Yuehong Shu, Junjun Lei, "Development for Control Strategy of ISG Hybrid Electric Vehicle Based on Model," in *Proceedings of the FISITA 2012 World Automotive Congress,* 2013, pp. 333-342, Springer.

[7]   O. Erdinc, B. Vural, and M. Uzunoglu, "A wavelet-fuzzy logic based energy management strategy for a fuel cell/battery/ultra-capacitor hybrid vehicular power system," *Journal of Power sources,* vol. 194, no. 1, pp. 369-380, 2009.

[8]   R. Rajamani, *Vehicle dynamics and control.* Springer Science & Business Media, 2011.

[9]   E. Kim and E. Choi, "Estimates of critical values of aggressive acceleration from a viewpoint of fuel consumption and emissions," in *2013 Transportation Research Board Annual Meeting,* 2013.

[10] K. Ahn, H. Rakha, A. Trani, and M. Van Aerde, "Estimating vehicle fuel consumption and emissions based on instantaneous speed and acceleration levels," *Journal of Transportation Engineering,* vol. 128, no. 2, pp. 182-190, 2002.

PERMISSIONS

LIST OF CONTRIBUTORS

Sandrine Delpeux-Ouldriane, Mickaël Gineys, Nathalie Cohaut and François Béguin
CNRS-Université d'Orléans, ICMN 1B Rue de la Férollerie, Orléans, France

Sylvain Masson, Laurence Reinert and Laurent Duclaux
Université Savoie Mont Blanc, LCME Chambéry, France

Amanda L. Ciosek and Grace K. Luk
Department of Civil Engineering, Faculty of Engineering and Architectural Science, Ryerson University 350 Victoria Street, Toronto, Ontario, Canada

Danielle Camenzuli and Damian B. Gore
Macquarie University, Department of Environmental Sciences, North Ryde 2109, Sydney, Australia

Scott C. Stark
Australian Antarctic Division, 201 Channel Highway, Kingston 7050, Hobart, Australia

Ju Zhang, Jianbing Li, Ronald W. Thring and Guangji Hu
Environmental Engineering Program, University of Northern British Columbia, 3333 University Way, Prince George, British Columbia, Canada

Lei Liu
Department of Civil and Resource Engineering, Dalhousie University, 1360 Barrington St., Halifax, Nova Scotia, Canada

Sanoopkumar Puthiya Veetil, Guy Mercier, Jean Francois Blais and Emmanuelle Cecchi
INRS-ETE 490 rue de la couronne, Québec, Canada

Sandra Kentish
The University of Melbourne, Chemical and Biomolecular Engineering, VIC 3010, Australia

Raymond Oriebe Anyasi and Harrison Ifeanyichuku Atagana
Department of Environmental Sciences Institute for Science and Technology Education University of South Africa, 1, Preller street, Muckleneuk Ridge, Pretoria, South Africa

Davis Amboga Anzeze and John Mmari Onyari
Department of Chemistry, School of Physical Sciences, College of Biological and Physical Sciences, University of Nairobi, Nairobi, Kenya

Paul Mwanza Shiundu
Department of Chemical Sciences and Technology, The Technical University of Kenya, Haile Selassie Avenue, Nairobi – KENYA

John W Gichuki
Environmental Protection Department (EPD), Big Valley Rancheria Bond of Pono Indiana, 2726 Mission Rancheria road, Lakeport CA 95453, USA

Zikun Xing
School of Civil and Environmental Engineering, Nanyang Technological University 50 Nanyang Avenue, Singapore

Lloyd H. C. Chua
School of Engineering, Deakin University Geelong Waurn Ponds Campus, 75 Pigdons Road, Waurn Ponds VIC 3216, Australia

Jörg Imberger
Centre for Water Research, University of Western Australia M023, 35 Stirling Highway, Crawley 6009, Western Australia, Australia

Mihyun Jo, Taeyuel Kim, Sirim Choi, Jongpil Jung, Hee-il Song, Hyunjin Lee, Gyoungsu Park and Jogyo Oh
Gyeonggi Province Institute of Health and Environment, North branch, Drinking Water Inspection Team 1 Cheongsaro, Uijeongbu-si, Gyeonggi-do, Republic of Korea

Jai-young Lee
University of Seoul, Environmental Engineering 163 Seoulsiripdaero, Dongdaemun-gu, Seoul, Republic of Korea

Nicolas Abatzoglou and Benoit Legras
Université de Sherbrooke, Department of Chemical and Biotechnological Engineering 2500 boul. Université, Sherbrooke, Quebec, Canada

Benoît Courcelles
Polytechnique Montréal, Department of Civil, Geological and Mining Engineering CP 6079, Succ. Centre-Ville, Montréal, Qc, H3C 3A7, Canada

Gopal Chandra Saha, Jianbing Li, Siddhartho Shekhar Paul and Ronald W Thring
Environmental Engineering Program, University of Northern British Columbia, 3333 University Way, Prince George, British Columbia, Canada

Afamia Elnakat
Environmental Science & Engineering, The University of Texas at San Antonio One UTSA Circle, San Antonio, Texas

Gleiciane Fernanda de Carvalho Blanc, Bernardo Lipski, Juliano José da Silva Santos, Karime Dawidziak Piazzetta and Juliane de Melo Rodrigues
Institutos Lactec, Department of Environmental Resources Centro Politécnico da UFPR. Rodovia BR-116, km 98, n° 8813, Curitiba, Paraná, Brasil

Luciana Leal
Copel Distribuição S.A. Rua José Izidoro Biazetto, n° 158, Curitiba, Paraná, Brasil

Ikuo Tanabe
Nagaoka University of Technology, Department of Mechanical Engineering 1603-1 Kamitomioka, Nagaoka, Niigata, Japan

I-Hung Khoo and Tang-Hung Nguyen
National Center of Green Technology & Education, College of Engineering, California State University Long Beach, 1250 Bellflower Blvd, Long Beach, CA 90840, USA

Yoichi Ichikawa and Syunsuke Mukai
Ryukoku University, Faculty of Science and Technology 1-5 Yokotani, Setaoe-cho, Otu, Shiga 520-2194, Japan

Masahiro Nishimoto
Polytech Add, Inc. RBM Tsukigi Square 3F, 1-18-8 Shintomi, Chuo-ku, Tokyo

Hideaki Mouri
Meteorological Research Institute 1-1 Nagamine, Tsukuba, Ibaraki 305-0052, Japan

Akihiro Hori
Meteorological and Environmental Sensing Technology Inc. 1-1 Nagamine, Tsukuba, Ibaraki 305-0052, Japan

Shamshad Khan, Wu Yaoguo, Zhang Xiaoyan and Hu Sihai
Department of Applied Chemistry, School of Science, Northwestern Polytechnical University, Xi'an, 710072, China

Xu Youning and Zhang Jianghua
Xi'an Institute of Geology and Mineral Resources, Xi'an, 710054, China

Sabrina Chelli, Adnan Falah and Rami El Khatib
School of Environment and Health Sciences, Canadian University of Dubai, 1st Interchange, Sheikh Zayed Road, Dubai, United Arab Emirates

Shamshad Khan, Wu Yaoguo, Zhang Xiaoyan and Hu Sihai
Department of Applied Chemistry, School of Science, Northwestern Polytechnical University, Xi'an, 710072, China

Liu Jingtao and Sun Jichao
The Institute of Hydrogeology and Environmental Geology, Shijiazhuang, 050803, China

E. Stora, M. Horgnies, I. Dubois-Brugger and L. Dao-Castellana
Lafarge Centre de Recherche, 95 rue du Montmurier, Saint Quentin-Fallavier, F-38291 France

Saadia Nousir, Tze Chieh Shiao, René Roy and Abdelkrim Azzouz
Nanoqam, Department of Chemistry, University of Quebec in Montreal, Montreal, Qc, Canada

Andrei-Sergiu Sergentu
Laboratory of Catalysis and Microporous Materials (LCMM), University of Bacau, Romania

Chin-Feng Chang and Jen-Hsien Weng
Department of Food Science, China University of Science and Technology, Nankang, Taipei 11581, Taiwan

Kao-Yung Lin
Department of Living Science, National Open University, Luzhou, New Taipei City 24701, Taiwan

Li-Yun Liu
Department of Food Science, Nutrition and Nutraceutical Biotechnology, Shih Chien University, Taipei 10464, Taiwan

Shang-Shyng Yang
Department of Food Science, China University of Science and Technology, Nankang, Taipei 11581, Taiwan
Department of Biochemical Science and Technology, National Taiwan University, Taipei 10617, Taiwan

Nicholas Hytiris and Panagiotis Fotis
Glasgow Caledonian University, School of Engineering & Built Environment (SEBE), Glasgow G4 0BA, United Kingdom

Theodora-Dafni Stavraka
Glasgow Caledonian University, School of Engineering & Built Environment (SEBE), Glasgow G4 0BA, United Kingdom
Université Paris-Est, Institut de Recherche en Constructibilité, ESTP, 28 avenue du Président Wilson, 94234 Cachan, France

Abdelkrim Bennabi and Rabah Hamzaoui
Université Paris-Est, Institut de Recherche en Constructibilité, ESTP, 28 avenue du Président Wilson, 94234 Cachan, France

Simona Schwarz and Mandy Mende
Leibniz-Institut für Polymerforschung Dresden e.V., Dept. of Polyelectrolytes and Dispersion Hohe Straße 6, 01069 Dresden, Germany

Dana Schwarz
Charles University in Prague, Faculty of Science, Department of Organic Chemistry Hlavova 2030/8, 128 43 Prague 2, Czech Republic

Changwoo Park
Hanyang University, Department of Electrical Engineering 222 Whangsimni-ro, Seongdong-gu, Seoul, Republic of Korea

Hyeongcheol Lee
Hanyang University, Department of Electrical and Biomedical Engineering 222 Whangsimni-ro, Seongdong-gu, Seoul, Republic of Korea

Index

www.ingramcontent.com/pod-product-compliance
Lightning Source LLC
Chambersburg PA
CBHW080642200326
41458CB00013B/4705